Open Source Mobile Learning:

Mobile Linux Applications

Lee Chao
University of Houston-Victoria, USA

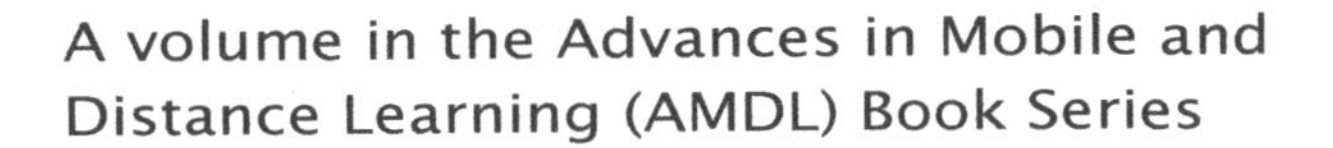

Senior Editorial Director:	Kristin Klinger
Director of Book Publications:	Julia Mosemann
Editorial Director:	Lindsay Johnston
Acquisitions Editor:	Erika Carter
Development Editor:	Julia Mosemann
Production Editor:	Sean Woznicki
Typesetters:	Keith Glazewski, Natalie Pronio, Milan Vracarich Jr.
Print Coordinator:	Jamie Snavely
Cover Design:	Nick Newcomer

Published in the United States of America by
Information Science Reference (an imprint of IGI Global)
701 E. Chocolate Avenue
Hershey PA 17033
Tel: 717-533-8845
Fax: 717-533-8661
E-mail: cust@igi-global.com
Web site: http://www.igi-global.com

Library of Congress Cataloging-in-Publication Data

Open Source Mobile Learning: Mobile Linux Applications / Lee Chao, Editor.
p. cm.
Includes bibliographical references and index.
Summary: "This book helps readers better understand open source software and its application in mobile learning, covering open culture and mobile learning in the open source setting and reviewing the pros and cons of various types of mobile network architecture, mobile devices, open source mobile operating systems, and open source mobile application software"--Provided by publisher.
ISBN 978-1-60960-613-8 (hardcover) -- ISBN 978-1-60960-614-5 (ebook) 1. Web-based instruction. 2. Linux device drivers (Computer programs) 3. Operating systems (Computers) I. Chao, Lee, 1951- LB1044.87.O64 2011
371.33'44678--dc22

2010040623

This book is published in the IGI Global book series Advances in Mobile and Distance Learning (AMDL) Book Series (ISSN: 2327-1892; eISSN: 2327-1906)

British Cataloguing in Publication Data
A Cataloguing in Publication record for this book is available from the British Library.

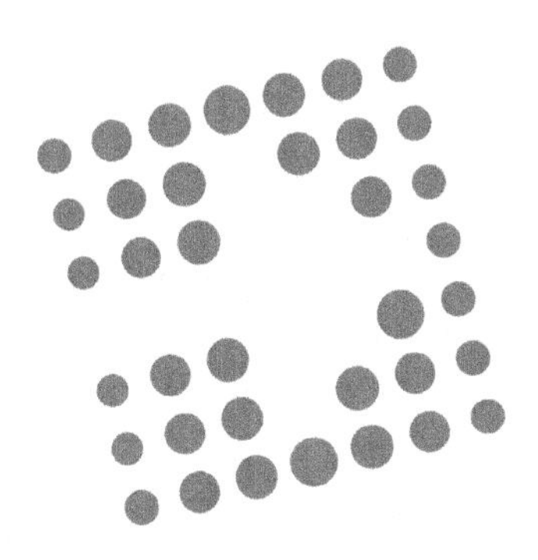

Advances in Mobile and Distance Learning (AMDL) Book Series

Patricia Ordóñez de Pablos
Universidad de Oviedo, Spain

ISSN: 2327-1892
EISSN: 2327-1906

Mission

Private and public institutions have made great strides in the fields of mobile and distance learning in recent years, providing greater learning opportunities outside of a traditional classroom setting. While the online learning revolution has allowed for greater learning opportunities, it has also presented numerous challenges for students and educators alike. As research advances, online educational settings can continue to develop and advance the technologies available for learners of all ages.

The **Advances in Mobile and Distance Learning** (AMDL) Book Series publishes research encompassing a variety of topics related to all facets of mobile and distance learning. This series aims to be an essential resource for the timeliest research to help advance the development of new educational technologies and pedagogy for use in online classrooms.

Coverage

- Cloud Computing in Schools
- Economics of Distance and M-Learning
- Educational Apps
- Ethical Considerations
- Lifelong Learning
- Managing Sustainable Learning
- Pedagogy & Design Methodology
- Tablets & Education
- Technology Platforms & System Development
- Ubiquitous & Pervasive Learning

IGI Global is currently accepting manuscripts for publication within this series. To submit a proposal for a volume in this series, please contact our Acquisition Editors at Acquisitions@igi-global.com or visit: http://www.igi-global.com/publish/.

The Advances in Mobile and Distance Learning (AMDL) Book Series (ISSN 2327-1892) is published by IGI Global, 701 E. Chocolate Avenue, Hershey, PA 17033-1240, USA, www.igi-global.com. This series is composed of titles available for purchase individually; each title is edited to be contextually exclusive from any other title within the series. For pricing and ordering information please visit http://www.igi-global.com/book-series/advances-mobile-distance-learning-amdl/37162. Postmaster: Send all address changes to above address.

Titles in this Series

For a list of additional titles in this series, please visit: www.igi-global.com

Outlooks and Opportunities in Blended and Distance Learning
J. Willems (Monash University, Australia) B. Tynan (University of Southern Queensland, Australia) and R. James (University of New England, Australia)
Information Science Reference • copyright 2013 • 358pp • H/C (ISBN: 9781466642058) • US $175.00 (our price)

Cases on Formal and Informal E-Learning Environments Opportunities and Practices
Harrison Hao Yang (State University of New York at Oswego, USA) and Shuyan Wang (The University of Southern Mississippi, USA)
Information Science Reference • copyright 2013 • 454pp • H/C (ISBN: 9781466619302) • US $175.00 (our price)

Refining Current Practices in Mobile and Blended Learning New Applications
David Parsons (Massey University - Auckland, New Zealand)
Information Science Reference • copyright 2012 • 334pp • H/C (ISBN: 9781466600539) • US $175.00 (our price)

Professional Education Using E-Simulations Benefits of Blended Learning Design
Dale Holt (Deakin University, Australia) Stephen Segrave (Deakin University, Australia) and Jacob L. Cybulski (Deakin University, Australia)
Business Science Reference • copyright 2012 • 454pp • H/C (ISBN: 9781613501894) • US $175.00 (our price)

Open Source Mobile Learning Mobile Linux Applications
Lee Chao (University of Houston-Victoria, USA)
Information Science Reference • copyright 2011 • 348pp • H/C (ISBN: 9781609606138) • US $180.00 (our price)

Models for Interdisciplinary Mobile Learning Delivering Information to Students
Andrew Kitchenham (University of Northern British Columbia, Canada)
Information Science Reference • copyright 2011 • 354pp • H/C (ISBN: 9781609605117) • US $180.00 (our price)

Architectures for Distributed and Complex M-Learning Systems Applying Intelligent Technologies
Santi Caballé (Open University of Catalonia, Spain) Fatos Xhafa (Open University of Catalonia, Spain) Thanasis Daradoumis (Open University of Catalonia, Spain) and Angel A. Juan (Open University of Catalonia, Spain)
Information Science Reference • copyright 2010 • 436pp • H/C (ISBN: 9781605668826) • US $180.00 (our price)

Cases on Online and Blended Learning Technologies in Higher Education Concepts and Practices
Yukiko Inoue (University of Guam, Guam)
Information Science Reference • copyright 2010 • 348pp • H/C (ISBN: 9781605668802) • US $180.00 (our price)

Table of Contents

Section 2
Mobile Learning in Open Source Setting

Detailed Table of Contents

Section 1
Mobile Learning and its Technical Foundation

The chapters in this section describe the technical and structural background of open source based mobile learning.

As an introduction, the goal of this chapter is to introduce mobile technology, mobile learning, and the open source approach. This chapter provides some basic knowledge for readers to make the reading of the later chapters easier. It first gives an overview of mobile technology. Then, it introduces the concepts and theories of mobile learning. The recent development in mobile technology and mobile learning will be briefly reviewed. Following the review of mobile learning, the next topic is about the open source approach. It discusses the advantages and disadvantages of the open source approach. It also briefly reviews the history of the open source approach. Next, the chapter investigates the open source approach in mobile learning. It discusses the possible usage of open source products in mobile learning. The last topic of this chapter is about the book framework. After going through the book framework, readers will have a general picture of what to be covered in this book. Lastly, the chapter wraps up with the conclusion and references sections.

This chapter discusses the availability of open technologies for mobile learning. The authors first introduce the issues and standards of adaptation mechanisms for mobile learning resources. Then, they describe some notable existing projects and software tools. At the end, the authors present some case studies and open trends about m-learning.

This chapter introduces mobile Linux operating systems that have been widely used by mobile devices. The author first gives a brief introduction to the history of mobile Linux. Then, he introduces the mobile Linux features which can be used to meet the mobile learning requirements. The last part of the chapter presents some strategies of selecting a suitable mobile Linux operating system for a particular mobile learning project.

In this chapter, the author discusses the role of openness, specifically standardization, in the mobile cloud - a cornerstone for an open, inter-operable realization.

This chapter reviews the model programs of mobile learning using mobile phones (mLearning/MPs) in Japan. The author first discusses the barriers, such as the psychological, pedagogical, and technological issues that mLearning/MPs had to overcome. Then, the author describes the findings obtained from four projects on teaching and learning the English language by mobile phones. At the end of this chapter, the author proposes open source-based mobile services as an effective model for English language learning using mobile phones.

This chapter introduces a worldwide popular mobile open source server project, Funambol. The author first discusses the background of the Funambol project and how it applies to mobile learning in three specific areas. Then, the author discusses major trends that are affecting mobile open source software for mobile learning, challenges that are confronting mobile software developers and content authors, and the benefits of using open source software for mobile learning systems.

Section 2
Mobile Learning in Open Source Setting

Each chapter in this section deals with a specific case of mobile learning in the open source setting.

This chapter provides a detailed account of developing a mobile network laboratory with a set of open source software that allows students to conduct labs either as an individual or as a group at anytime and anywhere.

This chapter introduces game based learning (GBL). The authors use an open source based simulation game platform to enhance the learner's experience. The Yet Another Telephony Engine (YATE) server is employed for the interactivity of the game platform. The effectiveness of the game platform is analyzed. The authors also explore new directions for further research in this area.

This chapter presents the Augmented Reality (AR) library developed for mobile phones. The authors demonstrate the use of such a system in teaching. They also present the results of the data analyses.

Foreword

How to make better use of open source technologies in mobile learning environment is a popular topic. It has been widely discussed in professional meetings and academic institutions. Though there are success stories, most of them are addressing one aspect of the whole picture. The challenge of effectively utilizing open source technology in the mobile learning environment still there for us to deal with, and a more systematic organization of such experience is both necessary and important to achieve such goals. Dr. Li Chao's *Open Source Mobile Learning: Mobile Linux Applications* is perhaps the first truly comprehensive and well-organized book to address this difficult task of planning, developing, and administering open source components for mobile learning.

This timely book organizes the content of mobile learning into three aspects: the technical foundation of mobile learning, the open source setting of mobile learning, and the open culture, as well as ubiquitous education. From these three sections, you could find almost all aspects of open source technologies in mobile learning environment s, and, moreover, many chapters discuss not just underlying instructional and foundational issues, but also the nuts and bolts of how to configure specific systems. The book provides a wealth of guidance and valuable information on the technologies that are available and what they can do, with a special emphasis on Linux mobile operating systems and related products, and how to develop and administer mobile learning systems.

Dr. Chao has authored and edited books on numerous topics in computer science field, including database, networking, and programming. He is the most popular computer science faculty in the university. Students like his courses because of his talents in teaching complex computer science theory and practice in an easy-to-follow way.

I believe this new book will become an instant hit in mobile learning research and teaching community. It is my pleasure to write the Foreword for this timely, well-organized, and highly relevant book, written by one of our most popular faculties.

Yun Wan

***Yun Wan** is an associate professor in Computer Information Systems at the University of Houston—Victoria, Texas. His research interests encompass intelligent software agents and e-commerce. His publications appeared in Communications of the ACM, IEEE Computer, IEEE Internet Applications, as well as other academic journals and edited books. He is the JECR special issue editor on comparison-shopping and recommendation agents and editor of a book about comparison-shopping agents. Wan received a PhD in Management Information Systems from the University of Illinois at Chicago.*

Preface

Mobile learning has become common practice among universities and schools. Numerous research results confirmed that learning and teaching can be extended to anywhere and anytime. Mobile learning goes beyond the limit of computer desks, classrooms, and campuses. Researchers show that mobile learning is very popular among students. More and more courses have adopted the mobile learning platform to enhance learning. A well established mobile learning infrastructure is the lifeline for situation based teaching and learning. For classes that require field trips, mobile learning is a platform that can get the job done. Mobile learning is the platform that can greatly enhance collaboration and group activities. Due to the special features that no other learning platforms can offer, mobile learning continuously attracts attention from educators.

The 3G mobile network has adopted the IP technology so that a 3G mobile device such as a smart phone can access the Internet. The forthcoming 4G mobile network will be the all IP based network. The trend indicates that the mobile network technology is moving towards to the all IP-based wireless network. This trend means that the computer technology will be more and more integrated into the mobile network. As the improvement of mobile networks and mobile devices, many new smart phones can have the functionalities usually handled by personal computers.

Due to its stability, flexibility, security, availability, and low cost, the Linux based mobile operating system has been gaining support from well-known technology companies. Google initiated the well known Android project, which is a combination of an operating system with the Linux kernel and an application development platform. Similarly, Intel introduced the Moblin project and Ubuntu introduced the Ubuntu Mobile project to the mobile computing community. There are a large number of open source tools and plug-ins available for the Linux based mobile operating system. For example, the open source Web 2.0 based mobile learning 2.0 technology has been widely accepted by the mobile learning society. These open source products have a significant impact on the mobile learning community.

Therefore, the open source approach in mobile learning has generated a great deal of interest. Researchers and practitioners in various academic fields have been using open source products in mobile learning for years. The open source products have been used for developing mobile learning course materials and for implementing mobile learning infrastructures. However, these experiences with the open source approach in mobile learning have not been systematically and comprehensively summarized and published for the mobile learning society to share. Although a large number of articles and books about mobile learning itself have been published in recent years, a few of them are about the open source approach in mobile teaching and learning. To enrich the mobile learning theory and practice, this book is aiming to collect the experiences of researchers and practitioners in open source based mobile learning. The book includes research studies about the mobile learning that is supported by open source software, the open source mobile technologies used to implement the mobile learning infrastructure, the class management with open source products, the implementation of mobile learning with open source games, and the development of mobile learning course materials with open source tools. The topics covered in this book will help readers better understand the open source solution in mobile learning.

THE CHALLENGES

Mobile network technology advances rapidly. When you look around, new ideas, new concepts, new theories, new devices, new software, new architecture, and new terminologies pop up every year or even every month. A mobile network system is not only updated rapidly, it is also a complex and versatile system. There are dozens of different types of mobile network systems. Different mobile networks follow different technology standards. Each mobile network system includes a telephony network, IP network, mobile devices, radio equipment, computers, operating systems, security protection, application software, and so on. It may consist of hundreds of or thousands of hardware devices and software products. The rapid growth in mobile technology as well as the complexity and versatility of mobile network systems create a great challenge to everyone who is involved in mobile learning.

Unlike personal computer systems where the software created for PCs are compatible with the computer systems produced by most of the computer manufacturers, the software developed for one type of phone is usually not compatible with other types of phones. Users may not be able to download the software on a Web site and run it on their phones. They may not be able to install and run the software created by themselves on their smart phones. The compatibility issue may cause concerns for some mobile learning projects. The course materials created for one type of smart phone may not work on other types of smart phones. It is difficult to require all the students to use one type of smart phone.

In most cases, the expenditure on mobile learning will increase dramatically. In addition to the complexity and compatibility issues, a mobile learning system that handles course material distribution, mobile learning service hosting, and mobile learning class maintenance can be very expensive. It costs money to upgrade and maintain the system. On the other hand, to keep up with the mobile industry's trends and to teach students the knowledge that is not obsolete, a mobile learning system needs to be upgraded frequently. However, the cost for adding additional hardware and software, reconfiguring the system, and providing training to users can take a significant portion of the annual budget. The shortage of funding often hinders the effort of improving a mobile learning infrastructure. The cost of developing a mobile learning infrastructure prevents many small universities from implementing their mobile learning projects. It can also be time consuming to upgrade and maintain the system.

Many educational institutions may not be aware of the open source solution for mobile learning. Some of them may not have the personnel with the knowledge and skills to support the open source based mobile learning system. Due the fact that there is little marketing effort on open source products, the open source products are less known among administrators, instructors, students, and mobile learning support staff members. Although many of the open source products are compatible with proprietary products, the decision makers may not be fully aware of these facts.

THE ANSWERS

Using the open source solution can be one of the solutions to overcome the difficulties in implementing mobile learning. The Linux based open source mobile operating system, Web based mobile service, and Java based mobile applications can potentially resolve the compatibility issues and reduce the cost.

The Linux kernel used by these mobile operating systems is designed to be compatible with a wide range of mobile devices and network equipment. This feature allows mobile devices to share the software. For example, over two dozens of Android based mobile devices produced by different manufacturers can share the software downloaded from Android Market. Java is the programming language supported by most of the application development platforms. The application developed with Java is known for its portability. The Web is an open source based service that runs on the Internet. It provides the interconnected hypertext which can be remotely accessed by mobile learners through their mobile devices. Documents provided by the Web generally can be handled by any 3G or 4G mobile devices.

By using an open source based mobile learning system, schools and universities can significantly reduce the cost of developing and supporting online classes while improving reliability and performance. Open source software and utilities such as those supported by a Linux based mobile operating system are free and powerful enough to handle tasks such as hosting mobile learning service and developing mobile learning applications. The open source solution is a great way to reduce the financial burden.

For a mobile learning system to keep up with the trends in mobile technology and to teach students the up-to-date knowledge, open source products provide a viable solution by allowing the user to add new components whenever they are available. With proprietary products, one has to wait for the new version to be made available on the market. The user of an open source product can update the mobile learning infrastructure as soon as the new product is available. Users also have the freedom to decide which part of the product should be upgraded instead of upgrading the entire package.

TARGET AUDIENCE

To teach courses based on mobile learning, instructors, mobile learning support staff members, and students need to understand and learn how to use new mobile technology. Instructors need to develop new pedagogies to adapt to the new teaching environment. They also need to create new teaching methodologies and new teaching materials that can utilize the new technology. It is also necessary for the instructors to come up with new course assessment methods to evaluate new teaching materials. New mobile technology creates new opportunities for the instructor to improve the existing teaching and research.

Students also need to get familiar with new mobile devices so that they can use them for learning. They should adjust their learning behavior and develop new ways to learn. They need to be more proactive, more collaborative, and more self-disciplined.

It is important for technicians to keep up with the new mobile technology. They need to develop and maintain the new mobile learning infrastructure. They need to know how to detect the vulnerability of a mobile network and enforce security measures. They must make sure that the qualified people can access the mobile course materials at anytime and from anywhere. They need to provide technical support for the hands-on practice in mobile labs. It is also required that they handle troubleshooting and training.

For the administrators of educational institutions, it is important closely follow the development of new mobile technology and new mobile teaching methodology so that they can make right decisions on when and how to implement the new curriculum. The administrators need to understand the cost and benefit of a mobile learning project. They should be actively involved in the development of mobile learning, which depends on their encouragement and financial support.

In response to the above needs, this book is designed for people who are involved in mobile learning. It is especially useful for the people who are involved in providing mobile learning services such as administrators, IT managers, instructors, m-learning support staff, and IT service personnel. This book also helps mobile learning decision makers on policies, planning, and strategies. Students can also benefit from this book on the open source mechanisms of accessing mobile learning services.

ORGANIZATION OF THE BOOK

Nineteen chapters are included in this book. The chapters are categorized into three main sections, Mobile Learning and Its Technical Foundation, Mobile Learning in Open Source Setting, and Open Culture and Ubiquitous Education. The following is a brief description of each section.

Section 1: The chapters included in this section describe the technical and structural background of open source based mobile learning.

Chapter 1 provides an overview of three key components of open source mobile learning, mobile learning, mobile technology, and the open source approach. Open source products can be used to implement the mobile infrastructure, support mobile learning, and create course materials. To help readers get a general picture of the book, this chapter provides a section about the book's framework.

Chapter 2 discusses the availability of open technologies for mobile learning. The issues and standards of adaptation mechanisms for mobile learning resources are explored. The authors provide some information about the existing projects and software tools that are available to mobile learning. Readers can also find some case studies and the analysis of open trends in this chapter.

Chapter 3 is about Linux based mobile operating systems. This chapter first investigates the requirements by a mobile learning system. To fulfill the requirements, this chapter introduces the Linux based mobile operating systems used by mobile devices. The architectures of various Linux based mobile operating systems are discussed. This chapter investigates what the Linux based mobile operating systems can do in mobile learning. It also provides recommendations on the selection of Linux based mobile operating systems for different mobile devices.

Chapter 4 reveals the role of openness, specifically standardization, in the mobile cloud technology. The topics such as mobile communications, global collaboration, transition from 3G to 4G mobile network technology, and mobile cloud are covered in this chapter. The author considers the mobile cloud the cornerstone for an open, inter-operable realization.

Chapter 5 introduces the mobile phone system (mLearning/MPs) which has been used in mobile learning in Japan. This chapter describes the barriers, such as the psychological, pedagogical, and technological issues that mLearning/MPs had to overcome. The author proposes open source-based mobile services as a way of overcoming barriers faced by mLearning/MPs and as an effective model for English language learning using mobile phones. This chapter demonstrates the implementation of four projects carried out in the teaching of the English language by mobile phones.

Chapter 6 introduces a popular mobile open source server project, Funambol. The chapter provides the background information about the Funambol project and how it applies to mobile learning in three specific areas, the impact of mobile open source software on mobile learning, the challenges that are confronting mobile software developers and content, and the benefits of using open source software for mobile learning systems.

Section 2: Ten chapters are included in this section. Each chapter in the section deals with a specific case of mobile learning in the open source setting.

Chapter 7 illustrates the development of a mobile network laboratory with a set of open source software. The mobile network laboratory allows students to conduct their hands-on practice in the networking class. The students can either work as an individual or as a group at anytime and anywhere.

Chapter 8 introduces game based learning (GBL). This chapter demonstrates the use of an open source based simulation game platform to enhance the learner's experience. The Yet Another Telephony Engine (YATE) server was used for the interactive mobile learning. This chapter provides the analysis on the effectiveness of the game platform. The authors suggest some new directions for further research in this area.

Chapter 9 presents the Augmented Reality (AR) library developed for mobile phones. The authors demonstrate the use of such a system in teaching. They demonstrate how ARToolKit can be used in mobile learning on Nokia N95 phones. The evaluation of the teaching with ARToolKit and games is also given in this chapter.

Chapter 10 discusses collaboration in mobile Learning Management Systems. The authors propose a collaborative game for students to improve their collaboration. This chapter provides details on the implementation of collaboration with open-source technologies.

Chapter 11 introduces the open source software, OpenLaszlo Presentation Server. Students can run the software on any device with the applications that blend to perfection a user-centered design. The software is used to facilitate the development of forms, menus and other components for a website as well as a full management back office or a trip booking site. As pointed out by the author, the open source software is comparable to proprietary software.

Chapter 12 exploits the potentialities of the 2.0 web tools and the advantages of the Open Source software to guide students towards effective linguistic competence and autonomy. In this chapter, the author describes the use of Web 2.0 in teaching to support the learning of English in an Italian secondary school. In the conclusion, the author presents the achievements and the drawbacks of the integration of e-learning 2.0 with classroom teaching.

Chapter 13 describes the use of mobile devices with the Web 2.0 technologies to improve collaboration, participation, knowledge sharing and construction. The author lays out the theoretical framework, which sustains learning with mobile devices. The author also describes the potential of Mobile Web 2.0 for the development of informal learning and the construction of personal learning environments. At the end of the chapter, the author presents mobile learning scenarios of using Web 2.0 tools, in particular, those made possible using Twitter and m-Flickr.

Chapter 14 explores the potential of Mobile Web 2.0 to enhance tertiary education today. It outlines both research-informed principles and case study examples which summarize Mobile Web 2.0 participants' experiences of using Mobile Web 2.0 within a pedagogical framework. It illustrates that the use of Mobile Web 2.0 has transformed both students' and lecturers' conceptions of teaching and learning.

Chapter 15 investigates the opportunity to teach computer programming courses on an open source based mobile learning platform. Based on the behaviors of novices in learning programming, the method of pair programming is adopted in the programming classes to improve the learning of computer programming. This chapter provides the information of the open source mobile technologies used to implement pair programming with mobile devices.

Chapter 16 discusses accessing remote laboratories through mobile devices. The authors present the main strategies for adapting a remote laboratory to mobile devices. The remote laboratory, WebLab-Deusto, is used to demonstrate the adaption. The strategies are analyzed and compared in order to decide which strategy is more suitable for which situation.

Section 3: Three chapters are included in this section to address issues related to open culture and ubiquitous education in general.

Chapter 17 aims to identify the theoretical and technological underpinnings for delivering mobile learning to the distance learner. It also discusses the possible learner communities that can benefit from mobile learning technology, with regard to their unique learning requirements and features.

Chapter 18 defines the virtual world concept and the m-learning scope. It distinguishes the different types of virtual worlds and makes a comparative analysis between them in order to bring out the features aimed at helping teachers to adopt them in their classes. This chapter also considers the choice of virtual world environments on open source platforms.

Chapter 19 explores a variety of issues, technologies and challenges associated with implementing mobile learning at open universities. The chapter begins with an investigation into open universities' common mandate and their nature. Then, the author explains the urgency and advantages for implementing mobile learning in their course and program delivery. This chapter also explores the technical requirements of mobile learning and presents some strategies for mobile learning implementation. The author also explores some of the challenges one may have to face when implementing mobile learning at an open university.

To help readers better understand the application of the open source solution to mobile learning with the open source approach, this book is designed to provide the information of open source products used in each phase of the mobile learning. Comparison of popular open source products and their usability in a mobile learning system is also given in selected chapters. In this book, readers will find design strategies and implementation methods of mobile learning systems. The book presents some possible solutions to the challenges encountered in the mobile learning system construction, management, and course material development.

Lee Chao
University of Houston – Victoria, USA

Section 1
Mobile Learning and its Technical Foundation

Chapter 1
Introduction

Lee Chao
University of Houston-Victoria, USA

ABSTRACT

The rapid development of mobile technology has created great excitement in e-learning. A new learning platform, mobile learning, has emerged. Mobile learning integrates the knowledge of e-learning and mobile networks. When properly implemented, mobile learning improves teaching and learning by providing better quality, lower cost, and a true anywhere-and-anytime learning environment.

Depending on the mobile technology and mobile service provider, mobile learning can be implemented with different approaches. Among various approaches, the open source approach has attracted a great deal of attention recently. This is due to the rapid development in mobile Linux and other open source products. Many well-known companies such as Google, Ubuntu, Intel, and Nokia have recently developed their own versions of open source operating systems and application development platforms. For readers to better understand these newly developed open source mobile operating systems and applications and to learn how to effectively use them in mobile learning, it is necessary to systematically introduce the technologies and demonstrate how these technologies can be used to support mobile learning.

As an introduction, the goal of this chapter is to introduce mobile technology, mobile learning, and the open source approach. This chapter provides some basic knowledge for readers to make the reading of the later chapters easier. It first gives an overview of mobile technology. Then, it introduces the concepts and theories of mobile learning. The recent development in mobile technology and mobile learning will be briefly reviewed. Following the review of mobile learning, the next topic is about the open source approach. It discusses the advantages and disadvantages of the open source approach. It also briefly reviews the history of the open source approach. Next, the chapter investigates the open source approach in mobile learning. It discusses the possible usage of open source products in mobile learning. The last topic of this chapter is about the book framework. After going through the book framework, readers will have a general picture of what to be covered in this book. Lastly, the chapter wraps up with the conclusion and references sections.

DOI: 10.4018/978-1-60960-613-8.ch001

INTRODUCTION TO MOBILE TECHNOLOGY

In a mobile network, signals are transmitted through radio waves (Stetz, 1999). Since there are only limited numbers of radio waves available to allow millions of people to communicate simultaneously, the region covered by a mobile network is divided into a number of small areas called cells. Within each cell, the mobile devices communicate with the base station through radio waves. Each base station is connected to a public switched telephone network (PSTN) or wireless service provider. The same frequencies of radio waves can be re-used in different cells. The size of a cell depends on the number of users within the cell. The more the users are within each cell, the smaller the cell size. The size of a cell can vary from 100m in diameter in a highly concentrated urban area to 30km in a rural area. In such a way, millions of users can contact each other simultaneously in the world.

In its short history, the mobile network system has improved significantly. It has progressed from an analog cell phone system to a high speed digital cellular system integrated with the high speed broadband Internet service.

The first generation (1G) mobile technology is an analog wireless cell phone system introduced in the 1980's (Harte & Bowler, 2003). Within each cell, 1G mobile technology such as Advanced Mobile Phone System (AMPS) developed by Bell Labs allows 56 users to use their cell phones simultaneously. The radius of a cell can range from 2km to 20km. The data transmission rate is 10Kbps. In today's standard, the transmission rate is a bit too low. In addition to the low transmission rate and the limited number of users that can use the mobile network simultaneously, the analog technology has some other significant weaknesses. It is easily interfered by noise and has no protection against eavesdropping. Due to these weaknesses, the 1G mobile technology has very limited use in mobile learning.

The second generation (2G) is the digital cellular system introduced in the 1990's (Harte & Bowler, 2003). Compared with the 1G mobile technology, the 2G mobile technology has improved significantly. It allows each radio wave to be dynamically shared by a number of users. Therefore, 2G can support more users operating simultaneously within a cell by using the radio waves more efficiently. Digital signals can be easily compressed to reduce network traffic. In addition, the digital mobile network provides encryption, and error detection and correction functionalities to improve the security and quality of data transmission. The 2G mobile technology has a data transmission rate ranging from 9.6Kbps to 1.2288Mbps. In addition to the cell phone service, the 2G mobile network includes data services such as text messaging. Currently, the 1G system has mostly been replaced by the 2G mobile technology. The 2G technology can meet some basic mobile learning needs. In fact, most of the early mobile learning reports are based on the 2G technology. Although the 2G technology can carry out some basic mobile learning tasks, it is not able to handle the rich multimedia course content posted on the Web.

Specified by the 3rd Generation Partnership Project (3GPP), the third generation (3G) mobile technology has made some significant improvement over the 2G technology by combining the cellular service with the Internet service (Harte & Bowler, 2003). The first 3G mobile network was released by Japan in 2001 for commercial use. Today, the 3G technology is available worldwide for commercial use. The early releases of the 3G technology included Universal Mobile Telephone Service (UMTS) which is implemented over the existing GSM (Global System for Mobile Communications) networks and CDMA2000 which is implemented over the existing CDMA networks. Later, the much improved UTMS-TDD was released. The 3G technology has been further improved by introducing the High Speed Packet Access (HSPA) family and the Worldwide

Interoperability for Microwave Access (WiMAX) IEEE 802.16e standard.

The International Telecommunication Union (ITU) defines the features of the 3G technology. The standard IMT-2000 specifies that the 3G spectrum ranges between 400MHz and 3GHz. The wide radio wave range allows more users to access the mobile network simultaneously. It also allows users to perform broadband based tasks such as video conferencing, video streaming, playing online games, watching TV, listening to music, and faxing documents. IMT-2000 also specifies that the minimum transaction bit rate be 348 Kbps in a moving vehicle and 2Mbps for stationary or walking users. IMT-2000 requires that the 3G mobile technology should be compatible with three channel access methods, FDMA, TDMA, and CDMA. 3G is also required to be backward compatible with the 2G mobile technology. By meeting the requirements specified by IMT-2000, the 3G mobile technology can now support multiple communication formats such as cellular, satellite, wired LAN, wireless LAN, and wireless WAN. It has better voice quality and enables worldwide roaming seamlessly. The 3G mobile technology also improves network security by allowing mobile equipment to authenticate the network to which the device is connecting and provides better encryption mechanisms.

By including the Internet service, the 3G mobile technology is ready to handle the rich multimedia course content provided by learning management systems (LMSs). Through the services provided by the 3G technology, students can make phone calls, send and receive e-mail, communicate through video conferencing, play online games, and surf the Internet anywhere and at anytime. The 3G technology allows the students to access universities' private networks from their mobile devices. In such a way, the students can use mobile devices to access the labs on campus for hands-on practice and collaboration.

Although 3G has provided a great platform for accomplishing many sophisticated mobile learning tasks. For some online lab activities or high definition movies, the performance of the early versions of 3G may not be adequate. The transmission bit rate of 348Kbps is not enough for these types of activities. The cost for subscribing 3G services is still too high in many parts of the world. It is difficult to require all the students to subscribe 3G services.

The fourth generation mobile technology (4G) is still under development (Glisic & Lorenzo, 2009). Often, some vendors claim that their products are the 4G technology. However, the term 4G is not well defined yet. There has been confusion about it. Commonly, we consider mobile technologies such as Long Term Evolution (LTE) and the 4G version of Worldwide Interoperability for Microwave Access (WiMAX) 802.16m as the 4G technology.

LTE is scheduled to be commercially available in the year 2010. It is designed to support an all-IP packet switching network. By doing so, the network structure of LTE can be significantly simplified, which means that less network equipment is needed and the data traffic latency can be significantly reduced. All-IP packet switching allows LTE to transmit a large volume of data. The LTE technology supports up to a 326.4 Mbps downlink data transmission rate and 172.8 Mbps uplink data transmission rate. The LTE cell boundary can be as far as 100km away from the center and still maintain reasonably good performance. There are two technologies used to boost LTE performance. The first one is the Orthogonal Frequency-Division Multiplexing (OFDM) scheme which is a multi-carrier Frequency-Division Multiplexing (FDM) method. OFDM offers high spectral efficiency, and it is robust against interference and less sensitive to errors. The second one is the Multiple Input Multiple Output (MIMO) technology. For data transmission, the MIMO technology uses multiple channels to send and receive data. By using multiple channels, LTE can transmit data several times faster than single channel transmission. The MIMO technology also improves transmission quality. LTE can handle mobile TV by including the Multicast Broadcast Single Frequency Network (MBSFN).

Another 4G technology is WiMAX II which is based on the IEEE 802.16m standard. The 802.16m standard is in progress. Similar to LTE, WiMAX II also includes technologies such as OFDM and MIMO. Besides OFDM and MIMO, WiMAX II supports real-time gaming, TV broadcasting, and video streaming over high definition screens. WiMAX II also enhances power saving management and supports the self management system. With the self management system, WiMAX II can manage the communication between base stations and mobile devices to reduce the interference and can improve the coordination of neighboring base stations. WiMAX II is designed to have an amazing 100Mbps data transmission rate for high mobility and up to 1 Gbps for low mobility. With these features, WiMAX II can meet the 4G requirements outlined by the IMT-Advanced standard.

4G is a desired technology for mobile learning that involves a lot of interactivity and requires high quality video. The 4G technology is especially useful for tasks such as hands-on practice in an online virtual lab or playing online multiplayer games.

In summary, some commonly used technologies for the different generations of mobile network systems are listed in Table 1.

Usage of Mobile Network System in Mobile Learning

During the past ten years, the number of mobile device users has increased dramatically. At the early days, only a small population was able access the first generation mobile network which is the analog cell phone system. By the year 2000, a majority of the population in developed countries were able to communicate through the second generation mobile network which is a digital cell phone system. Today, people communicate with each other through the third generation or fourth generation mobile network that can handle multimedia content and is integrated with the Internet technology. TAG (2008) reported that eighty five percent of the people in the United States have cell phones. According to Medford (2008), the demand for cell phones surged in Asia, Africa, and the rest of the emerging world. By the year 2008, more than 50% of the world population had cell phones. As the mobile network reaches the majority of the population and is able to handle multimedia content, it creates an ideal environment to carry out mobile learning.

In addition to the ability to reach out to the people, the mobile network system has other advantages in supporting teaching and learning. In summary, the following lists some of them.

- Mobile technology allows situation-based on-site teaching/learning. This means that teaching and learning can be extended to an outdoor environment. For example, the mobile network allows instructors to give a live demonstration on how to put a support beam in place on a construction site to students around the world.
- Mobile technology supports a truly ubiquitous teaching/learning environment. Teaching and learning can be carried out even when instructors and students are on the move.
- Mobile technology has greatly improved the communication among students and instructors. This makes collaboration and tutoring much more flexible.

Since mobile technology can accomplish so much in teaching and learning, it popularizes mobile learning in schools and higher education institutions. It also creates opportunities for the IT industry to open up new markets. On the other hand, mobile learning brings up some new challenges. In spite of the above advantages, there are some disadvantages. Learning becomes fragmented, which may not be good for certain course materials that require long-time deep thinking.

Table 1. Mobile Network Technologies

Mobile Technology			
Generation	**Standard**		**Description**
First Generation (1G)	AMPS, NMT, Hicap, CDPD, Mobitex, DataTAC		The first generation mobile technology is an analog mobile phone system. The analog mobile phone system can be easily interfered by noise. Its transmission rate is about 10Kbps.
Second Generation (2G)	Time Division Multiple Access (TDMA)	GSM	As a second generation mobile technology, GSM is a digital mobile phone system. The GSM transmission rate ranges from 9.6Kbps to 64Kbps. GSM is the most popular mobile phone system used by over 3 billion people.
	Code Division Multiple Access (CDMA)	IS-95	IS-95 is the first CDMA-based digital cellular standard. It is also a digital mobile phone system. The bit transmission rate for IS-95 is 1.2288 Mbps.
Third Generation (3G)	Universal Mobile Telecommunications System (UMTS) including UMTS W-CDMA, and UMTS-TDD		UMTS is a third generation mobile phone technology that combines the digital cellular system with the broadband Internet service. The bit transmission rate for the UMTS family ranges from 384Kbps to 16Mbps.
	Code Division Multiple Access (CDMA) 2000 including EV-DO, EV-DV, and 1xRTT		As a 3G mobile technology, CDMA2000 is also a system that combines the digital cellular system with the Internet service. Its bit transmission rate ranges from 0.144Mbps to 3.1Mbps.
Pre-Fourth Generation (4G)	High Speed Packet Access (HSPA) including HSDPA/HSUPA, HSPA+		The HSPA family is the upgrade of UMTS. The data download rate ranges from 14.4Mbps to 42Mbps and the upload rate ranges from 0.384Mbps to 11.5Mbps.
	Worldwide Interoperability for Microwave Access (WiMAX) IEEE 802.16e		Mobile WiMAX is used for portable devices such as Smartphones, game terminals, MP3 players, and so on. The WiMAX download bit rate can be up to 70Mbps at a short distance and 10Mbps at 10km.
Fourth Generation (4G)	Long Term Evolution (LTE)		LTE is a fourth generation mobile technology. It supports up to a 326.4 Mbps downlink data transmission rate and 172.8 Mbps uplink data transmission rate. The major 2G and 3G standards such as GMS, HSPA, IS-95, and CDMA2000 will upgrade to LTE.
	WiMAX II IEEE 802.16m		WiMAX II is designed to meet the IMT-Advanced standard so that it can be qualified as a 4G technology. It supports the data transmission rate up to 100 Mbps for high mobility and up to 1 Gbps for low mobility.

Students and instructors need to get used to the new teaching and learning environment, which may be a lot different from the environment they have been used to. Educators need to come up with new pedagogy to take advantage of mobile technology. Students need more discipline in their learning behavior and need to take care of their schedule in a more organized way. IT service staff members need to learn the new technology so that they can better manage the mobile network and support mobile learning. Mobile device companies and mobile service providers need to make more effort to meet the ever changing requirements from teaching and learning. There will be some additional cost to implement mobile learning. Administrators will face increasing demands for funding, technical support, new curricula, and skilled personnel.

Mobile Devices

Mobile devices are hand-held devices that can fit in a pocket, such as cell phones, GPS devices, multimedia players, and netbook computers. These devices usually have built-in wireless network components to allow users to communicate through a wireless phone or Wi-Fi network. As mobile technology improves, these mobile devices become more and more sophisticated. Today's mobile devices can be used to accomplish multiple tasks. For example, a cell phone can be used to make phone calls and to browse the Internet.

The following are some of the commonly used mobile devices.

- *Converged Mobile Device (CMD)*: A CMD combines a cellular phone with a Wi-Fi accessing device. Usually, a CMD also includes a digital camera, an MP3 player, a GPS, and a slot for adding a memory card. The commonly used CMD, Smartphone, such as the well-known Droid phone is powered by Android Linux created by Google and other Linux communities.
- *Netbook Computer*: A netbook computer is designed for Web surfing and e-mail checking. It is a small ultra light computer that usually weighs less than three pounds and has a LCD display size from 7 to 10 inches. Linux is one of the commonly used operating system on netbooks. Instead of using hard disks, some of the netbooks use solid state disks (SSDs) to store data. SSDs run faster than the commonly used hard disks. Some of the netbooks also include a mobile network adaptor or Wi-Fi wireless network adaptor.
- *Personal Digital Assistant (PDA)*: A PDA is a hand-held computer. The size of a PDA is slightly bigger than that of a Smartphone. Therefore, a PDA is also called a palmtop computer. It often integrates several functions such as computing, making phone calls, Web browsing, e-mail checking, networking, and document organizing. A PDA can also host various computer application software packages such as word processing and spreadsheet. In addition to hosting computer software, a PDA can also be used as a Smartphone, GPS device, digital camera, and media player.
- *Mobile Internet Device (MID)*: An MID is a multimedia-capable hand-held device designed for wireless Internet access. The size of a MID is between that of a PDA and a netbook. An MID is often built with various tools for handling high definition video and audio. With rich multimedia functionalities, it is also a good device for entertainment. In addition, it can be used as a communication device. It allows users to access mobile networks or Wi-Fi networks. Linux is the main operating system run on MIDs.
- *Portable Media Player (PMP)*: A PMP can be used to store and play digital media including the audio and video content. A PMP may include a user interface displayed on an LCD screen, a hard drive or flash memory for storing multimedia data, an FM radio, and a recording device.
- *Portable Navigation Device (PND)*: A PND may include a GPS which has a built-in map, a voice guidance system, a voice recording device, text-to-speech navigation, and so on. It is common that a PND also includes a graphic viewer, a digital camera, digital music, games, or a satellite radio.

As a computing device, a mobile device often includes a microprocessor and RAM to carry out computation. Intel Corporation is one of the semiconductor companies that produce processors for mobile devices. The Atom processor is specially designed for mobile devices. It is compact and power-efficient, and it is suitable for ultra-mobile PCs, MIDs, and Smartphones. There are other types of processors based on the Advanced RISC Machine (ARM) structure. Due to its very low power consumption nature, ARM processors are widely used by mobile devices such as PDAs, PNDs, cell phones, and game systems.

Mobile Services

To support mobile learning, a mobile network system provides various mobile services. Through the mobile network, these mobile services are accessible for mobile devices. The commonly supported mobile services for mobile learning are the podcasting service, Short Message Service (SMS), Common Short Code (CSC), Really Simple Syndication (RSS), mobile photo/video sharing, cell phone Skype calling, and so on.

Podcasting is one of the ways to access media files through the Internet. The term podcasting can be interpreted as personal on demand broadcasting. The service is designed for the Portable Media Player (PMP) type of mobile device platforms such as iTunes, Zune, Juice or Winamp. The podcasting service can be set up on a personal computer connected to the Internet. The podcasting service provides media files for subscribers to download. The software on a subscriber's PMP device can identify and download new media files automatically. The downloaded media files will be stored on a local hard drive for offline use. With the podcasting service, PMP mobile devices can be used to view class materials when they are available.

Short Message Service (SMS) is a service that allows mobile phones to exchange short text messages. Most of the digital mobile phones support SMS. Sending short text messages costs much less than making phone calls. In mobile learning, SMS can be used for event notification, short instructions, and exchanging messages among students and instructors. SMS is also a convenient service for group collaboration.

Another convenient mobile service is Common Short Code (CSC). The common short code is usually a five-digit number that represents a piece of text message such as a full length telephone number. CSC is particularly useful in online voting or survey. Each candidate or survey question is assigned a CSC number. The participant involved in the voting or survey process can quickly enter the code and send his/her choice. CSC is often required for the interactive SMS.

Really Simple Syndication (RSS) is a type of Web feed mechanism that is used to send frequently updated content to subscribers. A student can sign up for an RSS reader account and subscribe his/her favorite RSS service. Then, the RSS reader account will be automatically updated with new content from the RSS service.

A Skype user can use a Smartphone to make free calls. The free calls can utilize the Wi-Fi network or the cell phone carrier's cellular-data connection. First, the user's call is forwarded to the Skype server. Then, the message is delivered to another Skype user. This feature can significantly reduce the cost of mobile learning.

Some of the cell phone service providers provide photo or media file sharing services for customers. There are also some Web-based photo sharing services for mobile devices. These services offer customers free download client software. Once the client software is installed on cell phones, the cell phones can be used as the client for such services. These services also provide storage to host customers' media files. They make the media files viewable or downloadable by others if permitted.

Mobile services are offered by cell phone service providers, Internet companies, software producers, and consulting companies. For example, Google, Yahoo, and Microsoft all offer some free mobile services. The cell phone service providers such as Verizon, T-Mobile and AT&T also offer various mobile services to their customers. IT consulting companies such as Mobile Education and SunGard Higher Education develop and manage mobile services specially designed for universities and high schools.

Many of the mobile services depend on the Web, which requires mobile devices to be capable of dealing with IP data. Unlike the Web which is constructed with the open standard and is versatile, a cellular system has some constraints. Compatibility is the main issue among different cell phone service providers. Also, the relatively slow data transmission rate, high cost, small screen size, short battery life, and limited processing power for data management make the cell phone less appealing to mobile applications.

Utilizing open source software is one way to overcome these shortcomings. For example, the open source communication protocol, WAP (Wireless Application Protocol), is shared by many major cellular service carriers for Internet access related tasks such as e-mail, instant messaging, newsgroups, and Web browsing. Due to the superior adaptability, the open source approach is often used to develop universal mobile applications and services. For example, mobile applications developed with Java can run on various mobile devices universally.

Another way to improve the usability of mobile applications is to provide mobile services by using the cloud computing technology. Through cloud computing, data storage and data processing are kept on the Web. In such a way, a mobile device is an interface to the Web. The mobile device does not need to store and process the data. Better than the traditional thin client, cloud computing allows mobile devices to download and upload files, and do operation control tasks.

The open source approach and cloud computing can be very helpful in resolving the compatibility and data processing capability issues for mobile services and applications. Narian (2009) reported that, by the end of the year 2014, the revenue of mobile cloud computing will top 20 billion dollars. Well-known IT companies such as Apple, Google, and Microsoft all support their versions of mobile cloud service.

INTRODUCTION TO MOBILE LEARNING

Mobile learning can benefit from many features of mobile technology. Attewell (2005) summarized the benefits. Mobile learning encourages learners to be engaged in the learning process and therefore to raise the learners' self-esteem and self-confidence. Another benefit of mobile learning is that it is good for both individual and collaborative learning. Collaboration can be done through phone calls, text messaging, and exchanging photos and videos. Since the mobile phone is a device used by almost every student, there is very little resistance from students and instructors to adopt mobile learning. Schools in rural areas can particularly benefit from mobile learning.

By taking advantage of mobile technology, mobile learning has been adopted by a large number of schools and universities throughout the world. The research done by Ambient Insight (2008) indicated that, during the period from 2008 to 2013, the compound annual growth rate for mobile learning is 21.7%. By the year 2013, the mobile learning system can reach over 175 million learners.

As one of the rapid growing fields in education, mobile learning has been widely experimented and implemented. Quickly, mobile learning has become the research topic by many scholars. Researchers have illustrated how mobile technology changes the pedagogy and instructional design. They have also studied various approaches to improving the accessibility, security, and effectiveness in teaching and learning with mobile technology (Cheung & Hew, 2009).

Naismith, Lonsdale, Vavoula, and Sharples (2004) summarized the development in mobile learning. They provided information about mobile technology and mobile teaching/learning. They also discussed mobile-based collaboration and interactivity. They gave a review on mobile learning from the angle that reflected the activities involved in the mobile learning. They reviewed the activities in six categories: Behaviourist, constructivist, situated, collaborative, informal and lifelong, and learning and teaching support. Naismith, Lonsdale, Vavoula, and Sharples (2004) concluded that mobile technology supports a rich, collaborative and conversational learning environment.

Kukulska-Hulme & Traxler's (2005) book introduces mobile learning to educators. Through case studies, the book illustrates how mobile devices, such as hand-held computers, Smartphones and PDAs, are utilized to support teaching and learning. It also demonstrates how mobile technology impacts pedagogy and learners' behavior. According to the book, accessibility in mobile learning is superior to any other learning platforms. Learning materials, learning resources, and testing materials can be accessed by the learner from anywhere and at anytime; even the learning is on the move. The instruction can be carried out inside or outside the classroom. Students from anywhere can get the just-in-time course content. The book shows how communication and collaboration can be improved in the mobile learning environment. It also discusses how mobile learning changes students' learning behavior. It relates the issues related to the implementation of mobile learning such as lowering the cost, training users, and blending mobile learning with other learning platforms. The book examines how the application of mobile learning can be carried out by hand-held devices. The authors point out that more effort is needed to make mobile learning a life-long process.

Many researchers have studied the curriculum development for mobile learning. By following the theory and practice of the instructional design theory, various course materials have been created for classes taught in the mobile learning environment. Savill-Smith and Kent (2003) investigated PDA-based mobile learning. Then, Mitchell and Savill-Smith (2004) studied the usage of computer gaming in mobile learning. At Virginia Tech University, Dolittle, Lusk, Byrd, and Mariano (2009) studied the impact of iPods on mobile learning. Song (2009) did a research review about hand-held educational application. In the research conducted by Liu, Tao, and Nee (2008), mobile phones were used to engage the students in collaborative activities to improve their collaboration and performance.

Issues related to mobile learning implementation are also reported by many researchers. These researchers discus the strategies on making decisions about mobile network infrastructure, mobile devices, and services. Often, cost models are created to assist decision making. Mobile learning deployment and management are also the topics often discussed by the researchers. Training and professional development are usually included in the implementation plan. More specifically, Attewell and Savill-Smith (2004) conducted research on implementing mobile learning by using mobile phone networks. The online book edited by Herrington, Herrington, Mantei, Olney, and Ferry (2009) discusses new ways of teaching and learning with mobile technology and addresses education-related issues in mobile learning.

One of the main tasks during the implementation phase is to make sure that the mobile learning system is available to students and instructors whenever and wherever. This brings up mobile learning system support and management issues. Berger, Mohr, Nösekabel, and Schäfer (2003) addressed issues related to the efficient and effective ways to distribute learning materials to students.

Mobile devices are often used to play games. These devices provide a convenient and flexible environment for game based learning. Kuts, Sedano, Botha, and Sutinen (2007) discussed multiplayer games that were used as collaboration tools in learning. They reviewed the collaboration features of multiplayer games and provided recommendations on the use of these games in collaboration.

As a learner moves around, the learning environment and the learner's behavior change accordingly. Context-aware mobile learning has become the one of the widely studied subjects in recent years. For example, Riekki, Davidyuk, Forstadius, Sun, and Sauvola (2005) proposed the building context-aware architecture for mobile applications. Middleware plays an important role in handling context-aware computing. The author included the references of context-aware middleware studies done by other researchers.

INTRODUCTION TO OPEN SOURCE APPROACH

In the early days, most computation tasks are carried out by mainframe computers. There was little commercially available software. Professors and students wrote their own programs to accomplish the computing tasks. Often, the source code of these programs is shared by their colleagues and students. This may be considered the early stage of open source software used in teaching and research. Bull and Garofalo (2002) described the evolution from the traditional mainframe-based technology to today's Web-based technology. Since then, open source products have played a major role in every stage of the development in computer science.

- The first mile stone is the creation of the Internet in the early 1960s. The first Internet, Advanced Research Projects Agency Network (ARPANET), was the result of the collective effort of an open standard approach.
- The next significant development is the UNIX operating system. In 1971, the first UNIX operating system was released. Even though some versions of UNIX are now made proprietary by companies such as AT&T, Sun Microsystems, HP, IBM, and Santa Cruz Operation (SCO), there are still some versions of UNIX that are open source based. The well-known open source UNIX operating systems are FreeBSD which runs on Intel x86 PCs, NetBSD which runs as a starting point for portability, and OpenBSD which is a highly secured operating system. These open source UNIX operating systems are produced by the developers at University of California, Berkeley.
- The World Wide Web is another significant achievement under the open source approach. In 1991, the world's first Web site was presented online by the group led by Tim Berners-Lee. Unlike other commercial products, the World Wide Web has no copyright imposed on it. Today, the entire world benefits from the World Wide Web.
- Linux is another mile stone for the open source approach. In 1991, Linus Torvalds, a computer science student from Helsinki, Finland, introduced the GNU/Linux operating system which was designed to run on the Intel 80386 microprocessor. Now, GNU/Linux has been used by a wide range of computing devices, from embedded mobile devices to supercomputers. GUN/Linux has been widely supported by well-known IT companies such as Google, Ubuntu, Intel, Dell, Hewlett-Packard, IBM, Novell, and Sun Microsystems. Microsoft recently contributed device driver code to the Linux community using GPLv2 open source licensing (Marshall, 2009). With this driver code, while running virtual machines, GUU/Linux is able to handle the hardware that is only compatible with Microsoft products.
- Java is a programming language that can write once and run anywhere, which is a desired feature for today's Internet computing environment. In 1995, Java 1.0 was available to the general public. At first, Java is a free but a closed source product (Feizabadi, 2007). In 2007, Java was licensed under the GNU General Public License and became truly open source product. Nowadays, Java is one of the major programming languages for developing mobile applications and other Web-based applications.
- MySQL is an open source database management system. It is used to store and manage data for Internet applications. MySQL, developed by a Swedish consulting company called TcX, allows multiple processes work simultaneously without

costing much computing power. Therefore, it is suitable for working with Web-based service such as a learning management system (LMS). In addition to storing data, MySQL can also be used for user management and data management such as database backup and recovery. It also includes GUI tools to present data and perform data analysis tasks. MySQL provides a wide range of programming languages for developing database applications (Reese, Yarger, & King, 2002).

- For online teaching and learning, it is necessary to have a learning management system which can be used to manage student accounts, course materials, class schedules, and group activities. Moodle is one of the open source learning management systems. Created by Martin Dougiamas in 1999, Moodle is currently used in 199 countries to manage 2,251,536 online courses and 24,014,033 students (Moodle, 2009).

Open source products have become more and more popular in today's business operations. According to Asay (2009a), the research done by IDC showed that 24% of the software used in business was open source software in 2008, which went up from the 14% during the year 2007. The figures indicate that companies are confident to run their business with open source products. In another related article, Asay (2009b) stated that IDC predicted that the spending on Linux related products would increase 21% during the year 2009 and would grow with a 23.6 percent compound annual growth rate until 2013. Compared with the rest of the software industry which on average has about a 5% growth rate, the open source products' growth rate is far more significant. In addition to the fast growing open source software market, open source based service is another fast growing area. For example, as an enterprise service provider, Red Hat Linux has a growth rate two to three times faster than the broader software industry.

Introduction to Mobile Linux

Due to its stability, flexibility, security, availability, and low cost, Mobile Linux has been gaining favor from famous worldwide companies. For example, during the past few years, Google initiated the well-known Android project, which is a combination of a Mobile Linux operating system and an application development platform. Similarly, Intel introduced the Moblin project and Ubuntu introduced the Ubuntu Mobile project to the mobile computing community. As more mobile device manufacturers have adopted Mobile Linux operating systems and application software, some books began to cover Mobile Linux during the year 2009. The new development of Mobile Linux and other open source products will have a significant impact on the mobile learning community. Among the newly published books, one can find the book by Heuser (2009) that collects a set of documents on Mobile Linux installation and configuration of various mobile devices. For application development, readers can find the book by Lawrence and Belem (2009) quite informative for developing mobile applications for Ubuntu Linux. For readers who are interested in Android, they can take a look at the book by Ableson, Collins, and Sen (2009). This book provides instruction on Android and gives examples for developing multimedia applications for mobile devices. For more information about Moblin, one can visit its Web site (Moblin, 2009).

Usage of Open Source Products in Mobile Learning

Educational organizations have been one of the main sources that contribute to the development of open source software. They are also one of the largest users of open source software which is used in various academic fields such as biology, math, history, and so on. Open source software is also widely used in school administration and online teaching and learning. Chao's (2009) book discusses the application of open source products for online teaching and learning. Many of the high quality open source course materials are created and shared by universities and K-12 schools. The OpenCourseWare Consortium (2009) is such an organization that empowers people with open source course materials throughout the world. Over 200 higher education institutions are members of OpenCourseWare Consortium, including well-known universities such as MIT and UC Berkeley. The Open Source Education Foundation (2009) is an organization that promotes the use of open source software at K-12 schools. To promote the use of open source software in the federal government, Open Source for America (2009) was formed in 2009. The goal of the organization is to help the government to be aware of the advantages of open source software and to better use the open source software. Over 70 well-known companies have joined the organization. Among them, there are AMD, Canonical, Google, Novell, Oracle and Red Hat. Currently, the open source vendor Sunlight Labs is working with the federal government on a project that manages the data generated by government agencies. Also, the federal government has successfully applied some open source software such as SELinux for security management.

The Smartphone is a main mobile device used in mobile learning. On the Smartphone market, there are several well-known open source products such as Android, Palm Pre, and Moblin. Wallen (2009) lists ten reasons that open source makes sense for Smartphones.

- With open standards, it is easier to develop new software and load Web sites to meet specific needs. Also, hardware accessories will be more available.
- An open source Smartphone will accept more application software than that accepted by the proprietary counterpart.
- Since open source Smartphone operating systems are created on top of the Linux kernel, open source Smartphones in general have better security.
- Open source Smartphones give users more choice to customize their phones.
- It is relatively easier for an open source Smartphone to synch with other platforms such as OS X, Windows, and Linux.
- The total cost of an open source Smartphone is relatively lower than the cost of a proprietary Smartphone. The lower cost is due to no charge for the open source operating system.
- A Linux based open source Smartphone allows multitasking while some of the proprietary Smartphones do not have the multitasking capability.
- The open source Smartphone allows Internet mail such as Google mail to be pushed to the user's phone so that the user does not have to wait for e-mail downloads.
- The open source Smartphone is supported by the Linux community. A large number of developers can contribute to the system development and troubleshooting.
- By gathering the contribution from a large number of developers, the open source community is more creative on developing mobile services and network security tools.

According to Ambient Insight (2008), during the period between 2008 and 2013, the demand for mobile learning subject areas such as custom content, software-sold-as-a-service (SaaS), content hosting, development tools, and device-embedded learning will be significantly increased. Linux and other open source products can contribute to these areas in a way that no proprietary products can match. The following are some of the contributions from open source products.

- It is easy to customize open source products. In history, Linux and other open source products are the choice by customer-oriented service companies to serve highly dynamic customers such as petroleum exploring companies. Mobile learning is much more dynamic than the other learning platforms. The open source products can well meet the customized content requirement.
- As one of the business models, service companies often provide software to their customers as part of the service. The software provided to customers is often developed from open source software. For example, Red Hat Linux has two versions of Linux operating systems, Fedora and Enterprise Linux. Fedora is an open source based operating system which gets improved frequently by the open source community. On the other hand, Enterprise Linux based on Fedora is sold as part of the service to enterprise customers. In such a case, open source software allows service companies to provide the update-to-date software while keeping cost to the minimum.
- One of the features of mobile learning is that it can make interaction easier and it allows the content to be distributed dynamically. Content hosting is one of the key functions performed by a content management system (CMS) which is necessary for managing social network sites and personal blogs. An open source content management system is not only free but also simplifies the system upgrade.
- Traditionally, the open source Java Development Kit (JDK) has been widely used for developing Web applications. The later version of JDK has been significantly improved to support the development of wireless and mobile applications. In addition to Java, there are several open source application development platforms that are specifically designed for mobile application development. Among them, Google's mobile development platform is a well-known development platform used for developing applications for the Linux based operation system Android. Other mobile Linux distributions such as Moblin and Ubuntu Mobile also provide application development kits.
- Some of the mobile learning devices have embedded application software. The electronic translator is this type of device. The embedded Linux is important for many consumer electronics. Linux is well-known for being able to be embedded into small devices.
- The commonly used open source mobile application software includes browsers, file managers, e-mail, voice recorders, database clients, Web development tools, application development platforms, programming languages, and so on. These mobile applications can run on all major mobile phone platforms, such as Java ME, Google Android, Symbian, Windows Mobile, and iPhone.
- To develop and manage mobile learning materials, mobile learning system software is often used to accomplish these tasks. Georgieva (2006) compared 12 mobile learning systems. There are also some open source mobile learning systems. The open source learning system PIVOTE created by Daden Limited (2009) allows educators to create learning materials for iPhones and other sophisticated mobile devices.

The above reviews some usage of open source products. As indicated, open source products can make significant contributions to mobile learning.

In fact, an entire mobile learning infrastructure can be built with open source products. In later chapters, the authors will provide more information on the usage of open source products and how these products are applied in mobile learning.

BOOK FRAMEWORK

This book is about using open source products to support mobile learning. The authors of the book discuss research and case studies to demonstrate how mobile learning is implemented with open source products and how open source based mobile learning can improve teaching and learning. In the following, we take a look at how the book is organized. The chapters in this book are categorized into six main sections as below:

- *Introduction*: This section has one chapter that introduces the three major topics of this book, mobile technology, mobile learning, and the open source approach.
- *Open Source Mobile Learning*: There are eight chapters included in this section which covers the strategies, analyses, and evaluations of mobile learning that is supported by open source products.
- *Mobile Learning Technology*: There are seven chapters in this section which covers open source based mobile technology for the implementation of the mobile learning infrastructure.
- *Mobile Learning with Web 2.0*: This section includes four chapters. Each chapter illustrates how the open source Web 2.0 technology is used to support mobile learning.
- *Mobile Learning with Games*: The two chapters included in this section illustrate how open source games are used in mobile learning.
- *Development of Applications and Services with Open Source Products*: Three chapters are included in this section. These chapters illustrate how to use Android and OpenLaszlo to develop applications and services for mobile learning.

This book first reviews research studies on open source mobile learning. To better understand how open source products are used to support mobile learning, the book describes several types of open source mobile technology and their implementation on the mobile learning infrastructure. The next few main sections of this book focus on applying open source mobile technology to course material development, and class control and management.

CONCLUSION

With research papers and case studies, the authors in this book have made significant contributions to enrich the mobile learning theory and practice. To better understand their fields of study, this chapter introduces the background knowledge in three related areas, mobile technology, mobile learning, and the open source approach. By reviewing mobile technology, we can see that it is sophisticated enough to carry out mobile learning tasks. The recent development of mobile learning indicates that mobile learning has become common practice among universities and schools. Researchers have presented research analyses and evaluations. Their findings of mobile learning are positive, and there are lessons learned from the mobile learning practice. Another main topic introduced in this chapter is the open source approach. The history of the open source approach indicates that this approach has strong vitality. This chapter introduces some open source based products and services that have been widely used in a variety of mobile learning practice.

REFERENCES

Ableson, F., Collins, C., & Sen, R. (2009). *Unlocking Android: A developer's guide*. Greenwich, CT: Manning Publications.

Ambient Insight. (2008). *The US market for mobile learning products and services: 2008-2013 forecast and analysis*. Retrieved July 16, 2009, from http://www.ambientinsight.com/Resources/Documents/AmbientInsight_2008-2013_US_MobileLearning_Forecast_ExecutiveOverview.pdf

Asay, M. (2009a). *Up to 24 percent of software purchases now open source*. Retrieved July 16, 2009, from http://news.cnet.com/8301-13505_3-10238426-16.html

Asay, M. (2009b). *IDC: Linux spending set to boom by 21 percent in 2009*. Retrieved July 16, 2009, from http://news.cnet.com/8301-13505_3-10216873-16.html

Attewell, J. (2005). From research and development to mobile learning: Tools for education and training providers and their learners. In *Proceedings of mLearn 2005*. Retrieved December 20, 2005, from http://www.mlearn.org.za/CD/papers/Attewell.pdf

Attewell, J., & Savill-Smith, C. (2004). Mobile learning and social inclusion: Focusing on learners and learning . In Attewell, J., & Savill-Smith, C. (Eds.), *Learning with mobile devices: Research and development* (pp. 3–12). London, UK: Learning and Skills Development Agency.

Berger, S., Mohr, R., Nösekabel, H., & Schäfer, K. (2003). Mobile collaboration tool for university education. In *Proceedings of the Twelfth International Workshop on Enabling Technologies: Infrastructure for Collaborative Enterprises* (p.77).

Bull, G., & Garofalo, J. (2002). Proceedings of the Fourth National Technology Leadership Summit: Open resources in education. *CITE Journal, 2*(4), 530-555.

Chao, L. (2009). *Utilizing open source tools for online teaching and learning: Applying Linux technologies*. Hershey, PA: IGI Global Publishing.

Cheung, W. S., & Hew, K. F. (2009). A review of research methodologies used in studies on mobile handheld devices in K-12 and higher education settings. *Australasian Journal of Educational Technology, 25*(2), 153–183.

Daden Limited. (2009). *PIVOTE launches an open-source learning system for virtual worlds, the Web and iPhone*. Retrieved February 12, 2010, from http://www.prlog.org/10197856-pivote-launches-an-opensource-learning-system-for-virtual-worlds-the-web-and-iphone.html

Dolittle, P. E., Lusk, D. L., Byrd, C. N., & Mariano, G. J. (2009). iPods as mobile multimedia learning environments: Individual differences and instructional design. In H. Ryu, & D. Parsons (Eds.), *Innovative mobile learning: Techniques and technologies* (pp. 83-101). New York, NY: Information Science Reference.

Feizabadi, S. (2007). *History of Java*. Retrieved August 6, 2009, from http://ei.cs.vt.edu/book/chap1/java_hist.html

Georgieva, E. (2006). *A comparison analysis of mobile learning systems*. International Conference on Computer Systems and Technologies - CompSysTech' 2006. Retrieved August 6, 2009, from http://ecet.ecs.ru.acad.bg/cst06/Docs/cp/sIV/IV.17.pdf

Glisic, S., & Lorenzo, B. (2009). *Advanced wireless networks: Cognitive, cooperative & opportunistic 4G technology* (2nd ed.). Indianapolis, IN: Wiley. doi:10.1002/9780470745724

Harte, L., & Bowler, D. (2003). *Introduction to mobile telephone systems, 1G, 2G, 2.5G, and 3G technologies and services*. Fuquay Varina, NC: Althos.

Herrington, J., Herrington, A., Mantei, J., Olney, I., & Ferry, B. (Eds.). (2009). *New technologies, new pedagogies: Mobile learning in higher education*. Wollongong, Australia: University of Wollongong.

Heuser, W. (2009). *LINUX on the road - the first book on mobile Linux*. Scotts Valley, CA: CreateSpace.

Kukulska-Hulme, A., & Traxler, J. (Eds.). (2005). *Mobile learning: A handbook for educators and trainers*. London, UK: Routledge.

Kuts, E., Sedano, C. I., Botha, A., & Sutinen, E. (2007). *Communication and collaboration in educational multiplayer mobile games*. In IADIS International Conference on Cognition and Exploratory Learning in Digital Age (CELDA 2007) (pp. 295-298).

Lawrence, I., & Belem, R. C. L. (2009). *Professional Ubuntu Mobile development*. Indianapolis, IN: Wiley Publishing.

Liu, C., Tao, S., & Nee, J. (2008). Bridging the gap between students and computers: Supporting activity awareness for network collaborative learning with GSM network. *Behaviour & Information Technology*, *27*(2), 127–137. doi:10.1080/01449290601054772

Marshall, D. (2009). *Microsoft releases Hyper-V Linux drivers as open source*. Retrieved February 9, 2009, from http://www.infoworld.com/d/virtualization/microsoft-releases-hyper-v-linux-drivers-open-source-205

Medford, C. (2008). Cell phone market soars despite recession. *Red Herring*. Retrieved July 16, 2009, from http://www.redherring.com/Home/24181

Mitchell, A., & Savill-Smith, C. (2004). *The use of computer and video games for learning: A review of the literature*. London, UK: Learning and Skills Development Agency. Retrieved July 16, 2009, from http://www.m-learning.org/archive/docs/The%20use%20of%20computer%20and%20video%20games%20for%20learning.pdf

Moblin. (2009). *Moblin*. Retrieved July 16, 2009, from http://www.moblin.com/moblinen/AboutUs.aspx

Moodle. (2009). *Moodle statistics*. Retrieved August 6, 2009, from http://moodle.org/stats

Naismith, L., Lonsdale, P., Vavoula, G., & Sharples, M. (2004). *Mobile technologies and learning*. Retrieved August 6, 2009, from http://www.futurelab.org.uk/resources/publications-reports-articles/literature-reviews/Literature-Review203

Narian, D. (2009). *ABI research: Mobile cloud computing the next big thing*. Retrieved August 6, 2009, from http://ipcommunications.tmcnet.com/topics/ip-communications/articles/59519-abi-research-mobile-cloud-computing-next-big-thing.htm

Open Source Education Foundation. (2009). *Welcome to the Open Source Education Foundation website*. Retrieved August 6, 2009, from http://www.osef.org

Open Source for America. (2009). *Open Source for America*. Retrieved August 6, 2009 from http://www.opensourceforamerica.org

OpenCourseWare Consortium. (2009). *OpenCourseWare Consortium*. Retrieved August 6, 2009, from http://www.ocwconsortium.org/home.html

Reese, G., Yarger, R. J., & King, T. (2002). *Managing and using MySQL*, (2nd ed.). Sebastopol, CA: O'Reilly Media, Inc.

Riekki, J., Davidyuk, O., Forstadius, J., Sun, J., & Sauvola, J. (2005). Context-aware services for mobile users. In *Proceedings of IADIS Distributed and Parallel Systems and Architectures Conference as Part of IADIS Virtual Multi Conference on Computer Science and Information Systems (MCCSIS 2005)*.

Savill-Smith, C., & Kent, P. (2003). *The use of palmtop computers for learning: A review of the literature*. London, UK: Learning and Skills Development Agency. Retrieved July 16, 2009, from http://www.m-learning.org/docs/the_use_of_palmtop_computers_for_learning_sept03.pdf

Song, Y. (2009). Handheld educational application: A review of the research . In Ryu, H., & Parsons, D. (Eds.), *Innovative mobile learning: Techniques and technologies* (pp. 302–323). New York, NY: Information Science Reference.

Stetz, P. (1999). *The cell phone handbook: Everything you wanted to know about wireless telephony*. Middletown, RI: Aegis Publishing Group.

TAG. (2008). *An introduction to mobile technology: Information and marketing in 2008*. Retrieved July 16, 2009, from http://www.tagonline.org/articles.php?id=269

Wallen, J. (2009). *10 reasons why open source makes sense on smart phones*. Retrieved August 6, 2009, from http://blogs.techrepublic.com.com/10things/?p=808

KEY TERMS AND DEFINITIONS

CDMA2000: It is a third generation digital mobile phone system implemented over the existing CDMA networks.

Code Division Multiple Access (CDMA): It is a second generation digital mobile telephony system using the code division multiple access (CDMA) multiplexing scheme.

Global System for Mobile Communications (GSM): It is a second generation digital mobile telephony system using the time division multiple access (TDMA) multiplexing scheme.

High Speed Packet Access (HSPA): It is a family of high speed 3.5 generation digital mobile phone systems created on top of the GSM system.

Long Term Evolution (LTE): It is a fourth generation digital mobile phone system which is an upgrade of the UMTS, CDMA2000, and HSPA systems.

Universal Mobile Telephone Service (UMTS): It is a third generation digital mobile phone system implemented over the existing GSM networks.

Worldwide Interoperability for Microwave Access (WiMAX): It is a telecommunications technology for the broadband Internet connections and cellular phones.

Chapter 2
Adaptation Technologies in Mobile Learning

Paola Salomoni
University of Bologna, Italy

Silvia Mirri
University of Bologna, Italy

ABSTRACT

The explosive growth of Internet services and the widespread diffusion of mobile devices are offering ubiquitous and time-independent access to a huge amount of online resources. In particular, educational experiences on non-conventional contexts have been made possible and the use of mobile terminals has created a new form of e-learning, called mobile learning (or "m-learning"). The availability of open technologies has participated in the diffusion of such a trend. Needless to say, standards, metadata, and adaptation mechanisms have to be applied to the m-learning environment and content, in order to meet learners' needs and preferences, as well as device characteristics.

This chapter will introduce issues and standards for adaptation mechanisms for mobile learning resources, and it will describe some notable existing projects and software tools. Finally, it will present some case studies and open trends about m-learning.

INTRODUCTION

The term m-learning is meant to refer to learning by using mobile devices, both in anywhere/anytime context and in traditional classrooms (Clothier, 2005). During the last decade, mobile learning has grown from a research topic to a plethora of concrete projects in schools, museums, cities, and workplaces. This transformation has been facilitated by the widespread diffusion of mobile devices and connectivity. Moreover, the availability of open source technologies in this field strongly supports the diffusion of mobile learning. More recently, Web 2.0 technologies and

DOI: 10.4018/978-1-60960-613-8.ch002

applications have increased users' capabilities to collaborate and to share content and knowledge through mobile devices, constructing kinds of the so-called "collective intelligence" (Levy, 1997).

The diversity of mobile device capabilities, learners' preferences (and needs), supported content formats, and contexts of use has called for standards, metadata, and adaptation mechanisms. They have to be applied to the learning environment and content, in order to provide the best educational experience to learners. In particular:

- device capabilities and learners' needs have to be profiled on the basis of standards; open standards are strategically used in such a context;
- original learning contents have to be described on the basis of metadata and standards;
- learning contents have to be transcoded through adaptation mechanisms on the basis of device capabilities, learners' preferences, contexts of use, and original learning content characteristics.

This chapter will motivate the need of standards and adaptation mechanisms in the mobile learning field and it will describe related standards, open source projects, and libraries. Finally, it will present how Learning Content Management Systems face m-learning instances and it will introduce m-learning open issues and trends.

The remainder of the chapter is organized as follows. The section named "Mobile device profiling standards" presents main profiling mobile device standards (W3C CC/PP and OMA UAProf) which describe mobile device capabilities and drive learning content transcoding operations. This section also introduces open source repositories of mobile device descriptions (such as WURFL) compliant to the W3C Device Description Repository Standard (DDR). The section titled "Mobile learning content adaptation mechanisms" describes main concepts related to content adaptation and transcoding, by introducing architectural approaches, standards, and formats which apply adaptation mechanisms, open source projects and libraries in charge of operating adaptations to the whole learning content or to single media which compose it. The section named "Mobile issues in LCMSs" introduces main strategies to face mobile context issues applied by two of the most widespread and well-known open source Learning Content Management Systems: Moodle and ATutor. Finally, the section titled "Open issues and future research directions" concludes the chapter, by describing some open issues and future trends. In particular, it presents the growing need of tools and applications in order to let learners collaborate with each other, and create and share mobile learning content. Moreover, the relationships between mobile learning and accessibility topics are introduced in this last section.

MOBILE DEVICE PROFILING STANDARDS

In order to meet the needs of learners who are equipped with mobile devices, the adaptation of didactical materials has to be driven by device characteristics. Hence, profiling such characteristics is fundamental in mobile learning. In fact, standards which describe mobile device capabilities allow the identification and the definition of the characteristics (i.e. formats, sizes, etc.) the learning contents should have after the transcoding operations. There are currently two main standards which are devoted to perform device profiling: W3C CC/PP (Composite Capabilities/Preferences Profile) and OMA UAProf (User Agent Profile). Both of them are based on Resource Description Framework (RDF). The RDF format implies the document schemas are extensible.

As the diversity of devices increases, the device capability and preference for content negotiation and adaptation must be known. The goal of CC/PP (World Wide Web Consortium, 2004) and UAProf

(Open Mobile Alliance, 2007) profiles is letting client devices tell servers their capabilities. The CC/PP and UAProf data formats are based on RDF models and describe device capabilities with two-level hierarchies, consisting of components and attributes. Whenever these profiles are parsed, RDF represents an abstraction level over XML, so it must validate both XML and RDF. CC/PP and UAProf are useful for device independence, content negotiation and adaptation, as they allow different devices to specify their capabilities in a uniform way.

W3C Composite Capabilities/ Preference Profile (CC/PP)

The Composite Capabilities/Preference Profile (CC/PP) provides a standard way for devices to transmit their profiles when requesting Web content (World Wide Web Consortium, 2004). Servers and proxies can then provide adapted content which is appropriate to a particular device. A CC/PP vocabulary is defined by using RDF and specifies components and attributes of these components used by the application to describe a certain context. The three main components specify the hardware platform, the software platform, and the browser user agent, in particular:

- *Hardware Platform*: This component defines the device (mobile device, personal computer, palmtop, tablet PC, etc.) in terms of hardware capabilities, such as displaywidth and displayheight (that specify display width and display height resolution), audio (that specifies audio board presence), imagecapable (that specifies image support), brailledisplay (that specifies Braille display presence), and keyboard (that specifies keyboard type).
- *Software Platform*: This component specifies the device software capabilities, such as name (which specifies the operating system name), version (which specifies the operating system version), tool (which specifies present assistive tools), audio (which specifies supported audio types), video (which specifies supported video types), and SMILplayer (which specifies present SMIL players).
- *Browser User Agent*: This component describes the browser user agent capabilities, such as name (which specifies the user agent name), version (which specifies the user agent version), javascriptversion (which specifies javascript versions supported), CSS (which specifies CSS versions supported), htmlsupported (which specifies HTML versions supported), mimesupported (which specifies mime types supported), and language (which specifies languages supported).

The protocol for transmitting CC/PP profiles is based on an experimental HTTP extension framework. Many existing servers do not support this protocol, so developers have to adjust it to make it compatible in some way. There are two key problems related to device independence which are beyond the CC/PP working group scope:

1. The CC/PP profile does not provide a standard vocabulary for Web clients to communicate their capabilities to servers.
2. It does not describe the type of adaptation methods that servers should perform on behalf of devices based on their capabilities.

Such problems need to be solved so that the protocol can be used in practice.

OMA User Agent Profile (UAProf)

The second profiling standard we are going to introduce is UAProf, which has been defined as living between Wireless Application Protocol (WAP) devices and servers (Open Mobile Alliance, 2007). UAProf was born to be used for better content adaptation on different types of WAP devices. UAProfile also describes the next generation of WAP phones. The advantage of UAProf is that it defines different categories of mobile device capabilities:

- *HardwarePlatform Component*: as the related CC/PP component, this category provides information about the hardware capabilities of a mobile device, such as color capability (by using ColorCapable and BitsPerPixel attributes), model name of the mobile device (by using Model and Vendor attributes), text input capability (by using TextInputCapable attribute), screen size (by using ScreenSize and ScreenSizeChar attributes) and sound capability (by using SoundOutputCapable attribute).
- *SoftwarePlatform Component*: as the related CC/PP component, this category provides information about the software characteristics of the mobile device, such as audio and video encoders supported (by using AudioInputEncoder and VideoInputEncoder attributes), character sets accepted (by using CcppAccept-Charset attribute), Java capability (by using JavaEnabled, JavaPlatform and JVMVersion attributes), acceptable content types/MIME types (by using CcppAccept attribute) and the operating system name and version (by using OSName, OSVendor and OSVersion attributes).
- *BrowserUA Component*: as the related CC/PP component, this category specifies information about the browser of the mobile device. For example, the mobile browser name and version (by using BrowserName and BrowserVersion attributes), HTML version supported (by using HtmlVersion attribute), XHTML version supported (by using XhtmlVersion and XhtmlModules attributes) and JavaScript capability (by using JavaScriptEnabled and JavaScriptVersion attributes).
- *NetworkCharacteristics Component*: this category specifies information about the capabilities of the mobile device for network connection. For example, bearers supported (CSD, GPRS, SMS, EDGE, etc., by using SupportedBearers attribute) and encryption methods supported (WTLS, SSL, TLS, etcetera, by using SecuritySupport attribute).
- *WapCharacteristics Component*: this category provides information about the WAP features supported by the mobile device. For example, DRM (Digital Rights Management) capability (by using DrmClass and DrmConstraints attributes), maximum WML deck size (by using WmlDeckSize attribute), WAP version supported (by using WapVersion attribute) and WMLScript libraries supported (by using WmlScriptVersion and WmlScriptLibraries attributes).
- *PushCharacteristics Component*: this category specifies information about the WAP Push capabilities of the mobile device. For example, character encodings supported (by using PushAcceptEncoding attribute), character sets supported (by using PushAcceptCharset attribute), content types/MIME types supported (by using PushAccept attribute) and maximum WAP Push message size (by using PushMsgSize attribute).
- *MmSCharacteristics Component*: this category provides information about the MMS (Multimedia Messaging Service) capabilities of the mobile device, for example, the maximum MMS message size supported (by using MmsMaxMessageSize attribute), maximum image resolution supported (by using MmsMaxImageResolution attribute) and character sets supported (by using MmsCcppAcceptCharSet attribute).

The weakness of this standard is that it does not resolve how servers and proxies should use the UAProf profile, as well as the CC/PP one. Moreover, the UAProf production (and the CC/PP too) for a device is voluntary: for GSM devices, the UAProf is normally produced by the vendor of the device, whereas for CDMA/BREW devices, it's more common for the UAProf to be produced by the telecommunications company. Some drawbacks are listed in the following:

1. Not all devices have UAProfs.
2. Not all advertised UAProfs are available.
3. Real-time retrieving and parsing UAProfs is slow and can add substantial overhead to any Web request.
4. There is no industry-wide data quality standard for the data within each field in a UAProf.

All these issues and the huge amount of mobile device capabilities have made necessary the use of a repository of mobile device descriptions. This provides a quick and easy way to match learners' device types and their characteristics.

Device Description Repositories

The W3C MWI (Mobile Web Initiative) and the DDWG (Device Description Working Group), recognizing the difficulty in collecting and keeping track of UAProf, CC/PP and device profile information and the practical shortcomings in the implementation of UAProf and CC/PP across the industry, have outlined specifications for a Device Description Repository (DDR). DDR is supported by an initial core vocabulary of device properties (World Wide Web Consortium, 2008a). Implementations of the proposed repository are expected to contain information about Web-enabled devices (specially, but not limited to, mobile devices). By means of DDR compliant repositories, authors of Web content would be able to make use of the repositories to adapt their content to best suit the requesting device. This would facilitate the interaction and viewing of Web content and applications across devices with widely varying capabilities. Open and commercial implementations of the DDR Simple API are available (World Wide Web Consortium, 2008b). In particular, the open source "MyMobileWeb" (MORFEO Project, 2007) was presented by Telefónica I+D, as we will describe in the following section (in the subsection "Open source projects and libraries"). This implementation includes a Web-interface to the API implementation, which was developed in Java language.

An example of a device description repository is the Wireless Universal Resource File Library (WURFL), based on UAProf and CC/PP profiles (WURFL, 2010). The WURFL is an open source project that focuses on the problem of presenting content on a wide variety of wireless devices. The WURFL is an XML configuration file which contains information about device capabilities and features for a variety of mobile devices. Device information is contributed by developers around the world and the WURFL is updated frequently, thereby reflecting new wireless devices coming on the market. Currently, WURFL stores more than 500 capabilities related to thousands of mobile device models.

It is worth mentioning that, whenever an m-learning project takes into account mobile device profiling, it should provide integration with the learner's profiling system. In fact, learners' profiles have to be compliant to IMS Standards, such as the IMS Learner Information Profile (IMS Global Learning Consortium, 2002a) and the IMS Accessibility Learner Profile (IMS Global Learning Consortium, 2002b). The integration of such profiles guarantees a complete description of the learner's context, but it could cause some overlaps that the profiling system should manage. A solution has been proposed by Ferretti et al. (2009).

MOBILE LEARNING CONTENT ADAPTATION MECHANISMS

The growing diffusion of devices, coupled with the ability to deliver Web content and specifically online learning content anywhere at any time, has improved the learning process flexibility and quality of services (Anido, 2006). As a result, new techniques for delivering didactical materials according to device features and even specific languages emerged (Pandey et al., 2004). Adapting typical Web content and services (originally made for PCs) to small and mobile devices is still one of the content adaptation hot topics (Curran & Annesley, 2005). Information presentation on mobile devices needs to address the shortcomings of wireless appliances with small display sizes, different features for data input, limited graphics, etc.

The main obstacles to Web content interoperability are: possible application bugs; some devices don't support certain functions, such as new mobile phones that only support Java and non-standard proprietary markup language extensions. The final result is that the same online learning content might have a great variety of appearances and could run in several ways (or in no way), depending on the platform and device (Salomoni et al., 2008). Hence content adaptation and transcoding are necessary and should be based on device capabilities and preferences, on the network characteristics and on the strength of some application-specific parameters; therefore, Web learning content and applications should be generated or adapted for a better learner experience (Harumoto et al., 2005). Device independence principles set aside from any specific markup language, authoring style or adaptation process.

According to the W3C definition, content adaptation is the transformation and the manipulation of Web content (such as images, audio, videos, texts and multimedia presentations) to meet desired targets (defined by the terminal capabilities and the application needs) (Colajanni & Lancellotti, 2004; Harumoto et al., 2005). They include:

- media conversion (the conversion of content from its original form to another; it can be performed automatically, depending on the type of conversion: e.g., Text to Speech (TTS) or animation to image);
- format transcoding (e.g. XML to HTML, SVG to GIF, WAV to MP3);
- scaling (of images as well as video and audio streams; it can involve recoding and/or compressing specific media content and it has effects in terms of reduction of size, quality and data rate of contents);
- translation (from the original language to a different one, based on the user profile. This operation is only performed for textual and audio speech contents);
- re-sampling;
- file size compression;
- textual content fragmentation.

In particular, transcoding is the process of converting a media file or object from one format to another. This process is typically used to convert video, audio and image formats, but it is also used to adapt multimedia presentations and Web content to the constraints of non-standard devices, e.g. mobile devices (Laakko & Hiltunen, 2005). It is well-known that mobile devices have limited capabilities, such as smaller screen sizes, lower memory and slower bandwidth rates. Most existing multimedia presentations and Web content are created to be displayed on desktop computers and, usually, Web designers provide complex, detail-rich didactical content, with multimedia experiences. Thus, in mobile learning environments, transcoding must face the diversity of mobile devices. This heterogeneity imposes an intermediate state for content adaptation to ensure a proper presentation on each target device (Pandey et al., 2004; Curran & Annesley, 2005).

In the following of this section we are going to introduce architectural approaches, standards, open source projects and libraries, which can be useful in mobile learning content adaptation and transcoding.

Architectural Approaches

Due to different device capabilities, content adaptation and transcoding need to be implemented before the content is presented to the user. From an architectural point of view, four categories should be mentioned that represent the most significant distributed solutions for content adaptation (Colajanni & Lancellotti, 2004):

- *Client-side approaches*: The transcoding process is in charge of the client application. Client-side solutions can be classified into two main categories with different behaviours: (i) the clients receive multiple formats and adapt them by selecting the most appropriate one to play-out, or (ii) the clients compute an optimized version from a standard one. This approach suggests a distributed solution for managing heterogeneity, supposing that all the clients can locally decide and employ the most appropriate adaptation to them.
- *Server-side approaches*: The server (which provides contents) performs the additional functions of content adaptation. In such an approach, content adaptation can be carried out in an off-line (content transcoding is performed whenever the resource is created (or uploaded on the server) and a human designer is usually involved to hand-tailor the contents to different specific profiles. Multiple formats of the same resources are thus stored on the server and they are dynamically selected to match client specifications) or on-the-fly fashion (adapted contents are dynamically produced before delivering them to the clients).
- *Proxy-based approaches*: In proxy-based approaches, the adaptation process is carried out by a node (i.e. the proxy) placed between the server and the client. In essence, the proxy captures replies by the server to the client requests and performs three main actions: (i) it decides whether performance enhancements are needed, (ii) it performs content adaptations, and (iii) it sends the adapted contents to the client. To accomplish this task as a whole, the proxy must know the target device, the user capabilities (this information must be received from the client) and a "full" version of the original contents (this data must be received from the server). As a consequence, the use of network bandwidth could be intensive in the network link between the proxy and the server.
- *Service-oriented approaches*: The dynamic nature of adaptation mechanisms together with emerging opportunities offered by the new Web Service technologies now provides a new approach of service-oriented content adaptation. The philosophy at the basis of these approaches is fundamentally different from those previously discussed, since the transcoding and the adaptation activities are organized according to a service-oriented architecture. Indeed, the number of content adaptation typologies, as well as the set of multiple formats and related conversion schemes is still increasing. This dynamism is one of the reasons that make it difficult to develop a single adaptation system that can accommodate all types of adaptations; therefore, third-party adaptation services are important.

Standards

The diversity of the content presentation environments imposes strict requirements on multimedia applications and systems (Jannach et al., 2006). The emerging growth of mobile services (together with wireless technology such as Bluetooth, 802.11, GPRS and UMTS) defines more requirements for the content and service providers (Pandey et al., 2004). Content, terminal capabilities and underlying networks demand separate service creation processes, and mobile services require support for new billing and profiling mechanisms, based on the user and the service at hand. In particular, as these mobile devices are becoming more multimedia capable, one of the challenges is the multimedia content delivery on these embedded devices.

Hence, in this subsection we are going to introduce adaptation mechanisms applied by W3C Content Selection for Device Independence (World Wide Web Consortium, 2007), W3C SMIL markup language (World Wide Web Consortium, 2008c) and by MPEG-21 framework (MPEG Requirements Group, 2002).

Several attempts have been made to standardize the presentation environment and the presentation format for mobile service delivery. Markup languages such as the XML (Extensible Markup Language) and its applications like SMIL (Synchronized Multimedia Integration Language) developed by the World Wide Web Consortium (W3C) can be applied in modeling structured, document-like multimedia presentations.

W3C Content Selection for Device Independence

A more recent Candidate Recommendation for standardizing content adaptation was released in 2007 (World Wide Web Consortium, 2007). The Content Selection for Device Independence (DISelect) aims to provide the ability to select between different versions of content. DISelect has been designed to be used within other markup languages, and it was born to support content authors in the specification of different versions of the content they produce. The ability to select among different versions of the same content provides one important mechanism the authors can use in order to guarantee the adaptability of their materials, so that they could be enjoyed also by using mobile devices. DISelect is provided with a mechanism for the content selection which is to be expressed when adaptation takes place and which requires only modest computational capability. It defines two profiles of markup: the DISelect Basic and the DISelect Full. The former is a subset of the latter, and it is intended to contain only the necessary markup to be implemented on small-footprint DISelect implementations, running on mobile devices with limited processing power and memory. The latter is the complete set of defined DISelect markup. The main idea is to allow the use of DISelect elements and attributes within a host markup language by means of namespace mechanisms. In this module, attributes and elements for conditional processing have been defined, such as <if> (it defines a set of materials which has to be included in the final result if the associated expression has the appropriate value) and <select> (it encloses one or more sets of materials that are subject to conditional selection. It contains one or more <when> elements and an optional <otherwise> element. Expressions associated with the <select> and <when> elements control the conditions under which particular parts of the content are processed). Despite such a markup language and its related adaptation mechanism that seem to be very feasible and easy to use (also in creating suitable m-learning content), it is still a Candidate Recommendation and not yet a standard.

W3C SMIL

A stable and well-known W3C standard which applies client-side adaptation mechanisms is W3C SMIL (World Wide Web Consortium, 2008c). It plays the same role in a SMIL player that HTML plays in a Web browser (namely providing information on how to layout and format a page). A SMIL presentation can consist of multiple components of different media types (such as video, audio, text, and graphics) linked via a synchronized timeline. For example, in a slide show, the corresponding slide can be displayed when the narrator in the audio clip starts talking about it. SMIL 3.0 is the main representation of Web technology for describing timing and synchronization of multimedia presentations. Careful attention has been paid, in the design of SMIL, to modularity and extensibility of the recommendation, and three language profiles have been proposed. Most notably, a SMIL Basic profile is a collection of modules together with a scalable framework, which allows a document profile to be customized for the capabilities of the mobile device.

SMIL 3.0 is defined as a set of markup modules, which define the semantics and XML syntax for certain areas of SMIL functionalities. This specification provides different classes of changes to SMIL 2.1, covering the ten functional areas; in particular, new models are introduced, former SMIL modules are deprecated and replaced by new ones to allow differentiated features to be implemented in profiles, without necessarily requiring support for all of the functionalities of the former SMIL Modules, which are, in turn, revised allowing extended functionalities. Some of these changes are related to the use of SMIL through mobile devices and to the content transcoding mechanisms.

Several simple content selection mechanisms have been introduced in SMIL to provide greater flexibility and to meet different context of use constraints (including the ones that are related to mobile contexts). However, in most cases, SMIL adaptation is achieved at the client side. This implies that the client is adaptation-capable and that the profiles and the device capabilities are somehow set. In addition, adaptations do not necessarily belong to the same layer of a document presentation. One can start by designing a device-independent document layer and generate, once the profiles are identified, the SMIL content representation.

It is also possible to perform adaptation within a SMIL document instance beyond the mechanisms which are provided by the format and to modify the content to fit bandwidth and display limitations. In fact, SMIL language contains an "adaptation" or "alternate content" mechanism. By using the <switch> tag and the so-called "test attributes", it is possible to have a SMIL player choice between alternative content. Examples of attributes, which the player can use, are "systemBitrate" to select content that fits the current network bandwidth, "systemCaptions" to choose between video with or without captions, "systemLanguage" to select content in a given language, "systemScreenDepth", "systemScreenSize", etc. Besides the capabilities of specifying bandwidth, display limitations and learners' language for each medium, SMIL allows the declaration of author-defined variables to trigger the presentation of contents. The latter ones are a construction of the CustomTestAttributes (CTA) module. Language built-in and author-defined variables could be used by SMIL compliant players to present alternative multimedia items on the strength of their Boolean value. To enhance content adaptability, the SMIL CTA module extends this feature with the definition of author-defined custom test attributes. This module allows the attribute definition and setting in the header of the SMIL document, so that the author can set a default state for each custom test attribute and declare a default attribute which will be play out whenever no customization will be necessary. The customization is done by using a SMIL <switch> construct in the <body> of the SMIL document. It is used to select media for inclusion in a presentation depending on the values of the custom test attributes. The first object that contains a value true will be rendered. It is possible to set the last option so that it will always resolve true. In this way, it will be considered only if no other objects resolve to true. Hence, the CTA module could be used to offer alternative presentations to users who are equipped with mobile devices.

Another interesting SMIL 3.0 feature is the SMIL State Module. It provides mechanisms which permit the author to create more complex control media than the timing and content control modules could do. By allowing the use of variables, a document can have some explicit state along with ways to modify, use and save this state. So, such a module could be used to describe and/or define when different versions of the content should be played out, allowing also content adaptation to mobile learning contexts. It is worth noting that the improvement is the use of XPath expressions instead of boolean ones, which are used in the CTA module.

Finally, the improved SMIL 3.0 MetaInformation Module and the attribute label allow the definition of the alternative presentation and metadata within the document instead of in the <head> element of the presentation. In particular, the label attribute specifies the name of a SMIL file, which contains a description or an alternative version of the element. If it is selected, then the original presentation will be paused and a new document instance will be created to display the target SMIL file. The use of this attribute affects the original media synchronization and the provision of new versions of the multimedia presentation. Thus, by means of such an attribute, customized content could be provided to users equipped with mobile devices. Moreover, the new SMIL 3.0 Metainformation module permits the declaration of the element <metadata> as a child of media elements inside the <body> and not only in the header of the SMIL document.

These adaptation features enable a SMIL player to fit technical circumstances and some fairly static user preferences. This makes SMIL a standard able to provide adapted multimedia presentations to mobile learners.

MPEG-21

Another interesting standard is MPEG-21, which is an open standards-based framework for multimedia delivery and consumption by all the players (MPEG Requirements Group, 2002). It is the newest of a series of standards being developed by the Moving Picture Experts Group, after a long history of producing multimedia standards. The goal of MPEG-21 can thus be redefined as the technology needed to support users to exchange, access, consume, trade and otherwise manipulate Digital Items in an efficient, transparent and interoperable way. Interoperability is the driving force behind all multimedia standards. It is a necessary requirement for any application that involves guaranteed communication between two or more parties. Interoperability expresses the users' need for easily exchanging any type of information without technical barriers, supporting it also in mobile contexts (Burnett et al., 2003).

The basic concepts in MPEG-21 relate to *what* and *who* within the multimedia framework. *What* is a Digital Item, i.e., a structured digital object with a standard representation, identification, and metadata within the MPEG-21 framework. *Who* is a user who interacts in the MPEG-21 environment or uses a Digital Item, including individuals, consumers, communities, organizations, corporations, consortia, governments, and other standards bodies and initiatives around the world. The users can be creators, consumers, rights holders, content providers or distributors, learners, etc. There is no technical distinction between providers and consumers: all parties that have to interact within MPEG-21 are categorized equally as users. They assume specific rights and responsibilities according to their interaction with other users. All users must also express and manage their interests in Digital Items.

In practice, a Digital Item is a combination of resources, metadata, and structure. The resources are the individual assets or content. The metadata describes data about or pertaining to the Digital Item as a whole or also to the individual resources in the Digital Item. The structure relates to the relationships among the parts of the Digital Item, both resources and metadata. For example, a

Digital Item can be a video-lecture collection or a music album. The Digital Item is thus the fundamental unit of distribution and transaction within the MPEG-21 framework.

MPEG-21 is organized into several independent parts, primarily to allow various slices of the technology to be useful as stand-alone. This maximizes their usage and lets the users to implement them outside MPEG-21 as a whole, in conjunction with proprietary technologies. The MPEG-21 parts already developed or currently under development are as follows:

1. Vision, technologies, and strategy: this part describes the multimedia framework and its architectural elements with the functional requirements for their specification.
2. Digital Item Declaration (DID): this second part provides a uniform and flexible abstraction and interoperable framework for declaring Digital Items. By means of the Digital Item Declaration Language (DIDL), it is possible to declare a Digital Item by specifying its resources, metadata, and their interrelationships.
3. Digital Item Identification (DII): the third part of MPEG-21 defines the framework for identifying any entity regardless of its nature, type or granularity.
4. Intellectual Property Management and Protection (IPMP): this part provides the means to reliably manage and protect content across networks and devices.
5. Rights Expression Language (REL): this specifies a machine-readable language that can declare rights and permissions using the terms as defined in the Rights Data Dictionary.
6. Rights Data Dictionary (RDD): this is a dictionary of key terms required to describe users' rights.
7. Digital Item Adaptation (DIA): this identifies all the description tools for usage environment and content format features that might influence transparent access to the multimedia content (notably terminals, networks, users and the natural environment where users and terminals are located).
8. Reference software: this includes software that implements the tools specified in the other MPEG-21 parts.
9. File format: defines a file format for storing and distributing Digital Items.
10. Digital Item Processing (DIP): this defines mechanisms for standardized and interoperable processing of the information in Digital Items.
11. Evaluation methods for persistent association technologies: this part documents best practices in evaluating persistent association technologies using a common methodology (rather than standardizing the technologies themselves). These technologies link information that identifies and describes content directly to the content itself.
12. Test bed for MPEG-21 resource delivery: this last part provides a software-based test bed for delivering scalable media and testing/evaluating the scalable media delivery in streaming environments.

The seventh part of MPEG-21 specifies all the tools for the adaptation of Digital Items. One of the goals of MPEG-21 is to achieve interoperable transparent access to (distributed) advanced multimedia content by shielding users from network and terminal installation (by considering also mobile devices and contexts), management, and implementation issues. Achieving this goal requires the adaptation of Digital Items (MPEG MDS Group, 2003). A Digital Item may be subject to a resource adaptation engine, a description adaptation engine, or a DID adaptation engine, which produces the adapted Digital Item, according to the learners' context of use.

The usage environment description tools can describe the mobile device capabilities (in terms of codec, input-output capabilities, and device properties) as well as network characteristics (such as network capabilities and network conditions), the user (for example user info, usage preferences and usage history, presentation preferences, accessibility characteristics, including visual or audio impairments, and location characteristics) and the natural environment. In this context, the natural environment relates to the physical environmental conditions around a user such as lighting or noise levels, or circumstances such as the time and location (Vetro, 2004).

This part of MPEG-21 also includes the following specific items:

- Resource adaptability: i.e., tools to assist with the adaptation of resources, including the adaptation of binary resources in a generic way and metadata adaptation. In addition, tools that assist in making resource complexity trade-offs and associations between descriptions and resource characteristics for Quality of Service are also targeted.
- Session mobility: i.e., tools that specify how to transfer the state of Digital Items from one user to another. More specifically, the capture, transfer and reconstruction of state information.

Concluding, MPEG-21 is an ideal candidate to provide suitable and adaptable content in mobile learning environments.

Open Source Projects and Libraries

Different open source projects and libraries can operate adaptation to online learning content and to single media which compose it. In this subsection we are going to introduce some of them, in particular, FFmpeg (FFMPEG, 2010), ImageMagick (ImageMagick, 2010) and GAIA Image Transcoder (Gaia Reply, 2010). Finally, we will introduce an open source project that allows the development of mobile applications (MORFEO Project, 2007).

FFmpeg (FFMPEG, 2010) is an open source project which produces libraries and programs for managing multimedia data. The name of the project comes from the MPEG video standards group, together with "FF" for "fast forward". Such a project is composed by different parts, i.e., an audio/video codec library, an audio/video container mux and demux library and the FFmpeg command line program for transcoding multimedia files. FFmpeg is developed under GNU/Linux, but it can be compiled under most operating systems, including Apple Inc. Mac OS X and Microsoft Windows. There are two video codecs (FFV1 and Snow codec) and one video container invented in the FFmpeg project during its development. In particular, such a set of libraries allows different forms of audio and video transcoding, for instance, transforming video and audio files from a format to another one, adding subtitle or caption tracks to a video, re-sizing video width and height, disabling audio recording, and degrading audio quality.

ImageMagick (ImageMagick, 2010) is an open source software suite for displaying, converting, and editing raster image files. The software mainly consists of a number of command-line interface utilities for manipulating images. ImageMagick does not have a GUI based interface to edit images, as Adobe Photoshop and GIMP have, but instead it modifies existing images as directed by various command-line parameters. Many applications, such as MediaWiki and phpBB can use ImageMagick to create image thumbnails if it is installed. ImageMagick is also used by other programs for converting images. In fact, one of the basic and thoroughly-implemented features of ImageMagick is its ability to efficiently convert images between different file formats. In particular, ImageMagick can read and write over 100 image file formats. Another ImageMagick trancoding feature is the color quantization: the number of colors in an image can be reduced to an arbitrary number and this is done by intelligently weighing the most prominent color values present among the pixels of the image. Summarizing, such a library allows different kinds of image transcoding, such as format conversion, width and height resizing, image rotation, transparency, animation, insertion of text and comments.

The GAIA Image Transcoder (GIT) is an open source library which operates image transcoding for mobile applications (Gaia Reply, 2010). GIT is composed by two parts: a transformation and transcoding library which performs image adaptation to mobile devices (by using information retrieved from the WURFL file) and a JSP tag library which enables library utilization into a J2EE environment. The transcoding library works as a transformation pipeline, which is composed by a set of filters. Such filters could operate directly on the image body or indirectly on the associated meta-information. At the time the authors are writing, GIT supports filters such as re-sizing the image to conform it to the width and the height of the device screen, optimization of color depth of the image on the basis of device capabilities, transcoding of the image to supported formats, returning the associated image when a corresponding URI is given.

All the previous projects and libraries should be used by applications devoted to providing adapted content to mobile users. MyMobileWeb (MORFEO Project, 2007) is an open source project which allows the development of mobile applications with such features. It is based on Java and J2EE technology and on open-standards. By means of MyMobileWeb, it is possible to create applications which adapt their user interfaces according to the characteristics of the device and Web browser used. Such device capabilities are provided by a Device Description Repository (compliant to the W3C DDR Standard), such as the already described WURFL (WURFL, 2010). Content should be described and structured in a declarative language (based on Web standards). The applications developed by using MyMobileWeb automatically perform content fragmentation, when necessary. The application layout and all the appearance features are controlled through CSS stylesheets, and it is possible to define different stylesheets for different families of devices. Moreover, MyMobileWeb provides the selection of user interface parts, on the basis of the characteristics of the delivery context, allowing authors to specify adaptation policies by means of the selection attribute, as mandated by the W3C DISelect specification (World Wide Web Consortium, 2007). Finally, in order to adapt different media content (audio, video, images), selection and transcoding components are incorporated into MyMobileWeb.

MOBILE ISSUES IN LCMSS

In this section, we are going to present how some well-known LCMSs face mobile learners' instances.

Several open source projects are based on Moodle, a well-known open source Learning Management System (Moodle Pty Ltd, 2010). The MOMO (MObile MOodle) project is an add-on to Moodle, which is able to implement mobile learning scenarios with Moodle as a backend. Mobile learners may install the MOMO client on their mobile devices, and they can access m-learning courses. The MOMO client is a Java based application. Moreover, administrators are allowed to install the MOMO extension on the Moodle server in order to make content available for mobile environments. The MOMO project was designed to allow m-learning experiences by means of generic Java and Internet enabled devices. After some years it has been abandoned and substituted by the MLE-Moodle (Mobile Learning Engine-Moodle) plug-in. This is especially devoted to users who access m-learning content through mobile phones. Learners can either use the mobile browser or a special mobile phone application which was designed for learning on mobile phones (the MLE phone client). When a learner installs this special application on his/her device, he/she has to declare the device producer and model. Then, a customized version of the application is downloaded and installed. In such a context, there is no Device Description Repository. The MLE server side architecture also includes a media server, which is devoted to converting media (audio, video and images) to formats suitable for mobile phones.

Another widespread open source Learning Content Management System is ATutor (Adaptive Technology Resource Centre, 2010). ATutor has been designed and developed with accessibility and adaptability in mind. To reach such goals, ATutor supports a large number of accessibility, interoperability and e-learning standards (e.g., W3C WCAG, IMS AccessForAll, IMS ACCLIP, IMS Content Packaging, ADL SCORM, IMS QTI, IMS Common Cartridge), and its layout and interface are liquid and highly configurable by users. Thanks to the application of universal design principles, ATutor is easy to be enjoyed also by using non standard devices. Nevertheless, there is not yet a version explicitly dedicated to mobile learning environments. In order to overcome such a lack, the development team is designing a version of ATutor for mobile platforms (iPhone, Android, Blackberry). The main goal is to extend current Web services in ATutor to allow accessing learning content, network activity, communication tools, etc, by using a mobile device. Such a mobile application will be Java, JavaScript and AJAX based.

OPEN ISSUES AND FUTURE RESEARCH DIRECTIONS

This section will introduce some open issues and future trends in adapting technologies in mobile learning environments. A notable trend is related to the so called "E-learning 2.0". Such a term indicates the set of Computer-Supported Collaborative Learning (CSCL) systems and tools. At the moment, this is one of the most promising innovations to improve learning by exploiting current information and communication technologies. The main idea is to provide tools students can use to work together on learning tasks or in other words to collaboratively learn. Collaborative learning assumes that learners' knowledge is socially constructed through conversations and comparisons about learning content and idea sharing (Levy, 1997). Social networks, Web 2.0 applications and new communication technologies should be applied to m-learning in order to enhance such a more active role of learners. New features and tools should be adapted in order to be exploited in a feasible way by mobile devices with limited characteristics (computational capability, connectivity, small display, and so on). Collaboration among learners is an issue of the so-called constructivism, a theory of knowledge which argues that humans create their knowledge from their experiences. Constructivism is often associated with pedagogic approaches that promote active learning, or learning by doing. To apply such a theory, m-learning applications should provide mechanisms to allow learners' content creation, sharing and integration, according to mobile device capabilities.

Another interesting open issue is the accessibility of mobile learning environments. Their limited capabilities could affect the learning experience of people with disabilities (Barron et al., 2004). In fact, small screens, small buttons and keyboards, the lack of alternative input systems, and the restriction of configurable display options (i.e., text and background colors, text size, font) may represent barriers for learners with disabilities.

CONCLUSION

Content and application adaptation is fundamental in mobile learning contexts. It can be applied by exploiting different strategies, mechanisms, architectural approaches, device profiling standards and libraries. In order to provide a customized mobile learning experience, all these features and technologies have to be orchestrated. This way, learning content and applications could meet mobile learners' needs and other constraints which affect the m-learning context. In this scenario, standards and open source resources play a strategic role, because they significantly improve the tools to reach m-learning goals. In particular, the development of open source LCMS, the definition of profiling device open standards and repositories, and content transcoding open mechanisms and libraries are leading the field of adaptation in mobile learning environments. By means of such open technologies, m-learning designers and developers could effectively provide learners with adaptable and interoperable content and applications.

REFERENCES

Adaptive Technology Resource Centre. (2010). *A tutor learning content management system*. Retrieved March 21, 2010, from http://www.atutor.ca

Anido, L. (2006). An observatory for e-learning technology standards. *Advanced Technology for Learning*, *3*(2), 99–108. doi:10.2316/Journal.208.2006.2.208-0876

Barron, J. A., Fleetwood, L., & Barron, A. E. (2004). E-learning for everyone: Addressing accessibility. *Journal of Interactive Instruction Development*, *26*(4), 3–10.

Burnett, I., Van de Walle, R., Hill, K., Bormans, J., & Pereira, F. (2003). MPEG-21: Goals and achievements. *IEEE MultiMedia*, *10*(6), 60–70. doi:10.1109/MMUL.2003.1237551

Clothier, P. (2005). An introduction to m-learning: An interview with Ellen Wagner. *The E Learning Guild*. Retrieved March 21, 2010, from http://www.elearningguild.com

Colajanni, M., & Lancellotti, R. (2004). System architectures for Web content adaptation services. *IEEE Distributed Systems Online*, *5*(5).

Curran, K., & Annesley, S. (2005). Transcoding media for bandwidth constrained mobile devices. *International Journal of Network Management*, *15*(2), 75–88. doi:10.1002/nem.545

Ferretti, S., Roccetti, M., Salomoni, P., & Mirri, S. (2009). Custom e-learning experiences: Working with profiles for multiple content sources access and adaptation. *Journal of Access Services*, *6*(1&2), 174–192. doi:10.1080/15367960802301093

FFMPEG. (2010). *FFmpeg multimedia systems*. Retrieved March 21, 2010, from http://ffmpeg.org

Gaia Reply. (2010). *Gaia Image Trascoder* (GIT). Retrieved March 21, 2010, from http://gaia-git.sourceforge.net

Harumoto, K., Nakano, T., Fukumura, S., Shimojo, S., & Nishio, S. (2005). Effective Web browsing through content delivery adaptation. *ACM Transactions on Internet Technology*, *5*(4), 571–600. doi:10.1145/1111627.1111628

ImageMagick. (2010). *ImageMagick - convert, edit, and compose images*. Retrieved March 2010, from http://www.imagemagick.org

IMS Global Learning Consortium. (2002a). IMS Learner Information Profile (LIP). Retrieved March 21, 2010, from: http://www.imsglobal.org/specificationdownload.cfm

IMS Global Learning Consortium. (2002b). *IMS learner information package accessibility for LIP*. Retrieved March 21, 2010, from http://www.imsglobal.org/specificationdownload.cfm

Jannach, D., Leopold, K., Timmerer, C., & Hellwagner, H. (2006). A knowledge-based framework for multimedia adaptation. *Applied Intelligence*, *24*(2), 109–125. doi:10.1007/s10489-006-6933-0

Laakko, T., & Hiltunen, T. (2005). Adapting Web content to mobile user agents. *IEEE Internet Computing*, *9*(2), 46–53. doi:10.1109/MIC.2005.29

Levy, P. (1997). *Collective intelligence: Mankind's emerging world in cyberspace*. New York, NY: Plenum Publishing Corporation.

Moodle Pty Ltd. (2010). *Moodle*. Retrieved March 21, 2010, from http://moodle.org

MORFEO Project. (2007). *MyMobileWeb Project*. Retrieved March 21, 2010, from http://mymobileweb.morfeo-project.org

MPEG MDS Group. (2003). *MPEG-21 multimedia framework, part 7: Digital item adaptation*. (ISO/MPEG N5845). Retrieved March 21, 2010, from http://www.chiariglione.org/mpeg/working_documents/ mpeg-21/dia/dia_fcd.zip

MPEG Requirements Group. (2002). *MPEG-21 overview*. (ISO/MPEG N4991).

Open Mobile Alliance. (2007). *User agent profile v. 2.0 approved enabler*. Retrieved March 21, 2010, from http://www.openmobilealliance.org/Technical/ release_program/uap_v2_0.aspx

Pandey, V., Ghosal, D., & Mukherjee, B. (2004). Exploiting user profiles to support differentiated services in next-generation wireless networks. *IEEE Network*, *18*(5), 40–48. doi:10.1109/MNET.2004.1337734

Salomoni, P., Mirri, S., Ferretti, S., & Roccetti, M. (2008). A multimedia broker to support accessible and mobile learning through learning objects adaptation. *ACM Transactions on Internet Technology*, *8*(2), 9–23. doi:10.1145/1323651.1323655

Vetro, A. (2004). MPEG-21 digital item adaptation: Enabling universal multimedia access. *IEEE MultiMedia*, *11*(1), 84–87. doi:10.1109/MMUL.2004.1261111

World Wide Web Consortium. (2004). *Composite Capability/Preference Profiles (CC/PP): Structure and vocabularies* 1.0. Retrieved March 21, 2010, from http://www.w3.org/TR/2004/REC-CCPP-struct-vocab-20040115

World Wide Web Consortium. (2007). *Content selection for device independence* (DISelect) 1.0. Retrieved March 21, 2010, from http://www.w3.org/TR/cselection

World Wide Web Consortium. (2008a). *Device description repository core vocabulary*. Retrieved March 21, 2010 from http://www.w3.org/TR/ddr-core-vocabulary

World Wide Web Consortium. (2008b). *Device description repository simple API*. Retrieved March 21, 2010, from http://www.w3.org/TR/DDR-Simple-API

World Wide Web Consortium. (2008c). *Synchronized multimedia integration language* 3.0. Retrieved March 21, 2010 from http://www.w3.org/TR/2008/REC-SMIL3-20081201

WURFL. (2010). *Wireless universal resource file library*. Retrieved March 21, 2010, from http://wurfl.sourceforge

KEY TERMS AND DEFINITIONS

Accessibility: This term identifies the degree to which a product, device, service, or environment is accessible by as many people as possible. Accessibility can be defined as the "ability to access" some system or entity. This term is often used to focus on people with disabilities and their right of access, often by the means of assistive technology.

Content Transcoding and Adaptation: This term identifies the transformation and the manipulation of Web content (such as images, audio, videos, texts and multimedia presentations) to meet desired targets (defined by the terminal capabilities and the application needs).

Device Profiling: This term identifies the definition of a pre-determinate set of device capabilities and preferences, as well as the learner's profile; this group of information is fundamental in m-learning environments because it drives the adaptation of didactical materials.

Learner's Profiling: This term identifies the definition of a pre-determinate set of a learner's characteristics and needs. This information has to be profiled on the basis of standards and it is strategic in the adaptation of m-learning content, in order to provide a tailored and more effective learning experience.

Learning Content Management System (or LCMS): This term identifies an application (usually Web based) which allows the administration, provision, documentation, tracking and reporting of learning activities (e.g. classroom and online events, e-learning programs, training content, and so on).

Mobile Learning (or M-Learning): This term identifies learning activities which are available by means of mobile technologies and devices. In mobile learning environments, the learner takes advantage of learning opportunities without being bound in a predetermined location.

Open Standards: This term identifies a standard which is publicly available and has various rights to use associated with it. There is no single definition and interpretation because the terms "open" and "standard" have a wide range of meanings associated with their usage.

Chapter 3
Linux Based Mobile Operating Systems

Lee Chao
University of Houston-Victoria, USA

ABSTRACT

In today's mobile computing, Linux plays a significant role. The Linux kernel has been adopted by a variety of mobile operating systems to handle tasks such as device management, memory management, process management, networking, power management, application interface management, and user interface management. This chapter introduces Linux based mobile operating systems installed on various mobile devices. It first gives a brief introduction of the history of mobile Linux. Then, the chapter introduces the mobile Linux features that can be used to meet the mobile learning requirements. The last part of the chapter presents strategies on selecting a Linux based operating system for a particular mobile learning project.

INTRODUCTION

As an open source operating system, Linux plays an important role in mobile computing. This chapter first investigates the requirements by a mobile learning system. The requirement investigation to an operating system is carried out in three areas, the mobile learning infrastructure, the development of mobile learning course materials, and the mobile learning management. Traditionally, Linux operating systems have been widely used on servers, desktop and notebook computers to handle the tasks in mobile learning. This chapter focuses on the recently developed Linux based mobile operating systems used by mobile devices.

For mobile devices, it may not be feasible to use a full-blown version of a Linux operating system designed for servers and desktop computers. Instead, the Linux kernel has been modified for the mobile operating systems. In a mobile device,

DOI: 10.4018/978-1-60960-613-8.ch003

the Linux kernel handles tasks such as device management, memory management, process management, networking, power management, application interface management, and user interface management. This chapter describes the features and functionalities provided by the Linux kernel and illustrates how a Linux based mobile operating system can provide solutions to the requirements from mobile devices.

The architectures of various Linux based mobile operating systems are discussed in this chapter. The commonly available Linux operating systems such as Android, Palm webOS, Moblin, Ubuntu MID Edition, and Google Chrome are introduced. Based on the cost, usability, and flexibility, this chapter provides some recommendations on the selection of Linux mobile operating systems for different mobile devices.

BACKGROUND

Most of the mobile operating systems are UNIX based operating systems including Linux and iPhone (Robert Vamosi, 2009) which is derived from BSD UNIX. Linux distributions integrate the Linux kernel with other open source software to build various versions of Linux based open source operating systems. As the kernel of many mobile operating systems, Linux is known by its excellent flexibility and usability. More and more mobile device manufacturers commit to the open source approach. Mobile devices can benefit from Linux in the following areas:

- *Cost:* Since Linux is free for mobile device manufacturers, the cost to develop a mobile operating system is minimal. The manufacturers do not need to start their mobile operating system development process from scratch. To fit the needs of their mobile devices, they can build their mobile operating systems around the free Linux kernel.
- *Usability:* Another strong point of Linux is its usability. Linux has been adopted by the PC industry for decades. The mobile devices powered by Linux can easily communicate with the PCs and network devices installed on a Linux network. Also, the Linux powered mobile devices can share the application software and system management utilities developed for Linux. The embedded version of Linux is designed for the devices with limited resources. Therefore, the embedded Linux is ideal to be integrated into the mobile devices. Records have shown that Linux has been very successful in running embedded mobile devices.
- *Flexibility:* A Linux based operating system can be built to fit in a variety of devices depending on the available resources such as the CPU and RAM. It can also be built to consume only a little electric power. This feature is suitable for mobile devices since they in general have limited system resources.
- *Application Development:* One of the reasons for a mobile device manufacturer to be successful is the availability of application software. If a Smartphone is supported by abundant application software, it will be appealing to various users. Application developers specifically like open source operating systems which can make their job easier. Therefore, for a mobile device manufacturer to have a competitive edge, adopting the open source approach is a winning strategy. Proprietary companies such as Apple have also realized the importance of the open source approach. In 2008, Apple launched the Apps Amuck project to post iPhone application source code every day for a month (Apps Amuck, 2009). In such a way, iPhone application developers can get a quick start for their programs. To assist application develop-

ers, some of the mobile Linux distributions provide their application development platforms which provide a working environment for developers to create application software for mobile devices. Another advantage of the open source approach is that the number of application developers for an open source product is way more than that of an individual company.

- *Support:* Linux is backed by a large community of Linux developers and users. It is convenient for a Linux user to get help from the community. There are a large number of troubleshooting tips published on the Web. Most of the Linux distributions also have their own support Web sites. Usually, users can get solutions for their problems from these support Web sites.

Due to these great features, Linux can be easily modified to meet the needs of mobile devices. As a result, various versions of Linux based mobile operating systems have been developed. Some of the Linux based operating systems are well known in the mobile device market. Android is such a product developed by Open Handset Alliance (OHA) (2009). The Android project is initiated by Google. Later many other companies joined Google to form the organization Open Handset Alliance. Google has been developing another Linux based operating system called Chrome (Google, 2009) for Web related applications on netbook computers. Similarly, another well known Linux distribution Ubuntu also has a mobile version of Linux based operating system called Ubuntu Mobile Internet Device (MID) Edition (Ubuntu, 2009). Intel has been developing its version of Linux based mobile operating system called Moblin (Moblin, 2009) which is designed for mobile devices such as netbooks, mobile Internet devices, and in-vehicle infotainment systems. Linux Mobile (LiMo) Foundation, organized by companies like Motorola, NEC, NTT DoCoMo, Panasonic Mobile Communications, Samsung Electronics, and Vodafone, provides a version of Linux based operating system as well as an application development platform for mobile devices (LiMo Foundation, 2009). Palm is another company that supports Linux based operating systems. Developed by Palm, webOS is the Linux based operating system used by the Palm Pre Smartphone. Maemo is a Linux based operating system and application development platform. Maemo is developed by Nokia for developing mobile application software and for mobile devices such as Internet Tables (Maemo, 2009).

Benefit from these advantages, the market share of Linux based Smartphones is growing rapidly. According to the mobile metrics report released by AdMob, the market share of Android based Smartphones is ranked third in July, 2009 (Taylor Wimberly, 2009) which is only 1% less than the second ranked RIM operating system. Another Linux based Smartphone operating system Palm is ranked fifth in the report. The Linux based operating systems are ranked second if one combines the market shares of the Android and Palm operating systems.

Although the market share of Linux based operating systems keeps growing, the implementation of mobile Linux operating systems in mobile learning has not caught up yet. Readers may find some research about mobile learning based on the iPhone (Saponas et al., 2008). However, few research studies have been done for mobile Linux and its usage in mobile learning. The features offered by Linux based mobile operating systems in this area may need more research. In this chapter, the possible usage of Linux based operating systems in mobile learning will be explored.

OPERATING SYSTEM REQUIREMENTS BY MOBILE LEARNING

The requirements for the implementation of mobile learning can be classified into four categories, the requirements for constructing the mobile learning infrastructure, the requirements for developing the mobile based course content, the requirements for managing mobile learning activities, and the requirements for mobile learning services. The following briefly describes these requirements:

Infrastructure Requirements: To carry out the mobile learning activities, it is necessary to have a well established mobile network system in place. For a learner to access the learning materials and communicate with the instructors and other learners, the mobile device used by the learner must be able to communicate with the base station in a region called cell. The mobile device needs to have a low-power transceiver which is powerful enough to communicate with the base station in the cell. As the learner moves around, the mobile device will be carried by the learner to a different cell. The communication between the mobile device and the current base station will be handed off to the base station in the next cell.

A base station is connected to a public switched telephone network (PSTN) or wireless service provider; its transmitter, receiver, and control unit should always function properly. Operating systems are responsible for the communication between a mobile device and a base station, and between a base station and a wireless service provider. Operating systems are also responsible for the security, radio signal processing, networking, and data storage. Through the PSTN or wireless service provider, a learner is able to contact another learner or access the learning materials hosted by a server. Operating systems are also responsible for managing and controlling the activities on the server side. During the learning process, instructors and learners are able to share the learning materials or upload the course materials to the server. Operating systems are used to handle the uploading from mobile devices, receiving and managing the course materials on the server side. During the mobile learning process, operating systems are used to handle phone calls and manage voice mail massages, and receive and send Short Message Service (SMS) messages or e-mail. When used on mobile devices, an operating system should be able to handle a wide range of hardware devices and to optimize the usage of the available mobile network bandwidth based on the environment and the needs of the user. While on the road, battery power can be critical for the mobile learner to continue the learning activity. The operating system on a mobile device should optimize the usage of battery power so that the mobile learner can have adequate learning time.

Course Material Development Requirements: The mobile learning materials may include text-only or multimedia materials. For the development of text-only course materials, an operating system should be able to run the word processing software and the software that is able to convert files from one format to another format. Multimedia course content may include audio, video, graphic, image, animation, and handwritten materials. To develop multimedia course materials, various multimedia content development software and Web authoring tools are used. Media player software is used to playback audio and video content. Operating systems are also needed to host audio and video devices to playback multimedia content. Web authoring tools are required to create Web pages with multimedia content. While hosted on an operation system, the Web authoring tools can be used for audio editing, video editing, image editing, chart editing, screen capturing, clip art managing, and word processing. An operating system is needed to support the Web browser to display the course content and to run Web-based applications such as Google maps, Google images, and Google voices. The Web browser is also used for sending requests to and receiving responses from the Web server. Operating systems can also host photography equipment and the image storage equipment.

Mobile Learning Management Requirements: In many mobile learning projects, a learning management system (LMS) such as Moodle is used for user management, class management, and collaboration support. An LMS is an application server and should be hosted by an operation operating that has the server functionality. To fully implement mobile learning projects, many other application servers such as database servers, Web servers, and FTP servers also need to be supported by server capable operating systems. The operating systems on mobile devices should allow learners to access their course Web sites. The cell phone itself is a great tool for collaboration among students. To improve collaboration, various collaboration enhancement tools are developed. The commonly used collaboration tools are the Web conference, virtual whiteboard, screen sharing, text messenger, and document camera. These collaboration tools post a challenge to the operating systems used by mobile devices. Not all of the mobile operating systems can handle these collaboration tools.

Education games can be another form of collaboration. Mobile devices such as the Smartphone are designed to allow users to play some simple games. Some of these games involve multiple players for competition. Mobile operating systems need to handle these simple games designed for mobile devices.

Mobile Learning Service Requirements: There are various mobile services that support mobile learning. The cell phone notification service is commonly used for emergency notification by sending out security and weather-related warnings. The notification service can also be used to send event reminders and announcements for community activities. The location-aware service provides directions and background information about a location. The operating systems on the server side are required to provide these services, and the operating systems on the mobile devices should be able to subscribe for these services.

The context-aware service provides users with the formation relevant to the context. This service is often used in mobile learning. It is especially valuable for field trips. Based on the learner's knowledge and surrounding environment, the teaching and learning process is redesigned to best fit the learner's needs. The context-aware computing requires an operating system to process signals from various sensors, support machine learning activities, and conduct computation-intensive statistical analyses. The operating system should be able to support the computation that can match the user interface with the learner's cognition and learning environment.

In the above, we have investigated the several requirements for operating systems in a mobile learning environment. As an open source operating system, Linux is capable of meeting all the requirements listed above. For computing tasks on servers, desktop computers, and laptop computers, the server version or desktop version of Linux operating systems can be used. For mobile devices, the Linux kernel needs to be redesigned to adapt to the mobile computing environment. Often, the X Window component should be removed from the mobile operating system. Also, the library in a Linux operating system should also be redefined to exclude the files that are not suitable for mobile devices. Due to these changes, a mobile Linux operating system may not be able to use certain applications that are designed for regular Linux operating systems. A mobile operating system may not be able to support a full scale application development platform such as Java SE. Therefore, the Java applications developed for mobile devices may not be compatible with the Java applications developed for desktop or laptop computers.

OPERATING SYSTEM FUNCTIONALITIES

In general, a mobile device such as the Smartphone may include the following components:

- *Central Processing Unit (CPU)*: The main task of a CPU is to carry out calculation for communication and for applications.

- *Memory*: Memory is temporary data storage used by a CPU to quickly get necessary data for calculation and to store the calculation results.
- *Mass Data Storage*: It is used to permanently store information such as e-mail contacts, music, or photo images. A mass data storage device can be a built-in hard drive or a solid-state drive. It can also be a subscriber identity module (SIM) card or an external USB flash drive.
- *User Interface Device*: A user interface device may include a LCD panel, camera, keyboard, microphone, or speaker.
- *Network Device*: A network device may include a radio frequency unit, Global System for Mobile Communications (GSM) unit, IP network service unit, and Wi-Fi network service unit. It also includes air interface utilities for call control, report management, and location update.

In addition to these hardware related components, a mobile device may also be installed with various applications. An operating system installed in a mobile device is used to control computer operations such as the computer process and networking. It is also used to manage hardware and software such as the monitor, keyboard, library files, address book, application interface, and user interface. It can handle new additions to the hardware and updates of the software.

In the Linux based mobile operating systems mentioned above, a Linux based kernel is included. These mobile operating systems use the Linux kernel to handle tasks such as device management, memory management, process management, networking, power management, and application interface and user interface management. When a Smartphone is used in a mobile learning process, multiple computing programs will be running on the Smartphone. For example, the Smartphone needs to process the input from the keypad, dial the phone number entered from the keypad, and display the result of the dialing process. It will also maintain the mobile network connection to the base station nearby. The Linux kernel manages these processes so that they can be processed by the CPU of the Smartphone in a proper order. The Linux kernel also allocates the memory to store the data and commands to be processed by the CPU and store the processing results.

Device Management: The Linux kernel controls the hardware drivers, CPU, memory, keypad, battery, and network devices. The communication between a piece of hardware and software is done through a driver. The driver translates the command from the Linux kernel to the binary code used to control the activities of a device. The driver also translates a request from the device to a message that the Linux kernel can understand. Drivers are often provided by hardware manufacturers. One of the tasks handled by the Linux kernel is the management of various drivers. When multiple requests from different devices have arrived, the Linux kernel arranges these requests in a queue for the drivers to handle. Then, the kernel moves the messages translated by the drivers to the CPU for processing. When the rate of input is faster than the CPU's processing rate, the Linux kernel saves the requests in a buffer. When the hardware in a mobile device is updated, the kernel makes sure that the mobile device can keep up with the changes by using the updated drivers.

Memory Management: During computation, the operating system fetches or stores data to or from the fast storage called memory. As we know, mobile learning requires multiple tasks to be run on the server and mobile devices simultaneously. Linux is the type of operating system that can handle these multiple tasks. To make sure that the multiple tasks do not interrupt each other and each task has enough space to run, the Linux operating system allocates adequate memory and set up the boundary for each task so that no task will be interrupted by the other tasks. When a mobile device is started, the Linux operating system first allocates enough Random Access Memory

(RAM) for itself, and then it allocates some RAM space for hardware drivers. After that, the Linux operating system allocates the RAM space for the application software. The application software will run within the fixed boundary of the RAM space allocated for it. For the frequently used data, the Linux operating system creates a cache which is a small amount of high speed memory near the CPU to store these data. Some of the application software may require more memory than the available RAM. In such a case, virtual memory created on a mass storage device such as a hard drive can be used to extend the memory space. Based on the requirements and schedules of the tasks, the Linux operating system manages different types of memory to make sure that these tasks can be accomplished.

Process Management: While running an instance of a computer program, the operating system creates a process. During mobile learning, if a class of twenty students are logging on to a learning management system (LMS), then twenty processes will be created by the operating system. Many other processes are also created to support the activities of the LMS and to support the operating system itself. To manage these processes, the operating system creates a process control block for each process. A process control block may include a process ID, a record of files opened by the process, a set of process states that are stored in the registers, and the priority of the process. Each process is assigned a unique ID. The processes are sequentially executed. To speed up the execution of the processes, the operating system can start the next process while waiting for the completion of the input/output activities of the current process. When switching from one process to another process, the register contents of the old process are saved and the register contents of the new process will be loaded. The status of a process indicates if a process is currently running or suspended. The suspended process can be saved to the mass storage device such as a hard drive. When there are multiple processes, the processes frequently communicate with each other. The output of one process can be inputted to another process. When too many processes are started, the operating system will spend a lot of computing time managing these processes, which can significantly slow down performance.

Networking Stack: Mobile devices communicate with one another through mobile networks. The Linux kernel used in the mobile operating systems provides a networking stack to handle networking tasks. The networking stack provides various protocols to manage telephone calls and Internet access. In general, the networking stack may include the application layer and the kernel components as described below.

- *Application Layer*: The application layer includes protocols for handling the communication between applications such as conference calls or remote database access. The protocols in this layer can be used to establish, manage, and terminate sessions which handle requests and responses between the hosts.
- *System Call Interface*: This kernel component handles the communication between an application and a service provided by the Linux kernel. With the system call interface, an application is able to access the hardware or the operating system components.
- *Protocol Agnostic Interface*: As a component of the Linux kernel, the protocol agnostic interface supports various types of ports, communication protocols, and physical devices. By using the protocol agnostic interface, a mobile device producer can minimize the development time and design resources.
- *Network Protocols*: They are responsible for implementing various network transport protocols such as TCP and UDP.
- *Device Agnostic Interface*: This kernel component handles the communication be-

tween different hardware devices and network protocols. It is a common interface to support a variety of hardware drivers. This interface allows hardware manufacturers to add their own device drivers to Linux without conflicting with other device drivers.

- *Device Drivers*: A driver is used to handle the communication between a hardware device and software. The commonly used device drivers are Ethernet drivers, Wi-Fi device drivers, Bluetooth device drivers, game console drivers, GPS device drivers, digital camera device drivers, and so on. By using the open source code of the Linux kernel, the design and development of drivers can be made easier.

As described above, the Linux networking stack manages and controls network traffic. For mobile learning, the networking stack built in the Linux kernel is essential. In addition, the Linux kennel may also include some utility modules for network management.

Power Management: The Linux kernel supports two power management packages, Advanced Power Management (APM) and also Advanced Configuration and Power Interface (ACPI). ACPI is set up in the operating system to intelligently manage power consumption of the LCD, CPU, memory, and mass storage. In the newer version of the Linux kernel, ACPI is supported by default. APM is an operating system independent BIOS controlled power management package.

Many of the operating systems also manage application development platforms which are often included in the operating systems. Instructors and students can create their own application software on the application development platforms. Although there could be hundreds of different types of mobile devices, the application software developed for one operating system can run on different mobile devices if the same operating system is installed on those devices. The operating system can also make the application software run on these mobile devices even if they are upgraded with new hardware and software.

When a mobile device boots up, the operating system is loaded to memory by a piece of software called bootstrap loader. After the operating system is loaded to memory, it will take over the control and management tasks from the bootstrap software. It checks the hardware such as the CPU, memory, and network devices to make sure that these hardware components are working properly.

In addition to the Linux kernel, a mobile operating system such as Android also includes a few more components such as Applications, Application Framework, Android Runtime, and Libraries. These components include some commonly used mobile applications and handle the development and execution of the applications.

LINUX BASED MOBILE OPERATING SYSTEMS

Linux is known to be the operating system on the server side. Linux operating systems have been widely used to host Web servers, network servers, print servers, database servers, and LMS servers. The Linux based mobile operating systems have also been adopted by many well known mobile device manufacturers. Among the Linux based mobile operating systems, Android has been widely supported by many mobile device manufacturers and mobile phone companies such as Motorola, NEC, Panasonic Mobile Communications, Samsung Electronics, T-Mobile, Intel Corporation, and so on. There are several other Linux based mobile operating systems that are designed specifically for certain types of mobile devices or a specific brand of Smartphones. The following briefly introduces the commonly used Linux based mobile operating systems.

Android: Initiated by Google, Android is currently managed by Open Handset Alliance, which has over thirty members from hardware, software, and telecom companies. Android is a

Linux based mobile operating system plus an application software development platform. As an operating system, Android has been adopted by a large number of mobile device manufacturers such as T-Mobile, Samsung, and Motorola. The application software development platform provided by Android provides the environment for application developers to develop application software for mobile phones. Mobile device manufacturers can use the platform to create code for their own hardware.

On top of the Linux kernel, Android adds components such as Applications, Application Framework, Libraries, and Android Runtime. The following briefly describes these components:

- *Applications*: This component includes the Web browser, map, email client, short message service (SMS), and other software. All applications included in Android are written in the Java code and are run on the virtual machine called Dalvik which is optimized for low memory requirements. The Java programming language is also included in Android to help users create utilities and services for their own mobile computing tasks. To further help users develop their own software, Android also include tools such as the device emulator and Eclipse IDE.
- *Application Framework*: This component provides a set of standard APIs so that application developers can get a quick start on creating their application software. The application developers may find components such as View for GUI development, Content Provider for data access, Resource Management for sharing external source files, Notification Manager for creating alert message, and Activity Manager for managing the life-cycle of application software and user navigation history.
- *Libraries*: Through the Application Framework, the application developer can access the libraries which include the code used for developing graphics, multimedia content, applications for telephone services, applications for wireless communication, and database applications.
- *Android Runtime*: Runtime is a collection of software used to support the computer code execution. Android Runtime includes a set of core libraries to support the application software to run in its own process with its own instance of the Dalvik virtual machine.

Android is built on Linux version 2.6 which is used to provide the core system services The Linux-based mobile operating system in Android is used to handle tasks such as user account and security management, file management, memory management, process management, network management, and service management. The descriptions of these tasks are given below.

- *User Account and Security Management:* Android relies on the Linux kernel to protect the system from hackers and viruses. User management on a mobile device is also one of the tasks done by the Linux kernel.
- *File System Management:* The Linux kernel in Android manages the file system which controls the data storage service for mobile devices.
- *Memory Management:* The Linux kernel manages memory by allocating and deallocating memory for the file system, processes, applications, and so on. It also manages the cache used by a mobile device.
- *Process Management:* During the execution of computer programs, a process is an instance of a computer program which is executed by the operating system. The Linux kernel starts, executes, and stops processes.

- *Network Management:* The Linux kernel in Android is used to manage network communication. It controls the networking stack, drivers, network adapters, as well as routing devices.
- *Service Management:* The commonly used services of a mobile device are system logging, Internet search, and voice communication. Service management is another task of the Linux kernel.

Android supports mobile and wireless technologies such as GSM/EDGE, CDMA, EV-DO, UMTS, Bluetooth, and Wi-Fi. For multimedia content, Android also supports various multimedia formats such as MPEG-4, MIDI, WAV, JPEG, PNG, GIF, BMP, and so on. Android supports various hardware technologies such as touchscreens, GPS, accelerometers, video/still cameras, and so on. Android also supports the multi-touch technology. For the application software, Android provides different e-mail client software, maps, and Web browsers. It also supports Adobe Flash, Microsoft Exchange service, and is compatible with the Microsoft Office applications.

webOS: In addition to Android, there are other Linux-based mobile operating systems. webOS developed by Palm is another one used by the Palm Pre Smartphone. Like Android, Palm webOS is built on Linux version 2.6. On top of the Linux kernel, webOS adds components such as Applications, Mojo Framework, UI System Manager, and webOS Services. The following briefly describes these components:

- *Applications*: This component includes software for social networking, entertainment, maps, calendar, music, navigation, finance, games, video, and so on. webOS is designed to enhance social networking and to support the Web 2.0 applications.
- *Mojo Framework*: The Mojo Framework provides a set of tools to assist application developers. With the Mojo Framework, the application developers can get a quick start on developing Web-based applications such as e-mail, contacts, calendar, and finger gestures.
- *UI System Manager*: This component is responsible for user interface management. It manages application launching, navigation, and lifecycle. It schedules and controls the running of applications. It also manages events and notifications. It demonstrates the system status and the search results on screen.
- *webOS Services*: The webOS services includes the application services for accessing core applications such as launching browsers and running multimedia content, system services for hardware management and location tracking, cloud services for Web application management such as publishing and subscribing instant messaging service through the Extensible Messaging and Presence Protocol (XMPP).

webOS provides rich Smartphone features. For example, its graphical user interface is designed to support touchscreens. It has the multitasking capability of allowing users to run several applications simultaneously. To help the management of personal information, it supports JavaScript, HTML 5, and Cascading Style Sheets (CSS) for document formatting and appearance. Through Palm Media Sync, one can run the multimedia content of different formats and on different devices. It allows users to access an online music store and social network service such as Twitter. webOS also provides location-based service for locating restaurants and theaters.

Moblin: Developed by Intel, Moblin which stands for Mobile Linux is a combination of a Linux based operating system and an application suite. The Moblin operating system is based on the Linux kernel and includes various device drivers provided by the Linux kernel. The application suite includes a set of typically required software applications such as the Web browser, e-mail, office product, and multimedia software.

As a product of Intel, Moblin is optimized for the Intel Atom processor which is designed for Mobile Internet Devices (MIDs), netbook computers, Smartphones, and embedded devices. Moblin is featured by its speedy performance and fast boot time. Since Moblin is optimized for the Intel Atom processor, other major Linux distributions would like to integrate Moblin to their Linux operating systems. For example, Ubuntu and SUSE Linux have both released their versions of Moblin operating systems.

To support application developers in creating software for mobile devices, Moblin provides User Interface (UI) services and APIs. The UI panel displays components such as multimedia, status, pasteboard, people contact, Internet, and applications. The UI is developed on the Clutter animation framework which is an open source platform for developing animation based applications. With the Clutter animation framework, the developer can create a complex UI on top of the built-in classes, so the developing work can be reduced to minimum.

The application suite is designed to handle multimedia content, Internet browsing, and social networking. Moblin supports a variety of Linux desktop applications via GNOME Mobile technology. It includes a connection manager, Mozilla-based Web browser, media player, Moblin Garage application store, Moblin Application Installer, and Moblin Image Creator (MIC). These applications are briefly described below:

- *Connection Manager*: The connection manager is used to manage the Internet connection. By using the plug-ins, it can connect to the Internet through various wired or wireless technologies.
- *Mozilla-Based Web Browser*: The Mozilla-based Web browser is optimized for mobile devices. It includes a finger-driven UI and MID UI integration. It also supports crucial plug-ins such as Adobe Flash.
- *Media Player*: The media player performs audio and video playback and photo viewing. It includes GStreamer multimedia frameworks for constructing graphs of media-handling components. It also supports the Universal Plug and Play (UPnP) through the GUPnP framework designed for connecting consumer electronics, intelligent appliances, and mobile devices.
- *Moblin Garage Application Store*: With the Moblin Garage application store, one can store the open source applications compatible with Moblin as well as the freely distributed proprietary applications.
- *Moblin Application Installer*: The Mobile Application Installer can be used to launch the applications stored in Moblin Garage application store. The applications are categorized for easy access.
- *Moblin Image Creator (MIC)*: By using MIC, a platform developer can build a file system specifically designed for a mobile device by selecting a group of files provided by Moblin. The group of selected files can be loaded to a mobile device through a USB storage device.

Due to these features, Moblin performs very well on netbook computers. According to Smart (2009), Moblin is changing the trend in netbook market.

LiMo: LiMo is short for Linux Mobile which is the project supported by the LiMo Foundation organized by companies such as Motorola, NEC, NTT DoCoMo, Panasonic Mobile Communications, Samsung Electronics, and Vodafone. Like Android, LiMo includes a Linux based operating system and an application development platform. Unlike Android, LiMo's goal is to assist mobile device companies in creating their own applications by providing the middleware for the management of drivers and modem interfaces. Adding the middleware, LiMo has three components, Linux kernel space, middleware, and application. Descriptions of these three components are given below.

- *Linux Kernel Space*: This component includes the Linux kernel, device drivers, and modem interfaces.
- *Middleware*: This component provides functionalities related to security, registry, conflict management, frameworks for networking, database, multimedia, messaging, and so on. It also includes the management module which is used to maintain the consistency amount partners of the LiMo Foundation. To help mobile device companies create their own applications, the middleware is built to support the phone handsets produced by the partners of the LiMo Foundation. Application developers can add their own functionalities to the objects provided by the middleware.
- *Application*: This is an application software development kit (SDK). The SDK provides an environment for application developers to create mobile phone applications such as the user interface and Web tablet.

Ubuntu MID Edition: As a Linux operating system producer, Ubuntu has developed the MID edition of the Linux operating system specially designed for Mobile Internet Devices (MIDs). Ubuntu MID Edition includes three components, Ubuntu Linux, GNOME Mobile, and Intel's Moblin. It is built to fully support the Intel Atom CPU. Unlike Android and webOS which are designed mainly for telephones, Ubuntu MID Edition is designed mainly for mobile devices such as MIDs and netbook computers. Its features are summarized below.

- Ubuntu MID Edition is built to support rich Web content. It is able to run a variety of Web 2.0 applications such as FaceBook, DailyMotion, MySpace, and YouTube.
- To support the development of rich Web content applications, Ubuntu MID Edition provides Asynchronous JavaScript and XML (AJAX) which is a Web 2.0 application development environment. It also provides strong support for creating and managing audio, video, and graphic materials.
- For application development, Ubuntu MID Edition can supply application developers with programming languages such as Java, Python, Flash, and HTML.
- The application developers are allowed to modify the source code provided by Ubuntu for better functionality, flexibility, and extensibility.
- Ubuntu MID Edition can provide the desired open source, free or commercial application software and services to assist individual mobile device manufacturers to achieve their market strategies.
- Ubuntu MID Edition is designed for handheld and finger-tip use. It supports a full range of touch screen applications. It also supports the physical keypad, keyboard, and on-screen keyboard.
- To simplify the access of mobile networks, Ubuntu MID Edition provides the utilities for connecting to various types of mobile networks such as Wi-Fi, WiMAX, HSDPA, HSUPA, Ethernet, and Bluetooth. It also supports real-time conversations through VoIP or instant messaging.
- To support the mobile devices that have limited computing power, Ubuntu MID Edition is designed to consume less electric power. It is optimized for mobile devices so that it can run on 256MB RAM with a 2GB flash drive.
- Ubuntu MID Edition supports various devices such as GPS, flash memory, hard disk, and digital cameras.

- Ubuntu MID Edition is designed to support many commonly used network services such as Internet TV, discussion groups, sites for blogging, and online video games.
- Ubuntu MID edition can be adapted to be used by a wide range of embedded systems such as a DSL router.

As the above features indicate, Ubuntu MID Edition has a wide range of usage. It can run on mobile devices as well as wired network devices. However, it is not designed for mobile devices like low-end cell phones.

Google Chrome: Chrome is another Linux based operating system developed by Google. Chrome targets at netbook computers. It is designed to handle cloud computing for netbook computers. Therefore, it is more like the enhanced version of a Web browser. Unlike other operating systems, Chrome boots up very quickly and performs extremely fast when handling Web pages. On the other hand, Chrome does not support desktop applications such as office software and photo editing on the local computer. All the desktop computing tasks have to be done at remote servers through the Internet. As pointed out by Google, data can be stored in cloud storage equipment, multimedia editing can be done through Internet audio and video services, office documents can be handled by Google Docs which is a Web-based word processing, spreadsheet, presentation, and form application.

The Linux based operating systems and application development platforms introduced above target at various mobile devices. These mobile operating systems have their own characteristics. Their usage may vary. The following is a brief review on the usage of these Linux based mobile operating systems in mobile learning.

SELECTION OF LINUX BASED MOBILE OPERATING SYSTEMS

Linux operating systems have been used for servers, and desktop and notebook computers for decades. Linux distributions such as RedHat, SUSE, and Ubuntu provide both the server and desktop versions of Linux operating systems. Debian, another well known Linux distribution, provides the Linux operating system that can be used on both the server and desktop computers. Since these Linux distributions have been examined thoroughly in other books, this section will focus on the Linux based mobile operating systems.

A Linux based mobile operating system is openly designed for a specific device or even a specific phone. While we implement mobile learning, selecting the Linux based mobile operating system and mobile device is very important. To help readers with their decision making, the following is a brief guideline on the selection of mobile devices and Linux based mobile operating systems. The factors that may impact the selection are cost, usability, and flexibility.

Cost: In general, the cost of a mobile device powered by a Linux based mobile operating system is relatively low when compared with proprietary mobile devices. The mobile device companies only need to modify the existing Linux system to fit the needs of their mobile devices. Therefore, the development of operating systems for their mobile devices does not need to start from scratch. Since most of the Linux based operating system are pre-built in the mobile devices such as the Smartphone, there is not much choice on which mobile operating system to use in terms of cost. For some mobile devices such as a netbook, Ubuntu Linux and Google Chrome are good choices. These operating systems are designed for mobile devices with more processing power and do not cost the user anything.

Usability: Linux is a widely used open source operating system. As a server or desktop computer operating system, Linux supports a wide range of open source application software. Linux can be easily integrated into the computation platforms used by consumers' electronic and automotive devices. According to Beyers (2009), Android had 20%, Palm webOS had 5%, of the U.S. Smartphone traffic during October, 2009. That is, these two operating systems alone had a quarter of the Smartphone traffic. Beyers' report shows that Android is only second to iPhone OS in terms of the Smartphone traffic. The Android operating system has been or will be used to power dozens of Smartphone handsets such as, T-Mobile myTouch 3G, Samsung Behold II, Motorola Droid, HTC Magic, and Huawei U8230, just to name a few. In addition to mobile device makers, giant IT companies such as Dell, Lenovo, and Acer will also produce Smartphones powered by the Android operating system. Android supports many Web-based services provided by Google such as Maps, Gmail, and GTalk. Not only is Android adopted by many computer and telephone equipment makers, it also provides over 20,000 applications by the end of 2009 (Kirk, 2009). The later versions of the Android operating system support some nice features such as the multi-touch screen and voice dial. For mobile learning, students can use the Android emulator to develop Android application. Based on the usability, Android is the first choice. Sprint customers can purchase Smartphones powered by another Linux based mobile operating system Palm Pre. For Mobile Internet Devices (MIDs), Ubuntu MID Edition and Intel Moblin are good choices.

Flexibility: Due to the availability of open source code, Linux based mobile operating systems, in general, are much more flexible than proprietary operating systems. For example, Android allows mobile device makers to customize the operating system by modifying the source code or adding the makers' own code to the Android operating system. Even the users of Android phones can edit the operating system source code. The Android users can download and install any of its software as they want without limitation. The installation of applications can be done through the Internet, a computer, or a SD card. The Android users can update the operating system to use the latest soft keyboard, phone book, ringtones, and themes daily without waiting for the new versions of the operating system. Android allows users to tether the mobile phone so that the phone can be used as an Internet access modem. The Android operating system can run multiple tasks simultaneously. With these features, Android is the first choice for Smartphone devices. For other mobile devices, Linux based operating systems should be a better choice than proprietary operating systems which have many restrictions in terms of system update.

As described above, mobile learners can benefit greatly from Linux based mobile operating systems. On other hand, the Linux based mobile operating systems have their own weaknesses. For example, although the Android application store has over 20,000 applications, the number is much smaller than that in the Apple application store which has over 100,000 applications. Linux based mobile operating systems do not support many of the widely used proprietary application software such as audible online books. These weaknesses should be considered when you select operating systems for your mobile learning projects.

CONCLUSION

Based on the requirements of a mobile learning system, Linux based operating systems can be used to handle computing tasks on mobile devices as well as on the server side which supports the mobile learning services. On the server side, Linux based operating systems can be used to host application servers such as the LMS, e-mail server, and Web server. On desktop and notebook computers, Linux operating systems can be used to develop mobile learning course materials. On mobile devices, Linux based mobile operating systems support

Web browsing, text messaging, accessing mobile learning services, and telephone calls.

This chapter introduces some commonly used Linux based mobile operating systems for mobile devices. It also investigates the mobile operating systems' architectures and functionalities. For the Smartphone, this chapter identifies Android as a better choice based on the cost, usability, and flexibility.

This chapter points out that, to meet to the needs of mobile devices, the Linux kernel used by a mobile operating system is a simplified version of a regular Linux operating system. The Linux kernel is used to handle tasks such as device management, memory management, process management, networking, power management, and application interface and user interface management. In later chapters, more usage of Linux based mobile learning systems will be discussed.

REFERENCES

Amuck. (2009). *Welcome to 31 days of iPhone apps*. Retrieved November 6, 2009, from http://www.appsamuck.com

Beyers, T. (2009). *Android is even bigger than you think*. Retrieved December 6, 2009, from http://www.fool.com/investing/high-growth/2009/11/23/android-is-even-bigger-than-you-think.aspx

Google. (2009). *Introducing the Google Chrome OS*. Retrieved January 1, 2010, from http://googleblog.blogspot.com/2009/07/ introducing-google-chrome-os.html

Kirk, B. J. (2009). *Android market surpasses 20,000 applications, T-Mobile picks top apps for customers*. Retrieved January 1, 2010, from http://www.mobileburn.com/news.jsp?Id=8419

LiMo Foundation. (2009). *Welcome to LiMo*. Retrieved November 6, 2009, from http://www.limofoundation.org

Maemo. (2009). *Maemo basics*. Retrieved January 1, 2010, from http://wiki.maemo.org/Maemo_basics

Moblin. (2009). *Create the mobile Internet future*. Retrieved November 6, 2009, from http://moblin.org

Open Handset Alliance. (2009). *Android*. Retrieved November 6, 2009, from http://www.openhandsetalliance.com/android_overview.html

Palm. (2009). *Palm Pre*. Retrieved November 6, 2009, from http://www.palm.com/us/products/phones/pre/index.html

Saponas, T., Lester, J., Froehlich, J., Fogarty, J., & Landay, J. (2008). *iLearn on the iPhone: Real-time human activity classification on commodity mobile phones*. (University of Washington CSE Tech Report UW-CSE-08-04-02).

Smart, C. (2009). Linux will regain lost market share, thanks to Moblin. *Linux Magazine*. Retrieved November 6, 2009, from http://www.linux-mag.com/cache/7559/1.html

Ubuntu. (2009). *Ubuntu Mobile Internet Device (MID) edition*. Retrieved December 9, 2009, from http://www.ubuntu.com/products/mobile

Vamosi, R. (2009). *The pros and cons of iPhone security*. Retrieved January 1, 2010, from http://news.zdnet.co.uk/security/0,1000000189,39287778,00.htm

Wimberly, T. (2009). *Android closing in on BlackBerry, taking share from iPhone*. Retrieved January 1, 2010, from http://androidandme.com/2009/08/news/ android-closing-in-on-blackberry-taking-share-from-iphone

KEY TERMS AND DEFINITIONS

Android: It is a mobile operating system with the Linux kernel. It is designed for mobile devices such as smart phones and tablet computers.

Central Processing Unit (CPU): The main task of a CPU is to carry out calculation for communication and for applications.

Mass Data Storage: It is used to permanently store information such as e-mail contacts, music, or photo images. A mass data storage device can be a built-in hard drive or a solid-state drive. It can also be a subscriber identity module (SIM) card or an external USB flash drive.

Memory: Memory is temporary data storage used by a CPU to quickly get necessary data for calculation and to store the calculation results.

Ubuntu: It is an open source Linux operating system. Ubuntu has server, desktop, and mobile editions.

User Interface Device: A user interface device may include a LCD panel, camera, keyboard, microphone, or speaker.

Chapter 4
Openness—Evolution of Mobile Communications: A Cloud View

Sebastian Thalanany
U.S. Cellular, USA

ABSTRACT

In this chapter, the author reveals the role of openness, specifically standardization, in the mobile cloud – a cornerstone for an open, inter-operable realization. The discussion of mobile communication, the 4G mobile network, mobile cloud, and the openness in the mobile cloud are covered in this chapter.

INTRODUCTION

Technology enhancement continues to enable widespread connectivity. Mobility is an integral component of consuming and producing content, promoted through a wide array of connectivity choices. A choice of experiences enables proliferation, with a diversity of mobile devices and multimedia services. Service experience, in a connected world, is elemental, and one that has universal appeal. In this web of connectivity, both mobile devices - user facing - and machine type devices - non-user facing - are the two broad categories of information transactions.

Openness is an essential ingredient that fosters an unencumbered flow of information across entities: user-facing and non user-facing. Standardization is an integral component of openness. Indispensable – for the realization, preservation, and expansion of the Internet of Things – a paradigm, where the tethered and untethered entities interact and transact to provide an attractive service experience, in the ubiquitous Internet. Distribution – ubiquitous creation and consumption of information and intelligence.

DOI: 10.4018/978-1-60960-613-8.ch004

Services of all sorts are rendered through the availability of resources of connectivity and the execution of applications of interest of a consumer or producer of information. Typically, the execution may either be localized to a device or separated through connectivity between a requesting device and a remote server. For allowing the execution and the experience of a service on a device, with limited resources – mobile device – independent of location, would be through connectivity to an operating system, in a network server. The relatively unconstrained resource availability and management capabilities, resident in a fixed network has the potential of rendering a consistent user-experience across disparate geo-locations of user connectivity. Experience consistency is likely to be more uniform across different devices since the execution environment for service rendering remains unchanged. Mobile cloud - an integral component - a part of the cloud paradigm (Hartig, 2009).

A consideration in the preservation of a consistent user-experience, in the cloud, is the notion of grid computing, where the performance of the network servers is sustained through the participation of additional servers on a resource demand driven basis. For example, if a popular service is launched by users, additional servers may be required to meet the computing demands for a sustained experience.

The compelling aspect of the cloud paradigm is the ability of any connectivity enabled device - mobile or fixed - to provide an attractive user experience, independent of location. Users are empowered to access their favorite services, independent of device type - mobile or fixed - enabling virtually unlimited usage scenarios. A few examples include homes, libraries, cafes, airports, trains, planes, automobiles, etc. Cloud mobility - remote processing and storage - paves the path towards widespread user-centric information accessibility.

This chapter reveals the role of openness, specifically standardization, in the mobile cloud – a cornerstone for an open, inter-operable realization.

MOBILE COMMUNICATIONS

Evolution: A Shifting Paradigm: Circuit to Packet

The origins of mobile communications are rooted in the provision of voice services. The conversational, real-time, and the small information payload nature of voice media demanded and allowed the use of circuit-switched paradigm. Inherited from the fixed telephony world, it has worked well for decades, with widespread adoption. On the other hand, the explosive, ubiquitous, and continuing growth of the Internet has ushered in an era of multimedia services - a packet-switched paradigm – one that enables a distribution of transport paths and mixed-media (voice, data, and video). This shift allowed flexibility and adaptability in the formulation of architectural models towards larger capacities, mixture of real-time and non-real-time media, lower costs, and higher rates of information transactions.

Internet and Mobile Telephony

The Internet has been established as a global fabric, for the creation, delivery and consumption of multimedia services - a fabric that utilizes the distributed nature of the packet-switched paradigm. The pervasiveness of this model has influenced the evolution of mobile communications, through the adoption of the same vision. The IP (Internet Protocol) has been extended to the edges of the mobile network, beginning with 3G (Third Generation) systems – a vision that is being extended beyond 3G. Pivotal - extension enabling mobile broadband.

Mobile telephony, with voice-centric services, inherited a tightly controlled business and operational model from the much older fixed telephony ancestor. The high cost of licensed spectrum, the related, silo-oriented, switching infrastructure, operations, and customer support were accommodated by consumers, in the early days, for an untethered communication experience. A captive market, until lower cost packet-switched alternatives appeared, such as WiFi wireless modems in portable computing devices, such as laptops and smartphones.

Multimedia Aspects

The notions of the rich media experience over the Internet and mobility are a natural convergence - notions that require the incongruent cultures and technologies of the Internet and mobility to merge in a graceful manner. Using these notions, the nature of wireless connectivity is effectively mined, while embracing the ubiquity, flexibility, and openness of the Internet protocol. Such a framework enables a decoupling of service creation aspects from the nuances of the underlying transport technologies, through appropriate abstractions. The framework facilitates a promotion of lower operational costs, together with a simplified service creation environment, resulting in a wider participation of Internet application development communities – a collaborative union for the creation, delivery and consumption of multimedia services. This allows a seamless service experience - fixed or mobile - everywhere.

The technology enabled shift from a mobile telephony model to a mobile multimedia model has widespread ramifications across the value-chain. The former is characterized by a hierarchical, centralized, and rigid nature. The latter allows flat, distributed and flexible architectural models for the realization of a vast array of possibilities, for both multimedia service creators and consumers. Corresponding cultural changes demanded to meet the challenges of the ongoing shift will largely determine the effectiveness of mobile multimedia service evolution. A change that leverages the enabling benefits of technology evolution, while influencing appropriate business model shifts and regulatory reform.

Standardization: A Global Collaboration

Mobile technologies have been based on standards to allow a consistent and inter-operable implementation and deployment. Over time, these technologies have adopted the evolving standards to meet the demands dictated by the market. The radio interface standards evolution has generally preceded the market evolution, during the 2G and 3G timeframes. As the demand for multimedia services continues to grow - with the significant influence of the impact of Internet services and the declining voice service revenues - the evolution of standards above the radio layer is driven by expectations and trends in the market. In some cases, new capabilities have been adopted in the market, where there are adverse implications, such as limited inter-operability. These implementations are likely to stifle market expansion, through a projection of behaviors and expectations, particularly in the case of new capabilities, where inter-operability is restricted. On the other hand, proprietary capabilities at the service layer, where inter-operability is not restricted, may not require standardization, from a technology perspective. Historically, an open and collaborative environment, for the development of standardized specifications, lends itself to a generally more robust and lower cost product, through a healthy participation of interested parties.

The widespread participation fostered by the global standards bodies provides a forum for consensus across technological and market requirements that are often colored by regional variations. These challenges often take a longer time to settle and therefore require visionary leadership, with a long-term perspective, to steer and formulate viable strategies for the selection and initiation of candidate capabilities/features for standardization.

A systematic approach ensures inter-operability for market creation, market demands, and market expansion. With the rising complexity of communication systems, standardization plays a significant role in the management of inter-operability, cost, widespread implementation, knowledge transfer, and extensive collaboration across products and services. This is a sharp contrast with proprietary implementations, which may reach a market quicker but is severely limited in terms of interoperability, subject to monopolies, and a constrained market expansion potential. An open framework with standardized interfaces has a distinct advantage, both in terms of reducing costs, as well as promoting innovative implementations, while preserving inter-operability and extensibility.

This is the backdrop in the evolution of the next generation mobile system architectures - 4G mobile broadband. The packet-switched paradigm is the foundation of the LTE (Long Term Evolution) (Third Generation Partnership Project, 2010), WiMAX (World-wide interoperability for Microwave Access) (WiMAX Forum, 2010) and CDMA (Code Division Multiple Access) (Third Generation Partnership Project 2, 2010) systems, in the evolution towards 4G. These systems use an all IP core network environment for widespread inter-operability and inter-working across an assortment of underlying access technologies. To complement the evolution of the mobile core network, the radio access network is evolving towards spectrally efficient, cost efficient technologies that leverage OFDMA (Orthogonal Frequency Division Multiple Access) principles to support mobile broadband. A coordinated evolution trajectory towards the realization of the performance demands for a mobilization of the Internet, fostering a seamless multimedia service experience.

The features and capabilities in the 4G mobile ecosystem include support for a variety of media types – delay-sensitive, interactive, and non-delay sensitive. The notions of "always-on", "presence", "QoS", "policy and charging", and "roaming" require particular attention from a standardization perspective for consistent behaviors, expectations and experience. Standardization of the mobile broadband architectural models provides a framework of open interface specifications, which enable broad market participation across equipment vendors and service providers – essential elements for market evolution and expansion.

Towards 4G: Implications beyond 3G

The role and significance of standardization continues to grow proportionately with the complexity of the mobile ecosystem across the access, core, and service layers. The interfaces across each of these layers and also within each of these layers are required to be well characterized and defined to promote practical and robust implementations. Openness is the dominant theme, in the specification of the building-block functions and interfaces.

New techniques and optimizations are part of the ongoing research, to meet the demands of mobile broad, data-oriented service mobility. In the radio access layer, these improvements are approaching the limits that are dictated by the laws of physics - the Shannon bound. The noise-limited nature of a wireless link requires acceptable minimum signal energy per bit, relative to the noise floor for a desirable performance measure - a significant foundation that shapes the user-experience, via the mobile device. The form factor and the battery-life of the mobile device constrain the performance measure enhancements that are achievable through increased antenna size augmentation and/or increases in the transmit power. Advanced signal processing techniques to optimize the signal quality, such as interference cancellation, add computational complexities, which then have to be traded off with a corresponding impact on the power consumption, in the mobile device. These are some of the aspects being implemented as part of the ongoing research related to incremental enhancements.

Considerations for a Transition

The research suggests that the performance and capacity demands, related to the unprecedented growth of mobile multimedia services, require the exploitation of distributed radio links, in contrast with the traditional, vertically integrated radio links, which were sufficient to meet the demands of voice-centric services. Distributed radio links, inter-connected, via a harmonized core network is an attractive paradigm that offers possibilities to manage local traffic demands, while allowing mobility across disparate radio technologies. Some of these directions include the notions of smaller transceiver coverage areas, for macro coverage. Macro coverage is constrained in terms of capacity and data throughput, which are traded off for a large radio coverage area. On the other hand, local coverage is constrained in terms of the coverage area, which is traded off for a higher-capacity and data throughput. A co-operation between the macro radio access technologies and the local radio access technologies (e.g. Femto, WiFi access points) allows the crafting of an appropriate heterogeneous environment, to meet deployment specific trade-offs. This co-operation will pave a path forward to meet the demands of capacity, coverage, throughput, mobility, and cost. The inherent challenges in a co-operative distributed, wireless access environment include interference management and network management. A heterogeneous access network environment provides opportunities to utilize the distributed nature of local area radio access points to harness the benefits of higher capacity and data throughput, for both the radio access segment and the backhaul segment - Internet backhaul. This reduces the burden on the cost and topology constrained macro backhaul environment. The use of fixed wireless as an alternative for the macro backhaul environment has the downside of spectrum related constraints, as compared to the wired backhaul (e.g. Ethernet/IP), which has the capability to maintain capacity demands consistently, with attractive cost impacts.

The inherent cost-benefit advantage factors in a heterogeneous access network model provides a framework through the establishment of standardized functions and interfaces that invite a broad participation from vendors and service providers. The open and collaborative nature of such a direction will serve as a catalyst for a mass market adoption to meet the unique challenges of multimedia service mobility – a model which is in stark contrast with the traditional silo-oriented concepts of "proprietary" or "first-to-market" notions founded in narrow perspectives that constrain market adoptability.

The significance of the heterogeneous network model lies in accommodating the demands of data throughput and capacity - a foundation for orchestrating a seamless and consistent user-experience. The elements of a heterogeneous access framework include a variety of inter-working capabilities between the next-generation all IP access networks and the existing third-generation mobile macro access environment. The inter-working capabilities encompass multimedia, including circuit-switched voice. A necessary bridge to promote the adoption of islands of next-generation all IP access networks, while supporting the existing market of mobile users.

Migration Perspectives

Migration strategies require careful crafting to ensure that the customer expectations, with regard to the existing services are preserved, while standardized, open technologies evolve to meet the demands of the marketplace. Both standardization and the accompanying technology evolution must address the building-blocks that are essential to allow the formulation of practical migration strategies. At the same time, new capabilities and features that are envisioned must be sufficiently forward-looking, beyond the designs of proprietary implementation to both shape and influence the creation of new market potential.

The regulatory regime is a significant catalyst for evolution in the arena of wireless spectrum availability for supporting adequate bandwidths necessary to satisfy the capacity and throughput demands of mobile broadband. The use and availability of both licensed and unlicensed spectra are attractive avenues for the exploration, innovation, and implementation of novel architectures in the advent and promise of the mobile Internet - pathways to manage consumer expectations, operational costs, and the delivery of mobile multimedia services seamlessly across disparate access technologies. Wired or wireless – a converging communication landscape.

The dominant technologies in the evolutionary 4G landscape are LTE and WiMAX. Inherently, these technologies demand wider bandwidths to leverage the benefits of OFDMA, MIMO, noise cancellation, interference cancelation, spatial division multiplex techniques, etc. Aside from the wider wireless highway benefits and an enhanced utilization of spectrum, network layer enhancements for a robust security framework, inter-working and handover capabilities are critical to pave the way for a heterogeneous access environment. These cross-layer considerations require to be understood properly, from an operational perspective for access providers, for widespread adoption.

The motivation for a migration towards the evolving open standards and technologies to meet the anticipated market demand is complemented through a proper understanding of the corresponding cost and performance benefits, beyond the existing third-generation systems. The collaborative, consensus driven work in the open standards initiatives, around the globe, provides a platform for promoting an awareness and understanding of the potential attractiveness of evolving technologies. Significant intermediation is required to promote this understanding, between the nuances of the standards processes/specifications and the business leadership. The virtual nature of mobile multimedia communications is impacted by the non-technology aspects, such as consumer usage paradigms, perceptions and cultures. These aspects have significant diversity across the various market regions worldwide. The global open standards initiatives play a critical role in this dimension. A satisfaction of local market demands, while preserving global inter-operability.

MOBILE CLOUD

The evolution considerations, mentioned in the preceding section, are being developed with an awareness of the cross-layer protocol aspects, including the application layer. Notions of distributed architectural models, information hiding, and access layer independence are among the open standards considerations, capabilities, and features. These notions are directionally favorable, in terms of incremental steps towards a cloud-oriented mobile multimedia service environment.

The open nature of the ongoing global 4G standardization initiatives implies that multimedia applications should be available to consumers, independent of the type of technology access. This openness is being extended across all the layers of the protocol stack, including the application layer. Further, seamless communication experience is sought across the underlying platforms, such as hardware and software. Collectively, the notion of openness encompasses all the preceding elements to foster a rich, diverse, interoperable, flexible, scalable and configurable environment, for the creation and consumption of mobile multimedia services.

Cloud Paradigm

Simply stated, Cloud (Amazon Web Services, 2010) computing is a paradigm that enables virtualization. The building-block capabilities that enable this virtualization are connectivity and computing resources. Virtualization in this case refers to the execution of systems of hardware and software, or the rendering of applications and services on remote resources - the Cloud. The connectivity of a mobile device to the Internet enables access to the applications and services in the Cloud. This is in contrast with traditional virtualization where remotely hosted services and applications are executed, using hardware and software resources locally resident within the mobile device. Hybrid combinations of these models for service execution and delivery are possibilities as well.

The attractiveness of the Cloud computing model (Express Computer, 2008) is that the access to services is device and location independent. Service examples include a potentially unlimited variety of Internet-oriented applications - social networking, instant messaging, email, word-processing, collaborative software tools, presentations, file storage, etc. Beyond location and device independence, Cloud computing offers a simplified service environment for diverse consumers without being impacted by the nuances of the computing hardware and software locally resident in the device - mobile or fixed. The design variations in the locally resident resources such as operating systems and clients have an impact on the rendered behavior of services and applications - some noticeable to the consumer. In some cases, specialized configurations of the device may be necessary to optimize a service experience. This is a cumbersome, non-appealing constraint for consumers, not interested in spending the time to familiarize themselves with the necessary procedures. Further, with the enormous rate at which new and diverse devices appear in the marketplace, knowledge of the device dependent nuances is not always portable - a constraining model from a widespread market adoption perspective.

Among the various benefits, such as an enhanced ease of interaction for the consumer and for service acquisition, presence management for the consumer is appealing since the Internet is founded on the principle of "always-on" and "anywhere". From a service provider perspective, there are cost benefits since the remote computing resources, being distributed in nature, are better positioned to handle peak load conditions as well as load-balance through resource sharing. The requirements for a mobile device to access services in the Cloud include a minimized processing configuration to support connectivity and a user-interface to launch applications that are executed on remote application servers. The notions of software as a service, and a web-oriented operating system are utilized in the Cloud where updates in different parts of the Cloud can be accessed from other parts of the Cloud. The use of DNS (Domain Name System) enables applications in the Cloud to access the required resources, such as files located in remote servers, in a location-agnostic fashion. This simplifies the operation and maintenance of applications and services. A very compelling and attractive model for an increasing number of nomadic consumers, for business or personal services, where consistent service behaviors are an imperative, where access may be from cafes, libraries, airports, or from any location, while mobile. The preservation of profiles and interests provides consumers with enhanced capabilities to leverage and use information.

The Cloud computing model may be partitioned in a logical manner. The application server virtualizes the execution of software, where the results are then transferred to a presentation server. The presentation server manages the interactions with a mobile client connected to the Internet. The application servers and the presentation servers are hosted in the Cloud. The distribution of this logical model provides robustness and scalability, while at the same time providing services in a consumer device platform independent manner. Local device backups of user-specific data are unnecessary and are an option driven by user-preference demands – a paradigm with implicit inter-operability, while accommodating a variety of user-interface choices.

Since a web-oriented operating system is resident in the Cloud, in contrast with a traditional operating system resident locally in a mobile device, the benefits of redundancy and availability of services are inherent in the Cloud. With the reduced local processing demands in the mobile device, the benefits include extended mobile power availability. In the case of netbooks or laptops, the storage of information in the Cloud offers a minimization of potential information loss resulting from local hardware/disk failures or theft. The universal nature of a web-oriented operating system allows a consistent experience for services accessed via a diverse range of mobile device hardware/software platforms. A benefit that accrues from this model is improved security since the servers are fewer, as compared to the potentially vast numbers of mobile devices, and therefore less vulnerable to security threat scenarios.

The Cloud computing (Armbust et al., 2009) resources facilitate a simultaneous handling of a large and scalable volume of distributed mobile applications. A grid of computing resources interlinks geographically distributed computing resources to support the demands of a Cloud overlay of application servers. This approach provides the flexibility and the computing power to handle processing intensive real-time, near real-time, and non-real-time applications – a paradigm that provides mobile users with options to optimize their preferences, in terms of data management and experience, and of the services rendered by the Cloud. It leverages a distributed file system for the management and the storage of information across the computing nodes defined by the system – an information hiding model that veils the complexities of parallel programming, for mobile users. The grid manages the availability of computing nodes while also allowing a mobile user to partition information, where there may be local information copies acquired from the distributed file system. This is an essential capability in the Cloud for load balancing as well as fault management – an implicit self-optimizing capability, through information replication and a replenishment of computing resources, in the event of failures. The grid facilitates parallel processing through the use of artifacts, such as tables where each row is an independent record of column-oriented information elements that are organized according to a schema governed by a specified template.

The two aspects that are central to the theme of a mobile Cloud, containing the mobile computing function (Punithavathi & Duraiswamy, 2008), are: a) A robust underlying SON - Self Optimizing Network, for the processing and transport of multimedia application information, and b) A mobile device or a fixed device, with multimode (wireless broadband - LTE, WiMAX, WiFi, wired broadband - Ethernet cable) access capabilities, respectively. The evolution of the smartphone family of mobile devices is an example of mobile device types that are aligned with the required capabilities to leverage services in the Cloud. From the perspective of location independent service availability, the Web 2.0 paradigm (O'Reilly, 2005) provides a framework for the development, management, and deployment of multimedia services in the mobile Cloud. The pervasive information access and service availability over the fixed Internet has ushered in a significant interest in the extension of Web 2.0 to the mobile realm. The notion of Web 2.0 was crafted with the intent of providing web-oriented services for enabling an enhanced user-experience, in terms of both the personal and social context. Flexibility and choice related information transactions, collaboration, and applicability are some of the significant ingredients inherent in a Web 2.0 framework.

For both personal and business services, conveniences and intuitive user interfaces are critical. In an increasingly complex web of information exchange demands, in both spatial and temporal dimensions, the convenience and intuitiveness of applications are critical. Mobile commerce, context, content, social networking, and location provide a categorized structure of service types that shape critical service profiles for widespread appeal. The distributed computing resources in the Cloud provide content, context, and geo-spatial awareness, linked to user-profiles, which enable instant access to user-information history. This is particularly beneficial for users to recollect transactions or other storage information resident in the Cloud. This also facilitates convenient and dynamic changes to user-profiles and interest-preferences. In concert with the capabilities of the Cloud, the mobile device leverages sensors, locally resident in the devices, to present the user with an audio-visual and haptic experience – a rich virtual experience on the go.

Web Services: Cloud Enabled

The web-services environment provides an interoperable, extensible, distributed, and reusable framework for information processing in the Cloud – a loosely coupled framework enabling virtually unlimited possibilities for the creation and consumption of multimedia services. A separation of the interfaces from the implementation allows a decoupling of the services from a hosting platform - hardware and software - fostering an unencumbered application development model. The framework includes three types of entities: a) Service requester, b) Service registry, and c) Service provider. A service requester entity, which intends to acquire a service from a service provider, searches a service registry to acquire a corresponding service description. After the service description is acquired, it is bound to a corresponding service provider. When the binding is established, the service requester interacts with the service provider for the acquisition of the requested service.

The procedures that are necessary for the use of web-services may be classified as, a) Publication of service descriptions by the service provider in a service registry, b) Retrieval of a service description by the service requester, statically - application development phase or dynamically - application execution phase, from the service provider or the service registry, c) Binding of the service description with the service implementation, specific to a service provider which may be static - application development phase or dynamic - application execution phase. In the latter case, the binding may be to different service implementations, specific to one or more service providers, and after the binding procedure, the abstract service type is bound to an actual service resource, defined by the associated transport protocol and parameters, d) Invocation of service, where the service requester acquires the desired service from the service provider.

The standard network protocol for web-services is HTTP (Hyper Text Transfer Protocol). The web-service procedures are contained in XML (Extensible Markup Language) messages. This interface provides the description for the necessary web service interactions. The components of this interface include the transport protocol, message format, and service location. XML provides a standardized approach for content markup that is useful for information transactions. It is the foundation for the web-service framework. It fosters the notion of a loosely coupled Cloud environment, through platform - hardware and software - independence, for a ubiquitous multimedia service experience. Platform independence implies that service creation and consumption are independent of operating systems and access technologies. SOAP (Simple Object Access Protocol) messages use XML for the exchange of structured information across computing resources. HTTP provides the transport for SOAP messages.

The creation and experience of web-services hinge on multimodal access - heterogeneous capabilities, in concert with user-interface augmentation in the mobile device. A well-defined, standardized and open API (Application Programming Interface) regime is essential for the evolution of rich multimedia services. The benefits are enormous in term of enabling third-party application development communities to participate in the creation of new services, as well as mashups. A mashup of diverse data and functions, using open APIs, is particularly attractive. This allows the rapid creation of services to meet new and dynamically changing market demands. Various partnerships are also fostered across participating segments to leverage their strengths and to contribute to the overall value-proposition - namely, market creation and sustainability.

Protocols and APIs: Different Objectives

While protocols provide an interaction and communication across hierarchically oriented entities - mobile devices and network elements - categorized via the layers in the protocol stack, APIs provide application access to underlying resources - hardware and software - for entities at the same protocol layer. The abstract representation - an inherent characteristic of service-level APIs - enables innovative application-level development possibilities since the composition of related functionality is isolated from the details of the underlying resources. Another difference between protocols and APIs is that the former facilitates symmetric communications, while the latter operates in an asymmetric fashion – commands are sent to the underlying resources, and decisions are enforced by event triggers and responses from these resources. Protocols facilitate system-wide inter-operability across a multitude of entities with diverse implementations. On the other hand, service-level APIs, through various abstract representations, provide a vehicle for a widespread service realization, with a consistent user experience. Open standards-driven specifications for both protocols and APIs are critical for service realization in the Cloud.

The NGSI (Next Generation Service Interface) initiative, within the OMA (Open Mobile Alliance) standards development organization, is one where open API specifications are being developed. These open APIs are intended to address the challenges of an emerging multimedia service world where personal, business, social and government oriented interests and demands are being considered in the formulation of API specifications. The API environment provides an abstraction layer for applications to utilize the underlying computing resources and capabilities - independent of the access type - wireless or wired. The types of APIs being considered may be broadly classified in terms of large scale services - voice, messaging, location, etc. - and long-tail (Anderson, 2006) oriented user-driven niche services. The prominent benefits of the NGSI initiative are: a) Well-defined interfaces to access resources and capabilities, b) Awareness of mobile device capabilities, context, user-profiles, service subscription, dynamic configuration, and user-preferences, c) Utilize service creation and consumption capabilities for an enhanced service experience.

The service-level components and interfaces associated with the NGSI framework are the following:

1. Data configuration and management
    - Management and update of data stored in XML documents
2. Registration and discovery
    - Search and discovery of resources
    - Resource registration
3. Multimedia list handling
    - Management of lists containing media types, media identifiers, usage mode, etc.
    - Management of list conditions
4. Context management
    - Management of context information access
    - Management of context information associated with context processing entities
5. Multimedia call control and configuration
    - Management of media for the configuration of call handling procedures
    - Management of multimedia conference call configuration

6. Identity control
    - Management of identity information associated with functional entities
    - Management of a federation of identities, such as third-party identities

In the spirit of inter-operability - use of standardized protocols - and widespread service creation and consumption - use of standardized APIs, the need to satisfy both aspects is essential in the evolution Cloud services. The participation of a large and growing Internet application development community, for mobile services in the Cloud, is pivotal. In this regard, simplified API specifications are significant to encourage a widespread participation. The NGSI initiative within OMA has adopted the REST (Representational State Transfer) architectural style of protocol binding for API specifications. In this style, the operations are abstractions of the following operations: a) GET - resource state acquisition, b) PUT - resource state update, with a new state, an idempotent operation, c) POST - resource state update, with a new state, a non-idempotent operation, and d) DELETE - resource deletion. Since these operations are akin to the familiar Internet oriented HTTP operations and transport the API information as parameters in a URI (Uniform Resource Identifier) or as HTTP parameters, the model has an inherent appeal among Internet application developers. The REST model is therefore significant with respect to the growth of web-services in the mobile Cloud. The availability of API specifications with both SOAP and REST web-services binding provides flexible choices in terms of API implementations. The choice of protocol bindings does not impact the API behavior and is therefore attractive for the co-existence of current and evolving applications. The protocol bindings for the abstract APIs provide applications with the ability to communicate with resources in the grid, in the rendering of rich multimedia services, available in the Cloud.

Convergence: Way of the Cloud

The mobile Cloud computing environment is a confluence of virtualization, grid computing and a service-oriented architecture that encompasses web-services. The service-oriented nature of the Cloud provides mobile users with a consistent experience, while also minimizing the computing demands on the mobile device. The computing capabilities in the Cloud for mobile access may be partitioned, in terms of access point and remote computing resources. This allows a location dependent and a location independent distribution of resources.

The scalable and flexible nature of Cloud computing enables the availability of distributed computing resources. This is particularly attractive for mobile usage scenarios to harness the appeal of Web 2.0 services. The capabilities of Web 2.0 services include semantically intuitive interfaces, context awareness, social connectedness, and user-preference-oriented customization that are compatible with the architectural framework of the Cloud. Some of the intrinsic Web 2.0 characteristics are: a) Harnessing the statistical long-tail for service potential – Google AdSense, b) Trust building – Wikipedia, c) Widespread distribution/decentralization – BitTorrent, d) User content creation – YouTube, e) Flexible user feedback mechanisms – Amazon/eBay service rating, f) Crowd participation – Blogging, social networking, g) Folksonomy – RSS feeds, content tagging/classification – Webshots, Flickr, Caedes, h) Rich user experience – Multimedia applications. The Cloud facilitates these services capabilities on-demand. A synergistic combination where the Cloud is powered by an underlying scalable grid of distributed resources, to provide Web 2.0-like services "anywhere", "anytime".

The Cloud paradigm is a departure from the rigidly partitioned architectural models, prevalent in the existing mobile systems. The flexible and distributed nature of computing and access technology resources establishes both an enhanced utilization of resources and system availability. A model that allows dynamic linkages across information repositories to meet the mobile multimedia service demands for creation, delivery, and consumption.

TRENDS: OPENNESS IN THE MOBILE CLOUD

Among the various challenges in the evolution of next-generation mobile systems, the management and provisioning of adequate and scalable system capacity are significant. The highly distributed nature of the Cloud is pivotal to meet these demands. End-user customization, context awareness, content awareness, and location awareness provide a transparent framework –one which is supported by the inherent characteristics of the Cloud.

Social networks and the highly personalized nature of mobile communications offer a unique perspective, in terms of the related implications, as compared to fixed or wired communications. Mobile communications have a significant component of usage modes which are linked indistinguishably with user behaviors and interests. The technologies to support these unique requirements must adapt to these characteristics, in a non-intrusive fashion. The Cloud model complements these requirements, as a natural fit.

The web-oriented landscape of communications includes: a) Third-party API development; b) Mashups, such as Google Maps, which accommodate various levels of context-awareness. Inherently intuitive, and obviates the need to learn cumbersome and changing user-interface modes. A particularly useful capability, especially as new mobile devices appear in the market; c) Collaboration, to stimulate user participation, which promotes the dissemination of knowledge and information; and d) Growth of the long-tail for market expansion.

Long-Tail, Inter-Operability, and Self-Optimization

The long-tail nature of the evolution is a natural consequence of connectivity and services, provided by the Cloud. Social networking is an exemplification of user-generated information that is available in the Cloud. It is a participative model that ushers various and numerous possibilities, such as blogging, micro-blogging, e-commerce, and entertainment, to name a few. SONs (Self Optimizing Networks), specifically in the context of wireless access, are a part of the Cloud environment. It provides inter-operable capabilities, multi-vendor choices, self-configuration, and an enhanced operational and maintenance framework. The ability to self-configure to meet the demands of the long-tail is central for adapting to capacity and connectivity, in the presence of mobility. Open standards provide this critical function in the Cloud.

Inter-operability and service portability are significant components for the realization of the Cloud. Openness in the Cloud provides the framework for accommodating the needs of a variety of users. A federation of Cloud resources provides inter-operability and service portability since it avoids a rigid coupling to any specific domain of resources. Scalability is enabled through the notion of federation since a service can be acquired through any domain that hosts the intended service. Economies of scale are fostered through the leveraging of shared computing resources to manage utilization and the total cost of ownership.

In this direction, the emergence of 4G radio access networks, such as LTE and WiMAX, requires capabilities that leverage other models of communication, beyond the terrestrial wide area network models for mobile connectivity. Co-operation with other radio access technologies such as satellite and local area coverage (e.g. WiFi) is essential for distributed connectivity and capacity. Seamless handovers triggered by a spatial change of access, as well as service quality demands and load balancing, are significant components for mobile service access in the Cloud. Distributed connectivity is vital for both user-oriented communications and machine-oriented communications. Appliances of all sorts - entertainment systems, refrigerators, automobile systems, cameras, vehicular traffic control, location sensors, telemedicine, and climate control - are among a plethora of examples of the Internet of Things. The motivation is that the Internet - mobile or fixed - a part of the Cloud is an attractive medium for the management, provisioning and monitoring of these entities. The challenges include the need for enhanced security capabilities since vulnerabilities to threat scenarios are elevated, as a result of inter-linkages across the components in the Cloud.

A dynamic allocation of bandwidth and self-optimization capabilities, in a distributed heterogeneous radio access technology environment, is required to meet assorted multimedia service usage demands. Adequate spectrum and spectrally efficient radio access technologies are necessary for the practical realization of the associated information transport. Configuration parameters and policies are expected to shape the transport demands in a dynamic fashion.

The Cloud paves a way forward towards a proliferation of the creation, consumption and delivery of services in the long-tail – a market where there is a large demand for a variety of niche services. It fosters a participative and collaborative framework for service mobility. Openness and standardization will serve as catalysts in this emerging paradigm.

REFERENCES

Amazon Web Services. (2010). *Amazon elastic compute cloud* (Amazon EC2). Retrieved from http://aws.amazon.com/ec2

Anderson, C. (2006). *The long tail*. New York, NY: Hyperion.

Armbust, M., Fox, A., Griffith, R., Joseph, A. D., Katz, R. H., & Konwinski, A. ... Zaharia, M. (2009). *Above the clouds: A Berkeley view of cloud computing*. Retrieved from http://www.eecs.berkeley.edu/Pubs/ TechRpts/2009/EECS-2009-28.html

Express Computer. (2008). *Computing in the clouds*. Retrieved from http://www.expresscomputeronline.com/ 20080218/technology01.shtml

Hartig, K. (2009). What is cloud computing? *Cloud Computing Journal*. Retrieved from http://cloudcomputing.sys-con.com/node/579826

O'Reilly, T. (2005). *What is Web 2.0? Design patterns and business models for the next generation of software*. Retrieved from http://www.oreillynet.com/pub/a/oreilly/ tim/news/2005/09/30/what-is-web-20.html

Punithavathi, R., & Duraiswamy, K. (2008). *An optimized solution for mobile computing environment*. Paper presented at the International Conference on Computing, Communication and Networking, St. Thomas, VI.

Third Generation Partnership Project. (2010). *Third Generation Partnership Project*. Retrieved from http://www.3gpp.org

Third Generation Partnership Project 2. (2010). *Third Generation Partnership Project 2*. Retrieved from http://www.3gpp2.org

WiMAX Forum. (2010). *WiMAX Forum*. Retrieved from http://www.wimaxforum.org

Chapter 5
Mobile Learning Using Mobile Phones in Japan

Midori Kimura
Tokyo Women's Medical University, Japan

ABSTRACT

The past ten years has seen remarkable developments in mobile devices, especially mobile phones, and interest in the potential of using mobile phones in an educational setting has intensified recently. The author's working group, in cooperation with eLPCO (e-learning Professional Competency) at Aoyama Gakuin University in Japan, started a mobile learning project in 2002 to demonstrate model programs of mobile learning using mobile phones (mLearning/MPs), with the findings from all the experiments conducted over the past seven years contributing to the educational process. This chapter first discusses the barriers, such as the psychological, pedagogical, and technological issues, that mLearning/MPs had to overcome. Next, the author introduces findings obtained from four projects carried out on the English language by mobile phones, and then provides suggestions on essential conditions required for a good program for mLearning/MPs. The chapter proposes open source-based mobile services as a way of overcoming barriers faced by mLearning/MPs, and as an effective model for English language learning using mobile phones.

INTRODUCTION

There are teachers and researchers who are enthusiastic about using mobile technologies. They believe that providing the means for learners to study "anytime, anywhere" will encourage more frequent and integral use of learning technologies, as opposed to the more occasional use generally associated with computer laboratories (Roschelle, 2003). There are good reasons for establishing study programs using mobile phones in Japan.

DOI: 10.4018/978-1-60960-613-8.ch005

1. Most students already possess a mobile phone, but they may not possess a computer; therefore, using mobile phones for learning allows everyone to study under the same conditions.
2. The average round-trip commute is about two and a half hours, so it is beneficial to find something useful to do while commuting.
3. A mobile phone has faster access and is easier to use than a computer for mailing and browsing the Internet.
4. Mobile learning, in other words, "learning anytime, anywhere," is more feasible by mobile phones than by computers.

However, some caution has been raised with regard to assuming that mLearning/MPs will become the next generation of learning simply for the reason that most learners already possess mobile phones (see Australian Flexible Learning Framework, 2007: The MoLeNET Project, 2007). Levy and Kennedy (2005) also argue that the widespread acceptance of communication technologies in non-learning contexts does not necessarily mean that they will be effective or valued in educational contexts. In view of the controversy surrounding the value of mobile phones with regard to education, the four projects we carried out over a seven year period were very significant in the sense that our experiments and the results they produced always set out to provide models for mLearning/MPs, and promoted the effectiveness and the potential of mobile phones for learning.

Our working group started the mobile learning projects using mobile phones in 2002 in collaboration with eLPCO (e-Learning Professional Competency) at Aoyama Gakuin University in Japan. The results from all of the experiments we conducted in the four projects enabled us to contribute to the English language education process. In this chapter, we will discuss mLearning/MPs from the psychological, pedagogical, and technological points of view, based on the needs, learning attitudes, and learning strategies of the students. Furthermore, there will also be a short discussion of the feasibility of using mobile phones as a learning tool for improving language test scores.

When we started the mLearning/MPs project, it was believed that only computers were capable of being a useful tool for education in the 21st-century, and many people had not yet recognized the value of portable computers or mobile phones, or of mobile learning itself. However, many people in Japan, both university students and working people, spend hours commuting, and it would be to their advantage to find something meaningful to do while they are commuting, such as learning on the move. On the other hand, many people, including teachers, who advocated the more conventional styles of learning, were insisting that students who were serious about studying should sit at a desk in the library or in front of a computer in a laboratory, and that mobile phones were merely a bothersome distraction to learning. As a result, although e-learning has become more common, the amount of research done on mLearning/MPs has been limited, despite the extremely widespread use of mobile phones throughout Japan, especially among university students, and greater use of mobile phones over personal computers (PCs) (Kogure et al., 2006).

Over the last seven years, the circumstances have changed considerably. The data collected by our research group from 482 students from various universities in Japan in 2006 shows that for sending e-mail, 91% of the students used mobile phones, while only 3% used computers (Kogure et al., 2006). University students use mobile phones far more often than computers. This implies that mobile phones can be an even more powerful learning tool and a mainstream platform in the future, even more so than computers can in some way; therefore, from now on, teachers should pay more attention to the use of mobile phones for learning. Our research group has tried to show models for mobile learning using mobile phones (mLearning/MPs) and makes use of our findings from all the experiments conducted over the past seven years to contribute to better education.

The first project was to ascertain the effectiveness of preparing for the Test of English for International Communication (TOEIC) by mobile phones. The second and third projects focused on preparing and reviewing study materials for class by means of watching video programs on mobile phones with multimedia capabilities. For the third project, we developed a function that enabled us to add captioning to video clips viewed on mobile phones. For the fourth project, effective vocabulary learning contents for mobile phones were developed. In order to check the effectiveness of mLearning/MPs, surveys were conducted to find out the needs, lifestyles, and learning strategies of the students, and English proficiency tests were administered in most of the projects. Based on all the research results, an open source-based mobile service for mLearning/MPs was suggested as an effective model of English language learning using mobile phones in Japan. It is believed that it could be a good model for use in other countries as well.

PROBLEM ANALYSIS OF MOBILE LEARNING USING MOBILE PHONES

A number of obstacles that limit the use of mobile phones have been pointed out by many researchers (Thornton & Houser, 2006; Pęcherzewska & Knot, 2007; Stockwell, 2007; Kukulska-Hulme, & Shield, 2008; Nah, White, & Sussex, 2008). These obstacles will be analyzed from four perspectives: technology, content design, uniqueness of device, and psychological issues.

Technological Issues

When it comes to mobile phones as a study tool, commonly mentioned drawbacks include small screen and key pad, low screen resolution, narrow bandwidth, high cost, limited storage, battery capacity, and network connectivity. Among these obstacles, network connectivity seems to be the most challenging of all, especially for oral communication and multimedia approaches to mobile phone-based learning, which are very limited (JISC, 2005). For example, a complicated experiment was tried using wireless application protocol (WAP) sites for listening activities, but it did not work smoothly due to limitations inherent in mobile phones (Nah, White, & Sussex, 2008). City College, Southampton (JISC, 2005) set up a web-based media board, supplying students studying English as a second language with specific oral or visual information by way of their mobile phones, wherever the students happened to be. However, setting up this kind of media board was very expensive and difficult for institutions or teachers to actually use. Oral interaction was tried at Stanford University (Tomorrow's Professor Listserv, 2002) in California, where students used their mobile phones to take part in automated, voice-controlled grammar and vocabulary quizzes; however, the activity was abandoned primarily because of problems with the voice recognition software.

Content Design

Like the majority of research in the areas of materials development and activity types, mLearning/MPs also deals with design issues. The majority of activities using mobile phones appear to employ text messaging for vocabulary learning (Andrews, 2003; Levy & Kennedy, 2005; McNicol, 2005; Norbrook & Scott; 2003; Pincas, 2004), and quizzes and surveys (Tomorrow's Professor Listserv, 2002; Norbrook & Scott, 2003; Levy & Kennedy, 2005; McNicol, 2005). Interactive and collaborative activities, such as oral communication between a teacher and a student, or among students, seem very difficult due to technological problems encountered with the abovementioned technologies.

Unique Form of the Device

As access to wireless networks expands and ownership of devices capable of communicating with such networks increases, the use of mobile devices to support language learning becomes even more commonplace. Many initial studies viewed mobile devices as being the same type of tool as personal computers, without taking into consideration the unique form, fit, and function of mobile devices. However, it is important to understand the strengths and weaknesses of the technologies unique to mobile phones. MLearning/MPs is different from computer-assisted language learning (CALL) in its usage. Equally important is the content that is provided. For example, it is not possible to take a one-hour course by a mobile phone, as its capacity is much smaller than that of a PC or laptop computer. Mobile phones should not be viewed as a portable, miniaturized version of a computer. The form factor for mobile phones is different in such aspects as their size, screen, and keyboard, and their mobility adds a new dimension to the activities that are supported. To realize the full potential of mobile phone technologies, teachers should regard mobile phones as one type of tool for mLearning that can be used outside the classroom.

Psychological Aspects of Using Mobile Phones

Ellis (1994) states that the outcomes of the learning efforts of students are influenced by a variety of individual psychological variables, such as aptitude, personality, learning style, motivation, affective state, and belief; all of these elements are closely related to English language acquisition. Mobile learning is closely related to self-access, self-study, and autonomous learning. Autonomous language learning has long been associated with individualization (Geddes & Sturtridge, 1982; Brookes & Grundy, 1988) and the notion that learners each have their own preferred learning styles, capacities, and needs (Skehan, 1989). Therefore, research and data on these issues are essential for preparing good mLearning programs for use with mobile phones. For example, a study (Stockwell, 2007, 2008) showed that some students preferred studying in quiet places where they could concentrate for long periods of time, while others expressed an inability to concentrate if they tried to learn on a bus or train. Thus, Stockwell raised some concerns about whether learners were ready to start mLearning/MPs, as it doesn't necessarily fit everyone's learning style or study habits.

FOUR PROJECTS FOR ENGLISH LANGUAGE LEARNING USING MOBILE PHONES IN JAPAN

The four projects we undertook are introduced below. The controversial issues surrounding the projects are discussed by comparing and contrasting the results obtained from past research. The purpose of the first project was to ascertain the effectiveness of using mobile phones for drill-and-practice activities in preparing for TOEIC (Test of English for International Communication). The issues of the small screen and key pad, high connection fees, and whether the students are ready to start studying using mobile devices will be discussed. The second and third projects focused on investigating effective ways of utilizing multimedia to enhance listening comprehension skills with mobile phones to prepare and review study materials for class. In addition to the issues of the small screen and high costs, which were the focus of the second and third projects, there are discussions regarding low screen resolution, narrow bandwidth, limited storage, and battery capacity. In the fourth project, effective vocabulary learning contents for mobile phones were investigated. Network connectivity, high cost, and the readiness of students to undertake study by mobile phones were also key subject matters for this project. Lastly, suggestions are offered regarding an optimal vision of what mLearning/MPs should be like right now, and what the prospects are for developing technology-enhanced mobile phones to be used for learning in the future.

FIRST PROJECT: DRILL-AND-PRACTICE

We started the first project in 2002. It was a preparation study for TOEIC by making use of drill-and-practice exercises. There were two key reasons behind selecting TOEIC preparation and drill-and-practice exercises as the contents for this first project. The first reason was that TOEIC is regarded as the standard by which English language proficiency is measured for use in business or for making significant personnel-related decisions in Japan, and it is used everywhere from small businesses to multinationals, as well as givernment agencies. College students in Tokyo are under pressure to get high scores on English tests in order to earn college credits as well as to secure a good job, so the availability of the ubiquitous mobile phone service has been instrumental in helping people learn on the move. The second reason was that drill-and-practice activities are believed to be most effective when matched with the specific learning needs of individual students (Taylor & Gitsaki, 2005). The circumstances gave the students tremendous motivation to use their mobile phones to study in their spare time. Online services were provided that enabled the students to work on multiple-choice questions and receive feedback on the answers they selected. Included in the mobile Internet technologies was a learning management system (LMS), which monitored the learning progress of the students.

A pilot project was conducted with nearly 300 university students over a five-month period in 2002, and again in 2003, with 98 university students participating. Participants in the project took a TOEIC test as a pre-test at the start of the program, and a TOEIC test as a post-test at its conclusion. They were also asked to respond to a 40-item questionnaire regarding such things as their learning styles, personality, and motivation and attitude, because these factors are closely related to English acquisition. Practice exercises for TOEIC were uploaded on the web and delivered to the mobile phones of the students. The students were informed by e-mail of the results of their performance, as well as what assignments, which were provided on the website, they were to practice every day.

Students were divided into two groups according to their preferences of learning styles, with 41 students choosing to study in the mobile learning group (mobile group), and 57 students choosing the computer network learning group (PC group). This project made use of a service providing an online English language study program containing TOEIC-style multiple-choice questions with answers and answer keys. The students in the PC group could download and solve as many questions as they wanted and check the results themselves, both on campus and at home. For students in the mobile phone group, a set of three questions was sent each time they accessed the program. When the students clicked a button to select an answer to a question, their choice was sent via the Internet, marked, and the choice was determined to be either correct or incorrect, and the response was returned immediately with feedback. To make use of the features of this service, we designed the mobile language learning program around the TOEIC practice programs (Figure 1). In this study, we wanted to see statistically if the students would really study English by mobile phones in their spare time, and ascertain the effectiveness of mobile technologies in language learning.

Students showed great interest in mobile learning, and the frequency with which they studied English increased from once a week to several times a week. Interestingly, the places where they studied English shifted from inside the home and classroom to locations other than these places, such as train stations, parks, convenience stores,

Figure 1. Learning management system for TOEIC practice program

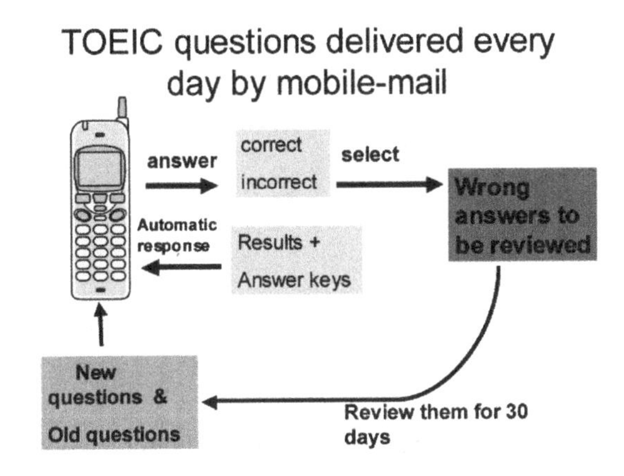

and the school cafeteria. Furthermore, the students became more proactive about the conditions of the experiment, requesting that they be allowed to choose the access time and frequency of study, as well as the quantity of practice materials. The results of the experiment are summarized here. We can conclude from the results shown in Figure 2 that the students in both the PC and mobile groups liked to study with IT devices, and they were interested in e-learning and m-learning.

Test results were compared between the two groups (Figure 3). We confirmed that the students in both groups improved their scores (full mark=50) and there was a significant difference between pre-test scores and post-test scores; PC group (0.0001), mobile group (P=0.007). Therefore, we found that the drill-and-practice format of the TOEIC practice program had the potential to be a good program for mLearning/PCs because the small screen and key pad of the mobile phone would not pose a problem. The program was effective for studying English, and learning on the move with a mobile phone fit the lifestyle and learning style of the students.

The students were ready to start studying mLearning/MPs. However, only one carrier out of the three major mobile phone companies was capable of enabling access to the program and running costs became very high by the end of the project. Today, however, there are many economical packages available for mobile phone service, although mLearning/MPs was yet not very prevalent in 2002 and 2003. For the first project, in particular, the high telephone bills and limited connectivity were serious drawbacks.

Second Project: Listening Practice Using Video Clips

The second project was conducted in 2004 and 2005 with the aim of investigating the possibilities that video learning offers by comparing the activities of mobile phone learners and computer

Figure 2. Interest in computer and mobile learning

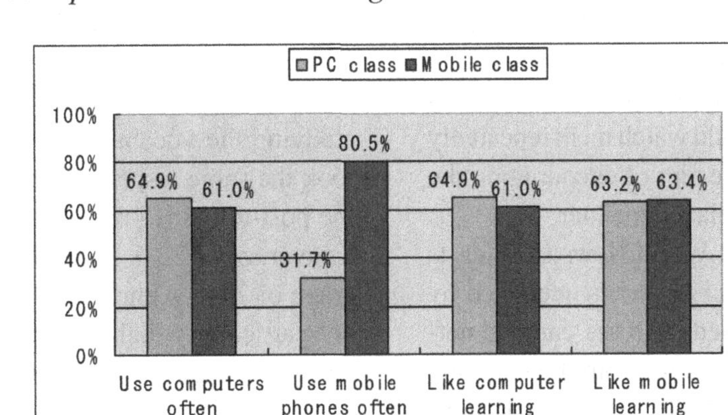

Figure 3. Change of TOEIC scores in the two groups

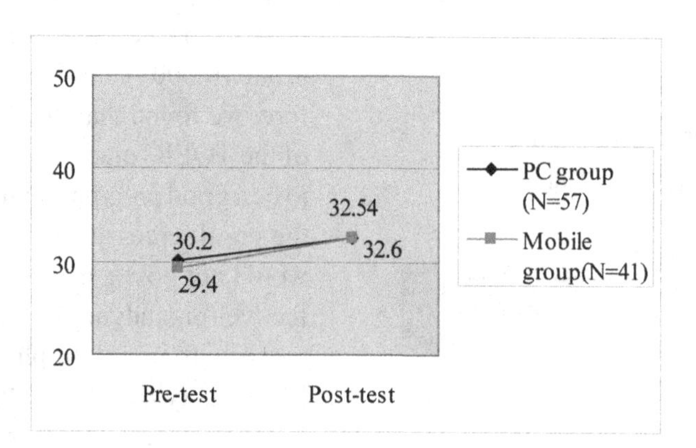

learners. Flowerdew & Miller (2005) state that video generally promotes the motivation to listen, provides a rich context for authenticity of language use, makes the paralinguistic features of spoken text become available to learners (more so than radio can), and aids in the understanding by learners of the cultural contexts in which the language is used. Multimedia materials also assist learners with different preferences regarding learning styles. Both visual and auditory learners learn better when lessons are presented in two or more formats, as opposed to a single format (Mayer, 2001). Gardner and Miller (1997) found that learners who watched movies in a relaxed atmosphere tended to give more personalized accounts of what they saw, whereas learners who watched under classroom conditions gave more factual and content-oriented accounts. Oftentimes, learners are unaware of the benefits they can derive from viewing videos for pleasure, and we expect learners to watch videos by mobile phones for fun. We made many short video clips out of one news program so students could watch them repeatedly in a short time, with the aim of encouraging the learners to study in a relaxed manner.

Video clips of ABC World News broadcasts contained in educational materials published by Kinsei-do were uploaded onto the campus network system with permission from the textbook company. The system, which requires an ID and a password to gain access, enabled students of both the PC and mobile groups to download clips to their preferred device and have the same opportunity to use the materials for study. However, only one of the three major mobile phone companies in Japan had the multimedia capabilities that would enable learning using video contents at the time of the experiment. As a result, the number of students, totaling only 11, in the mobile group was very limited. To check the effectiveness of using video clips, two types of tests were administered, a vocabulary test and a comprehension test. The number of participants in this sample of the second project was 38 university students, with 27 students in the PC group and 11 students in the mobile group.

1. Vocabulary test:

The students watched news clips three times and took a test (pre-test) to identify what vocabulary they did not know. After one week of studying the vocabulary from the pre-test, they took the same test (post-test). The full mark for the post-test was 46, with the PC group scoring an average of 22.6, and the mobile group an average of 25.4. Although the scores overall were not particularly high, the mobile group not only had a higher average score than the PC group, it

Figure 4. Vocabulary test on watching news

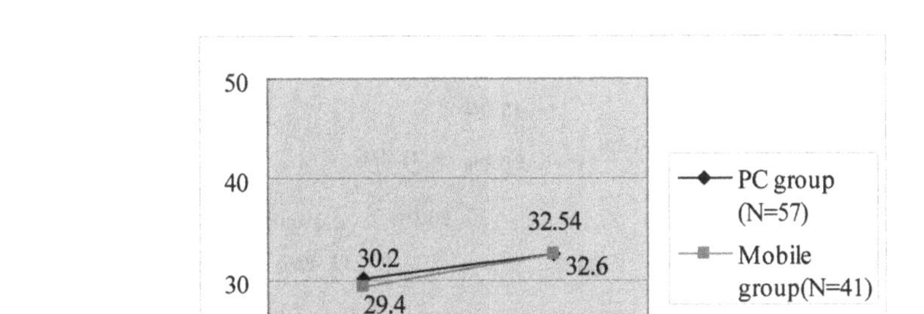

also showed greater improvement after one week (see Figure 4).

2. Comprehension test

Immediately after the vocabulary post-test, the learners took a comprehension test on the same video clips. The full mark was 9, with the PC group achieving an average score of 7.5 and mobile group an average score of 7.6, indicating that there was no significant difference between the two groups.

From the two test results, we could conclude that learning the news program by video clips can be useful for preparation and review for class. It was found that 73% of the students in the mobile group felt that the sound and images were clear enough, and watching video on a mobile phone was an effective way to learn. While resolution and bandwidth did not present any problems, it was found that the biggest reasons the students were reluctant to study the video clips by mobile phones were the high cost of telephone service and the need to frequently recharge the mobile phone battery. In 2005, most mobile phones in Japan did not have sufficient memory to hold effective learning engines.

Third Project: Assist Learning with Captioning

The third project, which was conducted in 2006, had two purposes. One was to assist English learning by using captioning, which provides a useful resource for self-study by a computer or mobile phone. The other was to overcome the problem of excessive battery consumption caused by downloading video clips, which was an issue in the second project. This problem in the second project was resolved by using a software application called *Moviegate* to save video clip files from the computer directly onto the mini-SD card contained in the mobile phones of the students. Video clips used in the second project were shortened versions of the original video tape, and the students studied a hard copy of the full transcript later on. For the third project, we tried something new by adding listening comprehension questions at the beginning of the each video clip, and putting key content-related words on the screen, similar to movie captions. Mayer (2001) insists that the learner's understanding is impeded when interesting but irrelevant words and pictures are added to a multimedia presentation. Therefore, for the third project, we focused on using simple captioning that consisted of only key words and phrases, and did not include the

Figure 5. How much the students' understanding changed. N=130

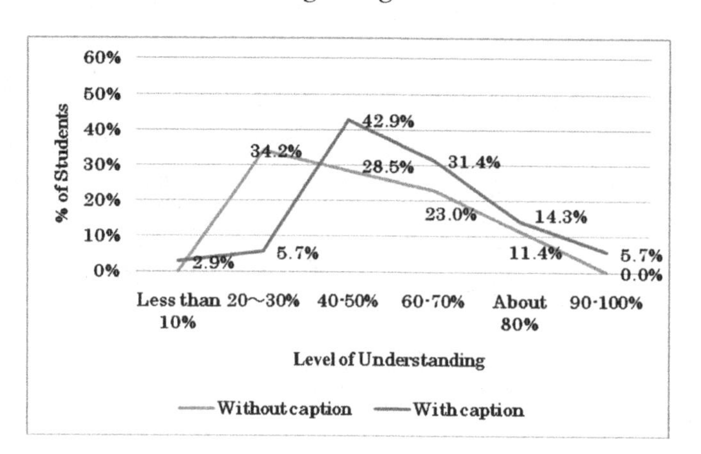

whole script. Another reason for using simple captions was that mobile phone screens are very small, so we worked out the optimal number of words that could appear on a single screen at one time, thereby enabling the students to listen and recognize the words at the same time.

The captioning we used was not ready-made, such as subtitles that are available for commercial movies, and had to be made by the teacher for each class. Key words and phrases were selected with the aim of helping the students focus more on listening. Students, particularly those with years of "classroom English" but little experience in actual use of the language, try to listen from the bottom up; in other words, piecing the meaning together, word by word. Moreover, key words and phrases are learned as recognition vocabulary elements when they are encountered in listening materials. The level of vocabulary is the most important element of L2 knowledge that all learners must develop, whether they are aiming primarily for academic or interpersonal competence. We investigated the effectiveness of captioning with respect to three points.

1. How the students' understanding changed.
2. How satisfied the students were with captioning.
3. How useful mobile phones were for watching video clips with captioning.

Two experiments were conducted to find the answer to the first point. In the first experiment, two types of video clips, one with captioning and another without captioning, were prepared for computers and mobile phones, to enable the students to use the video clips for preparation and review of class work. For the first experiment, the students watched a video clip titled "The purpose of informative speech." They watched this video clip for 50 seconds, first with captioning and then a second time without captioning, and marked the vocabulary they recognized after each viewing. Figure 5 shows the results of the experiment to see "How the students' understanding changed." The graph line indicates that the students recognized more vocabulary when viewing the video clips with captioning.

The second of the two experiments involved only mobile phones. For this experiment, video clips with captioning were used. This time, vocabulary lists and comprehension questions were included at the beginning of each video clip, and answers were provided at the end of each clip. The students using mobile phones to study video clips with captioning reported on how much of the video contents they were able to understand.

Figure 6. How much comprehension changed for students using mobile phones (self-report)

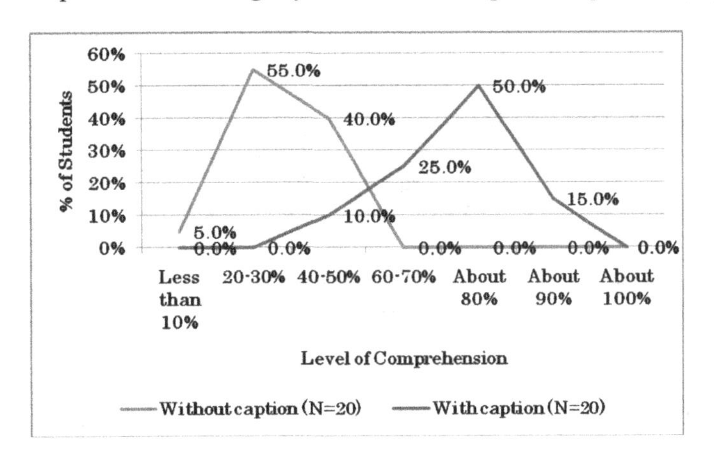

The students checked their understanding of the contents by taking a comprehension test, after which they replied to a questionnaire. Using this self-reporting method, the students were able to provide feedback on how much of the video they understood (Figure 6). This also provided verification of how helpful and useful it was to use mobile phones to study video clips with captioning.

Most of the students, including those using mobile phones, commented that captioning was helpful and they were satisfied with the results. Advantages and disadvantages of studying using mobile phones with captioning are summarized below based on the feedback from the students.

The advantages were:

- Visual images and audio were sufficiently clear enough and helped make the contents easy to understand
- Questions at the beginning of the video clip helped the students concentrate better on the contents

The disadvantages were:

- Details were difficult to pick up for fast-moving action shown on a mobile phone screen
- Reading captions was difficult when many words were scrolling or moving across the small screen of the mobile phone

Therefore, a small screen was a disadvantage when the students wanted to watch something moving on screen, such as images or text. The introduction of the mini-SD or micro-SD memory cards helped us to overcome such problems as the high cost of telephone access charges, frequent battery charging, and limited storage capacity.

Fourth Project: Ideal Vocabulary Learning Contents

The fourth project was conducted in 2007 in order to help students find a way to increase their vocabulary in a relaxed manner using mobile phones whenever the students had some spare time available. Expanding vocabulary is one of the most important issues for English learners in Japan, and vocabulary is the most important aspect of L2 knowledge for all learners (Harp & Mayer, 1998). With regard to short-term programs for learning, the data collected in our survey showed that about 80% of the students were more interested in using mobile phones than computers to learn vocabulary, and this led to our decision to develop contents for mobile phones that focused

Figure 7. Images of three different types of vocabulary learning by mobile phones

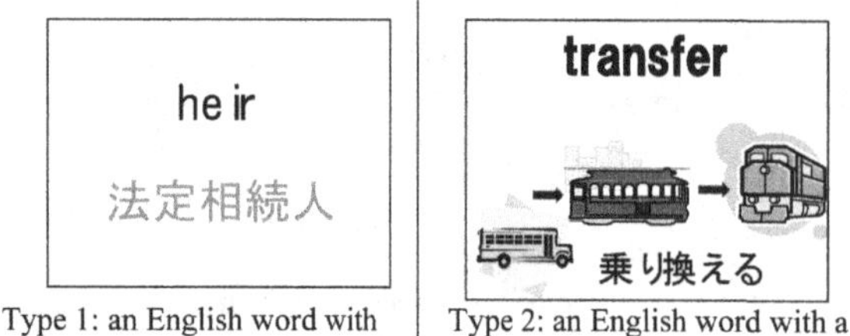

Type 1: an English word with Japanese translation

Type 2: an English word with a picture clue and Japanese translation

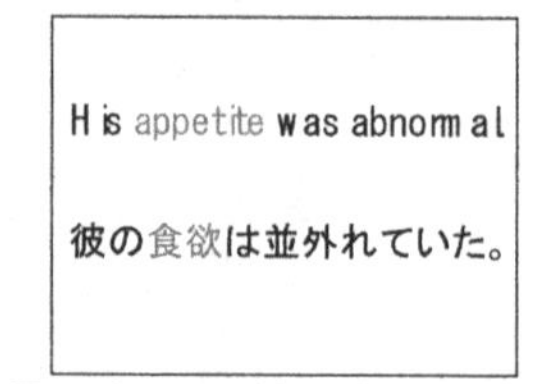

Type 3: a word in a short English sentence and Japanese translation

on vocabulary learning. According to Kadota and Ikemura (2006), presenting words in different forms motivates learners to actively engage in learning vocabulary. There are five formats: (1) translation, (2) illustration, (3) example sentences, (4) oral introduction with sound, and (5) the appropriate number of words. Another important element of this project was to develop learning contents from the perspective of the memory capacity of the learner. Memory is concerned with the encoding, storing, and retrieval of information. Illustrations and sentences that have been designed to provide equivalent amounts of information are found to make vocabulary easier to memorize as well as functional (Burns, Biswas, & Babin, 1993). Hatamura, Ikemura, & Togo (2006) found that studying English words together with sound is a very effective way of memorizing them. Mnemonics works by exploiting the feature of the brain that makes remembering easier if things are associated with other, previously remembered items (Wollen, Weber, & Lowry, 1972). Visual images have ample potential for transforming information stored in long-term memory into working memory. We chose 150 English words from a list of English vocabulary deemed essential for Japanese university students (Aizawa, 2005), and divided the vocabulary into three groups of 50 words each. Collating and organizing the above-mentioned information, our working members (seven teachers) collaborated and developed three different types of contents for the word groups, taking into consideration the distinctive features of mobile phones in Japan shown in Figure 7.

- **Type 1:** English words + sound links for the words + the meanings of the words in Japanese
- **Type 2:** Illustrations + English words + sound links for the words + the meanings of the words in Japanese
- **Type 3:** English words embedded in example sentences + sound links for the words + the meanings of the words in Japanese

Before we started the experiment, we carried out a background research on the views of the students regarding vocabulary learning, learning strategies, and learning style. We also administered a pre-experiment and post-experiment vocabulary test in order to check how much improvement was achieved. Feedback from the students was analyzed from the perspective of memory strategies and affective strategies (communication with others) (O'Malley & Chamot, 1990; Oxford, 1990).

A total of 135 students from six universities studying in six different majors participated in the experiment (Caretaking: 17, Economic: 13, Business: 24, English: 28, Nursing: 42, Pharmacology:11). The experiment took place over a six-week period. In the first week, a pre-experiment test consisting of 30 words out of the selected 150 words was administered, together with a pre-experiment questionnaire. In the sixth week, a post-experiment test and post-experiment questionnaire were conducted. The same words were contained in both the pre-test and post-test in order to evaluate the effectiveness of vocabulary learning by mobile phones.

The average pre-experiment test score of the participants was 6.4 (full mark=30), and average post-experiment test score was 18.6. The highest score was 28.4, achieved by an English major, and the lowest was 4.2, by a Caretaking major. There was a significant improvement in the scores between the pre-experiment test and post-experiment test, verifying that this vocabulary learning was successful. Moreover, the pre-experiment questionnaire found that many students studied vocabulary regularly, with a further study (Kimura & Shimoyama, 2009) finding that students in the low-scoring group (LG) had a tendency to study vocabulary mainly for mid-term and final exams. On the other hand, students in the high-scoring group (HG) did very well on proficiency tests as well as for overall English learning. The post-experiment questionnaire disclosed that almost 68% of the students studied during their commute, even before the project began, and the percentage increased by about 10% during the project. On the other hand, the time spent studying at a desk decreased from 61.6% to 41.8%. This is a firm evidence that the time spent commuting is a good time for mLearning/MPs in Japan. These findings confirmed that vocabulary learning is one of the best contents for mLearning/MPs. However, the data from the statistical analysis (Kimura & Shimoyama, 2009) showed an interesting trend that, although the HG students studied using mobile phones more than LG, many of them preferred studying at a desk while the LG students liked studying on the move.

The vocabulary content that proved to be the most popular among the students was words embedded example sentences (Type 3); however, the highest test scores produced from among the three types in the post-experiment test was words with Japanese translations (Type 1). A paired-test comprised of pre- and post-experiment test scores was conducted, with a comparison of the results between Type 1 and 2 vocabulary content ($p=0.0001$), and between Type 2 and 3 ($p=0.003$) vocabulary content, showing significant differences. The results showed that there was a discrepancy among the test scores, as well as between the preferences of the students, and the expectations the teachers may have that are based on Burns, Biswas, & Babin's statement (1993). The students commented that studying words in sentences helped them understand how to use the words in real-life communication situations. It was also found that the LG students tended to prefer studying words that were presented together with illustrations, more so than did the HG students.

The analysis of the strategies employed by the students found that prior to the start of the project; about 80% of them used the memory strategy of writing out words repeatedly. However, this percentage decreased to 27% during the project. Memorizing by writing out words is a typical Japanese strategy, and the HG students, in particular, preferred this strategy because they had developed the habit of memorizing Chinese characters by repeatedly writing them down when they were in elementary school. Their success with memorization from their early school years led them to continue using a strategy they were already accustomed to. Studying with friends, one of the affective strategies, motivates and encourages learners to study (Oxford, 1990). English and Nursing majors, who were members of the successful group, increased the use of this strategy during the project. On the other hand, Caretaking majors, who were part of the lowest scoring group, who used to use this strategy, tended to decrease its use during the project. In addition, later investigation of the involvement of teachers in this project discovered that Business majors were encouraged by their teacher to try mLearning/MPs, while Caretaking, Economics, and Pharmacology majors did not received such recommendations from their teachers. From these findings, affective strategies, such as encouragement from teachers and studying with friends, seemed to affect study outcomes in all types of learning.

Another challenge for mLearning/MPs in Japan is that materials and contents need to be formatted differently for different devices. However, we recently overcame the problem of different operating systems for mobile phones, enabling users subscribing to different carriers to learn the same contents. We have succeeded in delivering learning contents to every kind of mobile phone sold in Japan.

Although some students did not do well in the post-experiment test, 67% of the students said that they were motivated to learn vocabulary by this mLearning/MPs program, listing the following five features as reasons why vocabulary learning was an ideal program for mLearning/MPs: (1) study anywhere and anytime (lifestyle), (2) useful for review of classwork contents (purpose of the program), (3) words were augmented with example sentences (content type), (4) contents divided into proficiency levels (format of the program), and (5) periodic review tests (learning management system). Network connectivity, high connection costs, and student's readiness did not cause problems in this project, but it was necessary for us to take into consideration the lifestyles of the students, as well as their needs, language proficiency, and learning strategies to achieve successful mLearning/MPs.

Discussion: Solutions and Recommendations

The results of the four projects show that there is a shift from study inside the home and classroom to study in places other than in these conventional places, and confirmed from various perspectives the feasibility of mLearning/MPs. Our suggestions for successful mLearning/MPs will be discussed from four aspects: technological barriers, content design, unique form of the devices, and psychological issues related to the use of mobile phones.

Technological barriers, such as small screens, low screen resolution, and narrow bandwidth, as well as the high cost of telephone access and the devices themselves are not big problems today. In addition, we can expect the usefulness of mLearning/MPs academically to be on a par with that of studying with a PC. Although network connectivity still remains a challenging issue, receiving learning materials presents no problems whatsoever, unless synchronous collaborative activities are attempted. Delivering text message-type learning materials, including video clips, can be easily and safely provided by any mobile phone carrier. Today's mobile phones are equipped with highly-advanced technologies, and watching TV programs and video clips has become far easier compared with back when we began our experiment a few years ago.

The problems of storage, which can be solved by using mini- or micro-SD cards, in addition to limited battery capacities and small key pad no longer present a problem at all, unless the user has to type in many sentences. As a result of market competition among mobile phone companies, all of these companies offer attractive, economical fixed price packages to increase the number of users, making it possible for us to inexpensively use mobile phones and purchase devices reasonably.

As the results provided by our students suggest, there is tremendous demand for making use of mLearning/MPs for preparing and reviewing classwork contents, which can be prepared only by the teacher. These contents could consist of two parts, such as study materials and a small quiz to periodically test or review the materials. A drill-and-practice format would be suitable for the program. What we must keep in mind when making learning materials is to be aware that mLearning/MPs is different from CALL. We must find a unique way to take full advantage of the convenience and portability of the mobile device form factor, which is significantly different from that of a PC, to enable students to learn efficiently and effectively, as well as anywhere and anytime. For this purpose, learning materials should be short enough and organized simply enough to fit in a small screen, to ensure that students are able to start and finish their studies anytime and anywhere. In view of all these factors, it is clear that vocabulary learning is one of the best and easiest materials to study on mobile phones. Students can take advantage of whenever they have time available to study, and always click again to check learning contents using their mobile phones. Teachers should limit the time devoted to drill assignments to 10 to 15 minutes per day. This ensures that students will not become bored and that the drill-and-practice strategy will retain its effectiveness. In addition to vocabulary learning, listening to English programs and watching video clips are also good methods of study that support formal in-class lessons. Students can download learning materials from a computer and save them on the SD cards in their mobile phones in order to study them as many times as needed.

The next issue is who will make the learning materials for mLearning/MPs. Teaching is one of the most time- and labor-intensive jobs in our society, and it is a heavy burden for individual teachers to keep making new learning materials for different proficiency levels. Therefore, teachers should cooperate and develop learning materials and review questions together. Teachers can form a group, decide one common textbook, and assign each member several pages of the book from which to make learning materials. This is what we did for the fourth project. All of the created learning materials are then uploaded to the open source-based website, so the contents can be shared with students of other classes or other universities. This way, teachers can meet the needs of the students and provide a variety of contents that are always current and topical.

Last but not least, there is the issue of making programs that match the learning strategies of the students, with affective strategies, such as encouragement from the teacher, and studying with friends, which is important because it affects study outcomes in any kind of learning.

The majority of applications for educational use were developed by researchers and tested by students in a practiced and controlled environment. These applications were primarily developed to support the goals of teachers, and not those of students. Actually, teachers should customize their materials to match the different proficiency levels of students. Just as we did in the fourth project, teachers should cooperate and develop learning materials together, then upload them on the open source-based website, and share the contents with students of other classes or other universities. Through these activities, teachers can meet the needs of students and provide a variety of contents that are always current and topical. By taking the full potential of mobile devices, the convenience and portability, teachers can provide the students in the digital age with a better opportunity to learn, which matches their lifestyle. The key issue is that teachers must ensure that materials for use with mobile phones are appropriately designed for display on the small screen and be of a suitable length (from 30 seconds to a maximum of 10 minutes). mLearning/MPs is a form of informal study to assist and augment formal study in the classroom. Furthermore, a means for the students to perform self-assessments and obtain feedback is essential. By utilizing a learning management system, as we did in our first project, mLearning/MPs is capable of providing all these elements, making possible learner-oriented education.

There is a tendency among many teachers and researchers to overlook the psychological aspects. However, the study program must fit the needs, capabilities, and the study habits of students. A particularly interesting finding of this research was that the overwhelming favorite method among Japanese students for memorizing words was to write them down and read them aloud. In this sense, mobile phones do not seem to be perfectly compatible with their learning strategies. However, Japanese students are surprisingly good at typing text on mobile phones; therefore, we should try to find some way of incorporating this in the strategies. One of the most significant findings coming out of this project showed that self-access, such as this study, produced better results when teachers and students used affective strategies more effectively. These affective strategies included teachers giving encouragement to students, or students helping each other by studying together with their friends. Therefore, if we could add interactive elements to the vocabulary learning program, the results could be very promising.

mLearning/MPs is different from computer-assisted language learning (CALL) in how it is used. Moreover, how the content is delivered is also critically important. For example, it is not possible to take a one-hour course on a mobile phone because its capacity is much smaller than that of a PC or laptop computer. A mobile phone should not be viewed as a miniaturized, portable version of a computer. The form factor is different, with the mobile phone's size, screen, keyboard, and mobility adding a new dimension to the activities that can be supported. To appreciate the full potential of mobile phone technologies, teachers should regard mobile phones as one type of an outside-the-classroom learning device.

FUTURE RESEARCH DIRECTIONS

A number of limitations were encountered in the course of conducting this research, including the small size of the samples in the second and third projects, and the underdeveloped mobile technology infrastructure that existed at the time the study started. However, there is a very real need to continue the study of the application of mobile technologies to learning and teaching. Further studies with a wider range of participants and greater variety in contents would lead to better curriculum design and programs that are more learner-oriented.

Our next project is to develop nation-wide open source-based mobile services for students in Japan, in order to meet the needs of students to improve their competency in English, which is essential for scoring well on entrance examinations and for job hunting. The project is a combination of the first and fourth projects, integrating an online vocabulary learning program with drill-and-practice exercises in which students work on multiple-choice questions. Included in the project is an LMS (learning management system), which provides feedback on whether or not the students have provided the correct response, monitors the learning progress made by the students, and displays their proficiency level in comparison with all the other participants. The learning contents can be broadly divided into three parts: a list of vocabulary in sentences, exercises to check the memory of the students, and graded tests. The contents will be divided by proficiency level, in other words, by grade; and exercises and graded tests that include write-in vocabulary parts, which match the Japanese learning strategy. Participants will register themselves by the nicknames they have chosen as well as by their names, with the best thirty of each graded test being made public on the web, enabling the students to find virtual mates to study with.

CONCLUSION

Simply having students use technology does not improve their performance. How much impact is achieved depends on the ways the technology is used and the conditions under which the applications are implemented. Many educators look to educational research for evidence of technology's present and potential benefits; however, the process of integrating technology effectively into education requires substantial investments in technology infrastructure and teacher training (Ringstaff & Kelly, 2002). Roblyer (1990) is critical of many educators who try to promote dramatic improvements with technology, only to move on to the next fad when the improvements fail to materialize. This approach fails to solve real problems, and it draws attention away from the efforts to find legitimate solutions. Teachers must look for what has worked in the past to guide their decisions and measure their expectations in the present. Therefore, our research group has continued working on many projects and experiments, and investigated how technology can influence student academic performance.

Most of the drawbacks can be overcome by developing appropriate learning contents. Teaching is one of the most time- and labor-intensive jobs in our society. With so many demands on their time, most teachers cannot be expected to develop software or create complex technology-based teaching materials (Roblyer & Doering, 2010). Technically possible does not equal desirable, feasible, or inevitable. Therefore, we found that vocabulary study using drill-and-practice to prepare for proficiency tests could be one type of program that would be suitable for mLearning/PCs, as the small screen and key pads would not be a problem for this type of program. It would be effective for studying English, and learning on the move fit the lifestyle and learning habits of the students.

"New" methods often are old methods in a new guise. Our projects have led us to the conclusion that flash card activity using mobile phones is very effective for improving English vocabulary. This is the most basic type of drill-and-practice function, arising from the popularity of real-world flash cards. In this flash card activity, students see a set number of questions or problems, which are presented one at a time. The student makes a choice from multiple-choice items or types in an answer, and the program responds with positive or negative feedback depending on whether the student answered correctly. This function is believed to be most effective when matched with the specific learning needs of individual students (Taylor & Gitsaki, 2005).

One of the most significant findings coming out of this project showed that self-access study, such as this study, produced better results when teachers and students used affective strategies more effectively. Such strategies included teachers giving encouragement to students, or students studying together with their friends. Therefore, if we could add interactive elements to the vocabulary learning program, the results could be very promising. We hope that this kind of learning by mobile phones can be introduced to as many students as possible, including high school students, as it would enhance the communication capabilities of both students and teachers, and contribute to the achievement of educational objectives.

ACKNOWLEDGMENT

These four projects were supported by the Good Practice Project of Ministry of Education, Culture, Sports and Science in Japan. We wish to thank all the members of Task Force 26 at eLPCO at Aoyama Gakuin University.

REFERENCES

Aizawa, K. (2005). *JACET 8000-word list*. Kirihara Shoten.

Andrews, R. (2003). Lrn Welsh by txt msg. *BBC News World Edition*. Retrieved from http://news.bbc.co.uk/2hi/uk_news/wales/2798701.stm

Australian Flexible Learning Framework. (2007). *2017: A mobile learning odyssey*. Australia: Commonwealth of Australia. Retrieved from http://www.flexiblelearning.net.au/flx/go/home/news/feature/cache/bypass?sector_feature&id_277

Brookes, A., & Grundy, P. (Eds.). (1988). *Individualization and autonomy in language learning.* ELT Documents, 131. London: Modern English Publications in association with the British Council (Macmillan).

Burns, A. C., Biswas, A., & Babin, L. A. (1993). The operation of visual imagery as a mediator of advertising effects. *Journal of Advertising, 22*(2), 71–85.

Ellis, R. (1994). *The study of second language acquisition.* Oxford, UK: Oxford University Press.

Flowerdew, J., & Miller, L. (2005). *Second language listening: Theory and practice.* New York, NY: Cambridge University Press.

Gardner, D., & Miller, L. (1997). *A study of tertiary level self-access facilities in Hong Kong.* Hong Kong: City University.

Geddes, M., & Sturtridge, G. (1982). *Individualization.* London, UK: Modern English Publications.

Harp, S. F., & Mayer, R. E. (1998). How seductive details do their damage: A theory of cognitive interest in science learning. *Journal of Educational Psychology, 90*, 414–434. doi:10.1037/0022-0663.90.3.414

Hatamura, H., Ikemura, T., & Togo, T. (2006). *Goi no imi-gakushu ni hatsuon wa donoyouni riyo sareruka:Onin-riyo to gakushu-konnansei no yoin nitsuite no kento. (How pronunciation is useful for vocabulary learning? Investigation on the relationship between phonemes and learning difficulty, translated by the author).* Paper presented at Spring Conference of Japan Association of Language Education and Technology, Japan.

JISC. (2005). *Multimedia learning with mobile phones. Innovative practices with elearning. Case studies: Anytime, any place learning.* Retrieved from http://www/jisc.ac.uk/uploaded_documents/southampton.pdf

Kadota, S., & Ikemura, T. (2006). *Eigo goishido handbook.* Taishukan Shoten.

Kimura, M., & Shimoyama, Y. (2009). Vocabulary learning contents for use with mobile phones in education in Japan. In I. Gibson et al. (Eds.), *Proceedings of Society for Information Technology & Teacher Education International Conference 2009* (pp. 1922-1929).

Kogure, Y., Shimoyama, Y., Anzai, Y., Kimura, M., & Obari, H. (2006). *Study of mobile phone market for mobile-learning.* Paper presented at the meeting of the Japan Education Association of Technology, Niigata, Japan.

Kukulska-Hulme, A., & Shield, L. (2008). An overview of mobile assisted language learning: From content delivery to supported collaboration and interaction. *ReCALL, 20*(3), 271–289. doi:10.1017/S0958344008000335

Levy, M., & Kennedy, C. (2005). Learning Italian via mobile SMS . In Kukulska-Hulme, A., & Traxler, J. (Eds.), *Mobile learning: A handbook for educators and trainers* (pp. 76–83). London, UK: Taylor and Francis.

Mayer, R. E. (2001). *Multimedia learning.* New York, NY: Cambridge University Press.

McNicol, T. (2005). Language e-learning on the move. *Japan Media Review.* Retrieved from http://ojr.org/japan/wireless/1080854640.php

Nah, K. C., White, P., & Sussex, R. (2008). The potential of using a mobile phone to access the Internet for learning EFK listening skills within a Korean context. *ReCALL, 20*(3), 331–347. doi:10.1017/S0958344008000633

Norbrook, H., & Scott, P. (2003). Motivation in mobile modern foreign language learning . In Attewell, J., Da Bormida, G., Sharples, M., & Savill-Smith, C. (Eds.), *MLEARN 2003: Learning with mobile devices* (pp. 50–51). London, UK: Learning and Skills Development Agency.

O'Malley, M. J., & Chamot, A. U. (1990). *Learning strategies in second language acquisition*. New York, NY: Cambridge University Press.

Oxford, R. (1990). *Language learning strategies: What every teacher should know*. Boston, MA: Heinle and Heinle Publisher.

Pęcherzewska, A., & Knot, S. (2007). *Review of existing EU projects dedicated to dyslexia, gaming in education and m-learning*. WR08 Report to CallDysMProject.

Pincas, A. (2004). Using mobile support for use of Greek during the Olympic Games 2004. In *Proceedings of M-Learn Conference* 2004. Rome, Italy.

Ringstaff, C., & Kelly, L. (2002). *The learning return on our technology investment: A review of findings from research*. San Francisco, CA: WestEd RTEC. Retrieved March 20, 2009, from http://www.wsetedrtec.org

Roblyer, M. D. (1990). The glitz factor. *Educational Technology*, *30*(10), 34–36.

Roblyer, M. D., & Doering, A. H. (2010). *Integrating educational technology into teaching*. Boston, MA: Allyn & Bacon.

Roschelle, J. (2003). Unlocking the learning value of wireless mobile devices. *Journal of Computer Assisted Learning*, *19*(3), 260–272. doi:10.1046/j.0266-4909.2003.00028.x

Skehan, P. (1989). *Individual differences in second-language learning*. London, UK: Edward Arnold.

Stockwell, G. (2007). Vocabulary on the move: Investigating an intelligent mobile phone-based vocabulary tutor. *Computer Assisted Language Learning*, *20*(4), 365–383. doi:10.1080/09588220701745817

Stockwell, G. (2008). Investigating learner preparedness for and usage patterns of mobile learning. *ReCALL*, *20*(3), 253–270. doi:10.1017/S0958344008000232

Taylor, R. P., & Gitsaki, C. (2005). Using mobile phones in English education in Japan. *Journal of Computer Assisted Learning*, *21*, 217–228. doi:10.1111/j.1365-2729.2005.00129.x

Thornton, P., & Houser, C. (2006). Using mobile phones in English education in Japan. *Journal of Computer Assisted Learning*, *21*(3), 217–228. doi:10.1111/j.1365-2729.2005.00129.x

Tomorrow's Professor Listserv. (2002). *Message #289: Mobile learning*. Retrieved from http://sll.stanford.edu/projects/tomoprof/newtomprof/postings/289.html

Wollen, K., Weber, A., & Lowry, D. (1972). Bizarreness versus interaction of mental images as determinants of learning. *Cognitive Psychology*, *2*, 518–523. doi:10.1016/0010-0285(72)90020-5

ADDITIONAL READING

Beatty, K. (2003). *Computer-assisted language learning*. Pearson Education Limited.

Chapelle, C. A. (2001). *Computer applications in second language acquisition*. New York: Cambridge University Press.

Chinnery, G. M. (2006). Emerging technology: Going to the MALL Mobile assisted language learning. *Language Learning & Technology*, *10*(1), 9–16.

Collins, T. G. (2005). *English class on the air: Mobile language learning with cell phones*. Paper presented at the IEEE International Conference on Advanced Learning Technologies (ICALT'05), Kaohsiung, Taiwan.

Daniels, P. (2005). *Mobile learning tools for education.* Paper presented at the conference of Senshin IT katsuyo kyoiku in Kochi (Education Using the Advanced IT in Kochi). Japan.

Dias, J. (2002). MOBILE phones in the classroom: Boon or bane? *C@lling Japan, 10*(2). http://jalt-call.org/cjo/10_1.pdf

Dorneyei, Z. (Ed.). (2003). *Attitudes, orientations, and motivations in language learning.* Language Research Club, University of Michigan.

Fotos, S., & Browne, C. M. (Eds.). (2004). *New perspectives on CALL for second language classrooms.* London: Lawrence Erlbaum Associates, Publishers.

Gardner, R. C. (1985). The attitude/motivation test battery: Technical report (1985). http://publish.uwo.ca/~gardner/AMTBmanualforwebpage.pdf

Hanson-Smith, E. (1999). Classroom practice: Using multimedia for input and interaction in CALL environments . In Egbert, J., & Hanson-Smith, E. (Eds.), *CALL environments: Research, practice, and critical issues* (pp. 189–215). Alexandria, VA: TESOL.

Kamp, M. (2005, April). The role of information technology in our lives. *Mlearningworld.* http://www.mlearningworld.com/index.php?name=Articles&op=show&id=1421

Kiernan, P., & Aizawa, K. (2004). Mobile phones in task based learning. Are mobile phones useful language learning tools? *ReCALL, 16*(1), 71–84.

Kim, T. G. (2006, December 4). Cell phones revolutionize Koreans' lifestyle. *The Korea Times.*

Matcalf II, D. (2006). *mLearning. Mobile learning and performance in the palm of your hand.* HRD Press.

Merrill, Dd., Salisbury, D. (1984). Research on drill and practice strategies. *Journal of Computer-Based Instruction, 11*(1), 19–21.

Moreno, R., & Mayer, R. E. (1999). Cognitive principles of multimedia learning: The role of modality. *Journal of Educational Psychology, 90*, 358–368. doi:10.1037/0022-0663.91.2.358

Nah, K. C. (2008). *Language learning through mobile phones: Design and trial of a wireless application protocol (WAP) site model for learning EFL listening skills in Korea.* Unpublished Doctoral Dissertation, the University of Queensland, Brisbane, Australia.

Presky, M. (2005). What can you learn from a phone? Almost anything! *Innovate 1*(5). Retrieved March 22, 2007, from http://www.innovateonline.infor/ index.php?view=article&id83

Robinson, P. (2002). *Individual differences and instructed language learning.* Philadelphia: John Benjamins Publishing Company.

Salisbury, D. (1990). Cognitive psychology and its implications for designing drill and practice programs for computers. *Journal of Computer-Based Instruction., 17*(1), 23–30.

Sheerin, S. (1997). An exploration of the relationship between self-access and independent learning . In Benson, P., & Voller, P. (Eds.), *Autonomy & independence in language learning* (pp. 54–63). Longman.

Vygotsky, L. S. (1978). *Mind in society: The development of higher psychological processes.* Cambridge, MA: Harvard University Press.

Wang, S., & Higgins, M. (2006). Limitations of mobile phone 4 learning. *The JALTCALL Journal, 2*(1), 3–14.

Warschauer, M. (1997). Computer-mediated collaborative learning: Theory and practice. *Modern Language Journal, 81*(4), 470–481. doi:10.2307/328890

Wilen-Daugenti, T. (2009). *Technology and learning environments in higher education.* New York: Peter Lang.

Worthington, T. (2004). The Web on small screen: Mobile phones and Interactive TV. http://www.tomw.net.au/2004/wed/mobile.html

Zähner, C., Fauverge, A., & Wong, J. (2000). Task-based language learning via audiovisual networks: The Leverage project . In Warschauer, M., & Kern, R. (Eds.), *Network-based language teaching: Concepts and practice* (pp. 186–203). New York: Cambridge University Press.

KEY TERMS AND DEFINITIONS

Content Development: The goal of mLearning is to develop learning content that integrates with mobile applications and provides dynamic learning and activities anywhere, anytime.

Learning Strategies: These are strategies that help learners to study effectively. According to Oxford (1990), such strategies are broadly divided into two types: direct and indirect strategies. Direct strategies are further divided into memory, cognitive, and compensation strategies; and indirect strategies consist of meta-cognitive, cognitive, and affective strategies.

Learning Styles: These are an individual's natural, habitual, and preferred way(s) of absorbing, processing, and retaining new information and skills. These learning styles persist, regardless of teaching methods and content areas, and indicate the methods through which students are likely to excel.

mLearning/MPs: mLearning using mobile phones only.

Mobile Learning (mLearning): A form of learning that combines the technologies of mobile communications, such as laptop computers, PDAs, smart phones, and mobile phones, with eLearning.

Chapter 6
Open Source for Mobile Devices and Mobile Learning

Hal Steger
Funambol, USA

ABSTRACT

Open source software is increasingly used for many types of mobile apps and wireless devices. A leading example is Funambol, the world's most popular mobile open source server project. This chapter discusses the background of the Funambol project and how it applies to mobile learning in three specific areas: wireless device compatibility, data and content synchronization, and mobile device management. The chapter also discusses major trends that are affecting mobile open source software for mobile learning, challenges that are confronting mobile software developers and content authors, and the benefits of using open source software for mobile learning systems.

INTRODUCTION

Open Source software is increasingly being used on a diverse spectrum of wireless devices, for a broad range of purposes. Examples include Google Android, a major mobile operating system, WebKit, a layout engine that is the basis of several mobile web browsers, and Funambol, a mobile cloud sync server that keeps data and content in sync between mobile devices and other systems.

DOI: 10.4018/978-1-60960-613-8.ch006

This chapter describes Funambol open source and how it can benefit mobile learning. Funambol is the most popular mobile open source server project in the world. Funambol open source software has been downloaded more than four million times by 50,000 mobile developers in 200 countries. The open source project that Funambol is based on was started in 2001. Today, it has a vibrant worldwide community of developers and users who utilize open source for numerous projects on many types of wireless devices.

Funambol's experience as the world's leading mobile open source server project can shed light on the utility of open source for mobile devices and mobile learning in three primary areas: device compatibility, data and content synchronization, and device management.

BACKGROUND

The open source project behind Funambol, Sync4J, started in 2001. Its goal was to create an open source software server that could sync any type of data or content with any mobile device. It was called Sync4J because it was initially intended to provide synchronization for Java applications. Since then, although many components of Funambol software are written in Java, the software has expanded its scope to support billions of mobile devices and connected devices, as well as hundreds of backend systems, regardless of whether they use Java.

Funambol is the leading open source implementation of the SyncML (OMA, 2010a) standard and protocol that is built into more than one billion mobile handsets. SyncML is also known as the Open Mobile Alliance (OMA) Data Synchronization (DS) protocol. It is a data-neutral standard that is optimized to sync a wide array of data between mobile devices and servers. It is commonly used to sync mobile address books (contact information), calendars, and other personal information management (PIM) data (e.g. tasks and notes), although the standard supports practically any data. SyncML has also been used to synchronize email, files and rich media such as photos and videos. The standard was recently updated to support the syncing of large objects, such as video files, which can be important for mobile learning.

Beyond data synchronization, SyncML provides the ability to notify mobile clients when there is new information (using 'push' notifications), conflict resolution (e.g. what happens if the same data is changed on both a device and server at the same time), and disconnected (or offline) use.

SyncML was developed by a consortium of organizations to promote mobile data interoperability. They wanted to avoid reinventing the wireless infrastructure wheel to sync data with diverse mobile devices. This is why Sync4J was based on SyncML. It is also why Funambol is the most popular open source implementation of SyncML and why Funambol is for those pursuing mobile learning.

Funambol has played an important role in evolving the SyncML protocol over time. The OMA, which is the governing body for OMA DS (SyncML), also provides a standard specification for mobile device management known as OMA DM (OMA, 2010b). Funambol also provides open source OMA DM software, consisting of both client and server software that can be used to build systems that remotely manage mobile devices. Typically, device management capabilities include over-the-air (OTA) provisioning, firmware and software updates, and diagnostics. This can also be important for mobile learning systems as they are often used remotely by users, which necessitates remote setup, monitoring, and diagnostics.

The Sync4J project became the Funambol company in 2006 when its headquarters relocated from Italy to Silicon Valley and the company received its initial venture capital. The majority of the company's R&D remains in Italy although there have been numerous software contributions from Funambol's worldwide community of mobile developers. Funambol acquired another open source company, Zapatec, for its AJAX (Wikipedia, 2010a) web 2.0 software framework in 2009. Zapatec had several developers in Ukraine who are now part of the Funambol R&D team.

Today, Funambol provides an open source mobile cloud sync solution and platform for billions of mobile phones and connected devices. It has three primary components, as illustrated by Figure 1.

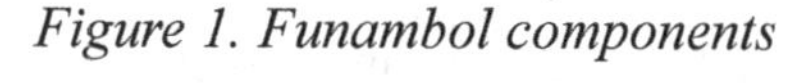

Figure 1. Funambol components

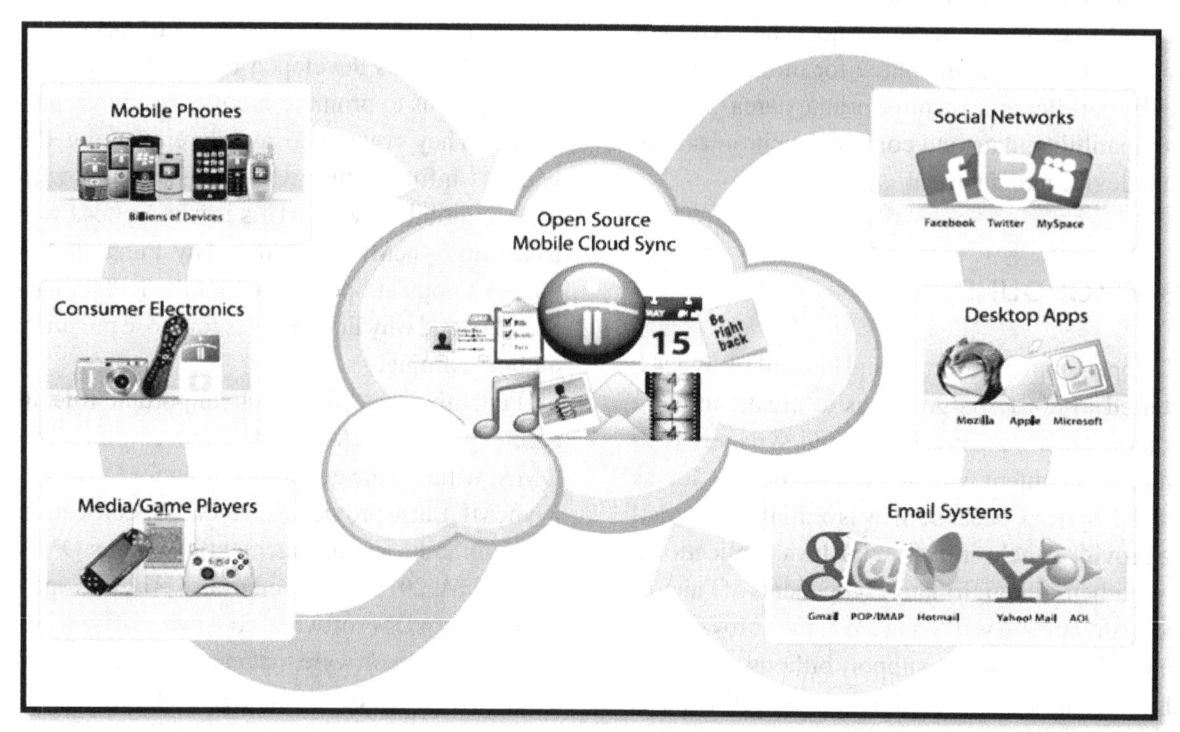

The Funambol server resides in the Internet cloud and provides data synchronization and device management capabilities. It syncs a wide range of data types, including PIM data, messaging and rich media. Funambol mobile client software, on the left side of the diagram, provides SyncML capabilities for a diverse set of mobile phones and devices. It includes mobile client software development kits (SDKs) to create software for virtually any wireless device. Funambol connectors, on the right side of the figure, enable the Funambol server to interface with backend systems such as social networks, content servers, desktop applications, and email servers.

As a commercial open source company, Funambol provides two versions of its software, which is relevant for people considering using open source for mobile learning systems. The Funambol Community Edition software is open source. It uses the AGPLv3 open source license that, in brief, allows Community Edition to be freely used and modified. Its primary requirement is that, if a system that uses Community Edition is deployed as a service (SaaS) for users outside of an organization (e.g. as part of a hosted deployment), the entire system must be made open source. This requirement is meant to ensure that the spirit of open source is preserved; i.e., people who use the open source software also contribute to its continued evolution.

Funambol Carrier Edition is commercial software that differs from Community Edition in a few important respects. It uses a commercial software license that allows organizations to forego open sourcing their code. It contains extra functionality that enables Carrier Edition to be deployed for large numbers of users, such as an AJAX end user web portal (which can be viewed and tried at http://my.funambol.com), over-the-air (OTA) configuration of devices (that eliminates manual set up), and performance enhancements such as scalability and fault-tolerance.

This "dual license" approach of offering an open source and commercial software has become popular among commercial open source companies. It is what enables the Funambol company to sustain itself financially so that the company can continue to support and enhance the Funambol open source project. In rough terms, the open source Community Edition software contains about 90% of the project code, while Carrier Edition includes the rest.

As a point of interest, the name Funambol is derived from the Italian word Funambolo, which loosely means tightrope walker. A tightrope walker symbol can be seen in the company's logo (Figure 2). The symbolism of a tightrope walker represents the actions of Funambol employees and community who sometimes walk a symbolic tightrope that requires balancing the needs of commercial customers with those of the open source community.

Figure 2. The symbol in the Funambol logo portrays a tightrope walker

TRENDS

Several trends are significantly impacting open source for mobile devices and mobile learning.

Evolution of Computing

Computing has undergone distinct shifts over the past decades. Figure 3 shows how it has evolved from personal computing (1980s) to client server computing (early 1990s), Internet computing (late 1990s), cloud computing (mid 2000s), smartphone computing (2008+), and now mobile cloud computing.

These shifts have culminated in a much richer form of computing on mobile devices. It allows mobile learning systems to be more distributed, where users can access a vast array of information and content.

Along with these changes in computing infrastructure has come the rise of open source software. Open source has seen significant development and user adoption in several areas. It has progressed up the computing technology stack, from server and desktop operating systems such as Linux, to application, web and database servers, and most recently, to mobile software. The effect that these advances in computing and open source have on mobile learning is described later in this chapter.

Proliferation of Smartphones and Other Wireless Devices

According to IDC, 1.2 billion phones (IEEE Communications Society, 2010) will sell in 2010, bringing the worldwide total to five billion (ABI Research, 2010). This equates to an average of 5

Figure 3. Evolution of computing

out of every 6 people on earth having a mobile phone.

Although mobile phones are the most common wireless device, there are many other types, such as:

- laptops and netbooks
- tablet PCs (such as the iPad and many others that are forthcoming)
- e-book readers (such as the Kindle and Nook)
- mp3 players (such as the iPod and many others)
- game consoles, digital cameras, printers, e-picture frames and Internet-based appliances
- vehicles and navigation systems

The list of mobile devices is rapidly expanding. In addition, there are many other types of wireless devices, such as those that use near field communications (NFC), Bluetooth, RFID and other forms of wireless connectivity.

The explosion of wireless devices impacts open source and mobile learning in several ways:

- Although phones are the most ubiquitous mobile device, and smartphones are being rapidly adopted, most mobile phones in existence in the next few years will continue to be less costly feature phones, especially in emerging markets
- There are many non-phone form factors that are more appropriate for mobile learning, due to their physical characteristics that are superior for reading and input, such as netbooks, tablets and e-book readers
- Creating content that works on a wide range of mobile devices is more challenging than ever, due to the fragmentation of the many makes and models of wireless devices. This is explored further below

Explosion of Mobile Rich Media and Content

Wireless devices are gradually progressing from being primarily text-oriented to becoming capable of rich media, with touchscreen interfaces, integrated audio and video, and desktop-like web browsers. Many devices can now play rich media and have capabilities such as text-to-speech that can aid mobile learning. At the same time, a wealth of rich content is increasingly being created for consumption on mobile devices.

Memory and Wireless Bandwidth Become Cheaper

An important consideration for mobile learning is that wireless devices are gaining increasing amounts of memory. As the cost of mobile memory drops, that makes it possible to store greater amounts of data and content. Further, the cost of wireless network connectivity is continuing to drop, which makes it more feasible to deploy mobile learning applications and content on wireless devices.

CHALLENGES

The aforementioned trends are laying the foundation for the rapid growth of mobile devices capable of being used for mobile learning. At the same time, there are several challenges that make it difficult for mobile developers and content authors to create mobile learning systems. These challenges are described below. In the subsequent section, the ability of open source to address these challenges is examined.

Device Limitations

Mobile devices have physical characteristics that distinguish them from personal computers and non-mobile devices, which affect their ability to be used for mobile learning systems. Here are some of their limitations which should be taken into account when considering the design of mobile learning systems:

- mobile screens, which vary greatly by size, color support, touch capabilities and technology, which affects how well things can be seen outside versus inside; in general, mobile screens are a fraction of the size of personal computer or fixed device screens
- battery life, which greatly affects the performance of software on mobile devices; in particular, if a mobile device is used for multi-tasking or frequent communication, it can drain the battery quickly, so it is important that a mobile learning application be aware of this, so it can conserve power requirements
- CPUs, which typically are less powerful than PC CPUs, although this is gradually changing
- memory, which as noted, is becoming cheaper but is still not as plentiful compared to PCs
- keyboard and input capabilities tend to be more limited, although newer smartphones have QWERTY keyboards or touchscreen interfaces; in general, however, it is better to assume that mobile devices are better for reading and not as useful for extensive writing and human input
- network access, which can range from slow and intermittent, to fast and continuous; however, as a general rule, the more radio or wireless connectivity used, the more power is consumed, so as noted, this needs to be a factor in the design of mobile learning systems

Device Fragmentation

Perhaps the most serious challenge facing software developers and content authors of mobile learning systems is device fragmentation. The diversity of mobile device characteristics, as well as the numerous mobile operating systems and environments (iPhone, Android, Symbian, BREW, Java, BlackBerry, Windows Mobile, Mobile Linux, and dozens of proprietary systems), and development tools, makes it difficult, time consuming and costly to develop and enhance apps and content that work on multiple device types. The situation is very different than developing for personal computers, as there are a relatively small number of platforms to support (e.g. Windows, Macs and various forms of Unix). Even then, for PCs, there are tools that make it possible to develop for one platform and deploy on others (such as Flash for animation). The state of cross-platform tools and technology for mobile is less advanced, for many reasons, including device fragmentation and limited physical capabilities of devices.

Native vs. Mobile Web Apps

A major decision facing mobile developers and content authors today regards building native mobile apps vs. mobile web apps. There are pros and cons to each approach. The advantage of native apps is that the resulting application performs better and can utilize more aspects of devices such as their physical attributes (e.g. touchscreen, GPS, accelerometer, specific keys). The downside of native apps is that they require specialized knowledge of each platform, including its development tools, which often involve a steep learning curve. Furthermore, when native applications are updated, on many platforms, there are not easy ways to update the software or content. In some cases, users must go through an app store to download or acquire software or content, which is not always convenient or natural, plus it places an additional requirement on the developer/author to work with an app store (imagine doing this with a dozen different app stores, which are now on the market).

On the flip side, mobile web apps that work across multiple phone platforms are no panacea. Although it is possible to create rather plain HTML-content that can work on many mobile devices, if the content is delivered in this fashion, it will likely be viewed as non-compelling or non-competitive with native apps and content. Although there are advances in HTML, the web browsers on mobile phones and devices are not standard and there is no guarantee that advanced HTML capabilities will be present on a user's device. The advantage of building mobile web apps vs. native apps is that, in theory, it is possible to build apps that work across mobile platforms; however, the usability of these apps often leaves much to be desired (although on low cost phones, this may be all that is possible). The decision to build native apps vs. mobile web apps depends on the specifics of the situation, such as the users being targeted, available resources, and more.

OPEN SOURCE ADVANTAGES

Open source provides several capabilities for mobile devices and learning.

Data and Content Synchronization

An important aspect of mobile learning systems is that they be easy to use. It is often helpful for applications to have some data and content local on the device, as this enables apps to perform better and have superior usability. It also means that devices can work in a disconnected mode. Funambol provides an open source platform to sync diverse data and content with a wide variety of wireless devices. This allows developers and authors to focus more on the content, interface and logic of the material, rather than reinventing the infrastructure wheel for mobile data synchronization.

Push Notifications

Some mobile learning systems require updated or dynamic information. In this case, it is useful for the wireless infrastructure to be able to notify mobile devices only when there is new information, as a means of conserving network bandwidth and device power. Funambol open source provides multiple push notification mechanisms to alert mobile devices when there are updates so that client software or the user can decide whether to download them. It is helpful to offer multiple methods for push notifications because mobile networks offer different capabilities, as well as charge differently for bandwidth, so having options for push notifications provides flexibility to adapt to different network attributes and cost models.

Device Management

Depending on the mobile device being used, it may be important for a mobile learning system to inspect or control different aspects of it remotely. A simple example is checking that the device meets the required hardware and software requirements, or that it has an up-to-date version of firmware or embedded software, and if not, to alert the user or update the device remotely. The ability to perform over-the-air OTA setup, provisioning, diagnostics, software updates and more falls into the category of mobile device management. There are some open source solutions, including Funambol, that provide a framework for performing remote device management capabilities.

Cross-Platform Development

Device fragmentation is a major challenge, as mobile developers and content authors must decide whether to create native or mobile web apps. There are some open source projects that provide development tools and frameworks to enable the creation of cross-platform mobile apps. Their benefit is that developers can create one version of an app and deploy it across multiple mobile platforms. It also means less maintenance, because in theory, a developer would only need to change one version of an app.

Funambol plans to release soon a free open source mobile web application framework that lets developers build one version of a mobile app and deploy it on multiple platforms. Funambol's approach allows developers to program in AJAX (Asynchronous JavaScript and XML), which is a higher level language that is familiar to developers who use it to build web apps. This allows a more common developer skillset to create mobile apps, as it does not require learning lower level, native, proprietary tools and code.

Funambol is not the only open source dev tool for creating cross-platform mobile apps. Another example is Appcelerator Titanium (Wikipedia, 2010b). However, Funambol is the only one that uses AJAX to bridge the web and mobile worlds. Other open source projects involve different languages and approaches, which have pros and cons. The pros are that the resulting applications may perform better and take advantage of the underlying hardware; the cons are that this requires yet another language to master. Which of these approaches are better for mobile learning depends on multiple considerations, such as the importance of performance versus platform coverage.

SyncML and Broad Device Compatibility

SyncML provides broad device support for syncing a wide variety of mobile data and content. If a developer uses open source software such as Funambol that is based on the SyncML standard, it provides a good foundation for syncing a wide variety of information with a large number of wireless devices.

Flexibility and Cost

An important requirement for many mobile learning systems is flexibility, as it is crucial to be able to respond to user feedback, support new devices, translate content for different languages, and so forth. When developers have access to the system's underlying source code, they have extensive control of what their system can do, in contrast to proprietary systems. This is not just about technology flexibility but licensing flexibility, as many open source projects offer free and unlimited distribution rights, albeit sometimes with basic requirements, such as copyright and attribution clauses. The details depend on the open source license but, in general, open source for mobile learning typically provides more flexibility and involves less cost than proprietary solutions.

Community Development, Support and Crowdsourcing

An important benefit of open source for mobile learning stems from the community effect of open source, which allows for and encourages the participation of many people to assist with a project. For example, if a mobile learning system was developed with open source, with the premise that it can be freely used and shared, perhaps this would encourage more people in an online community to help the project by volunteering to work on some aspect of it, such as software or content development, testing, documentation, website development, translations, device support, and more.

Further, one of the most beneficial aspects of open source is the support and assistance provided by people familiar with a project, who can help each other with technical questions and expertise. For example, crowdsourcing (Wikipedia, 2010c), which is similar to outsourcing (although in this case, the work is performed by a 'crowd'), can be used to involve a community for help. In the case of Funambol, which has a large community of project participants and volunteers, it is not unusual to get help from many project members all over the world.

Community interaction, support, and crowdsourcing enable a project to leverage the inspiration, talent, and ability of many people, not just those who work on a project as a full-time job or pursuit. Often, some of the best ideas and feedback for software come from those not directly involved in a project, as other people have a different perspective and think "out-of-the-box" beyond the primary actors in a project. Mobile open source is ideal for encouraging the flow of innovative ideas and work performed on mobile learning projects.

ADDITIONAL RESOURCES

There are many good sources of online information about mobile open source projects such as Android, WebKit, and Funambol. Links to these projects are listed below, where you can find developer forges that let you learn more about the software as well as download the software and documentation, look at demos and videos, and interact with members of their respective communities.

Google Android - http://www.android.com
WebKit - http://webkit.org
Funambol - http://www.funambol.org
Accelerator - http://www.appcelerator.com

REFERENCES

IEEE. Communications Society. (2010). *IDC, 1.2B mobile phones will be sold in 2010*. Retrieved from http://community.comsoc.org/blogs/ajwdct/idc-market-forecasts- mobile-broadband-and-lte

OMA. (2010a). *OMA data synchronization* V1.2.1. Retrieved from http://openmobilealliance.org/Technical/release_program/ds_v12.aspx

OMA. (2010b). *Device management Working Group*. Retrieved from http://openmobilealliance.org/Technical/DM.aspx

Research, A. B. I. (2010). *5B devices in 2010*. Retrieved from http://www.abiresearch.com/press/1684- Worldwide+Mobile+Subscriptions+Forecast+to+Exceed+Five+Billion+by+4Q-2010

Wikipedia. (2010a). *Ajax (programming)*. Retrieved from http://en.wikipedia.org/wiki/Ajax_(programming)

Wikipedia. (2010b). *Appcelerator Titanium*. Retrieved from http://en.wikipedia.org/wiki/Appcelerator_Titanium

Wikipedia. (2010c). *Crowdsourcing*. Retrieved from http://en.wikipedia.org/wiki/Crowdsourcing

Section 2
Mobile Learning in Open Source Setting

Chapter 7
Carrying That Ten Thousand Dollar Lab in a Backsack:
A Mobile Networking Laboratory with the Use of Open Source Applications

Dongqing Yuan
University of Wisconsin- Stout, USA

Jiling Zhong
Troy University, USA

ABSTRACT

In the past decade, with the development of wireless and other mobile technologies, including mobile computer, cellular phone, and GPS, educational practitioners have had the opportunity to develop a ubiquitous learning environment. This chapter provides a detailed account of developing a mobile network laboratory with a set of open source software (OSS) that allows students to conduct the labs either as an individual or as a group at anytime and anywhere.

INTRODUCTION

Computer networking is one of the most challenging subjects not only to students, but also to teachers. One of the most difficult issues teachers are facing is how to convey many of the theories when there are not so many labs right on the point text books can provide. On one hand, students may find the topic too technical and dry when presented; on the other, most IT instructors still primarily use lectures as the exclusive means to teach. As stated in IEEE/ACM IT Computing Curricula (IEEE/ACM, 2008), it is strongly recommended to incorporate hands-on lab components into teaching an undergraduate networking course as they help students apply the theory in solving real-world problems. As such, it will require significant investments on space, equipment and

DOI: 10.4018/978-1-60960-613-8.ch007

software in setting up such a dedicated laboratory that will lend support for the course. Unfortunately, the limitations of budget, space and facility do not allow for permanent institutionalization of such a physical laboratory. At the same time, current educational philosophy suggests that hands-on labs may be an effective and yet more efficient way to achieve the same goal. In this case, a mobile open source network lab will be a valuable solution.

The mobile open source lab is composed by mobile hard drives and laptops with a suite of open source software that, in aggregate, provide students with a means to experiment what they have learned in the lecture and at the same time to allow them gain real world hands-on skills. In this chapter, we describe the requirements for developing and implementing such a mobile open source network lab. The design process, focusing particularly on the mobility, usability and affordability issues, is addressed. We also discuss the impact provided by the mobile open source lab, such as how the lab enhances practical experience, engage active learning, and promote collaboration among students.

The chapter starts with a review of the related work, followed by a description of the mobile open source network lab including objectives, lab components, design, and implementation of the lab. Particular attentions are paid towards the use of OSS applications on the mobile lab. Finally, the evaluation of the mobile open source network lab is presented, and future work is discussed.

BACKGROUND

IEEE/ACM IT Computing Curricula (IEEE/ACM, 2008) recommends the incorporation of hands-on lab components into the undergraduate networking course as they help students apply the theories to solving real-world problems. Honey and Mumford (1984), and Kolb (1985) claim that students have different learning patterns, and science and technology students have the ability to learn primarily through hands-on experience. Denning (2003) also indicates that, ignoring application, computer education will end up like the failed "new math" of the 1960s- all concepts, no practice, lifeless, and dead. Dewey (1985) argues that "the first stage of contact with any kind of education, from children through adults, must be hands-on and experiential. Learning is a process of discovery and enactment and of wrestling with problems first hand" (p. 160). Having laboratory components and well-designed lab materials are essential to the success of networking education (IEEE/ACM, 2008).

However, most computer networking courses do not have laboratory components coupled with a modular curriculum, which allow students to practice the real-world problem-solving skills expected in the IT career field (Casado & Mckeown, 2005; Goyal, Lai, Jain, & Durresi, 1998; Kneale, Horta, & Box, 2004). Two factors contribute to this problem. One is the fact that it requires significant cost, space and time to set up a networking laboratory in an academic environment (Hughes, 1989; Lawson & Stackpole, 2006). It would cost more than $150,000 to set up a networking laboratory for 20 students initially just for the hardware and software (Gerdes & Tilley, 2007). Most schools do not have the budget, space and facility to create and maintain a "hands-on" learning environment. Another is that it would take many hours for the instructors to design a lab exercise that meets the objectives (Helps, 2006). To aggravate the problem further, to stay abreast with the rapid change of technology, the instructors, in addition to their busy teaching schedule, need to design and re-design the curriculum to explore the new technology (Helps, 2006).

The relevance of the study is further evidenced by *A Governor's Guide to Creating a 21st-Century Workforce* (National Governors Association, 2002). The guide states that the United States does not produce enough qualified graduates in science and technology to meet the specialized workforce demands. The skill gap is especially critical in high-tech professions. Universities and colleges have to produce higher quality graduates to fill the skills gap. The students need to learn integrated knowledge and know how to apply the knowledge to the real world problems rather than memorizing the facts.

However, the traditional "lecture-only" teaching style discourages the acquisition of deep understanding and higher-level skills in students. In addition, students also need non-technical skills such as writing a technical report, collaboration, negotiation and team work to be successful in the career (IEEE/ACM, 2008). IEEE/ACM IT computing curricula (IEEE/ACM, 2008) list ten qualities which are considered most important by employers in the IT industry. These qualities include communication skills (verbal and written), honesty/ integrity, teamwork skills, etc. As IEEE/ ACM IT computing curricula (IEEE/ACM, 2008) state, "that employers seek candidates with these general qualities underscores the importance of making professional practice a central component of the curriculum." (p. 43).

Over the years, a lot of studies have been done in developing a "hands-on" laboratory and curriculum. Hnatyshin and Lobo (2008), Morgan et al. (2007), and Wong et al. (2007) propose different virtual network labs. Minch & Tabor (2003) developed a class in which students took in charge of a real productive network as a lab class. McAndrew (2008), Nelson & NG (2000), and Wright et al. (2007) have used open source software to teach a variety of computer science courses. University of California (Wright et al., 2007) develops an open source lab learning facility to support a variety of computer science and engineering courses. Victoria University at Australia (McAndrew, 2008) implements an open source software based cryptography course. North Carolina State University (Meneely, Williams, & Gehringer, 2008) creates a Repository for Open Software Education (ROSE) which is a website containing three education-friendly open source projects for an undergraduate software engineering course. Moravian College (Corbesero, 2006) uses open source software to create a small lab for gaming classes. The above research studies create simulation learning tools with OSS, so students can apply the theories they have learned to the level in the simulation environment.

The problem of missing the hands-on lab components in teaching a networking course appears to be threefold: 1) students may not have a clear understanding of how to apply what they have learned in the classroom to solving real world problems (Guo, Xiang, & Wang, 2007; Morgan et al., 2007), 2) there will be gaps in the expectation between the IT industry and student skills (Minch & Tabor, 2003), and 3) it will be hard for the program to satisfy the requirements of Accreditation Board for Engineering and Technology (ABET) which proposes a set of objectives for the use of laboratories in IT education (Corter et al., 2007).

MOBILE OPEN SOURCE NETWORK LABORATORY

Solutions and Recommendations

This research addresses the above problem and extends the previous research on networking labs. The labs are designed according to the following criteria: mobility, affordability, compatibility, portability, and clarity. Mobility is achieved through installing Linux on USB flash drives. Installing a portable Linux operating system on a flash drive or USB key no larger than the thumb allows the Linux operating system to run from any computer that can boot from a flash device and allows students to bring the operating system with them. The lab curriculum will closely match the learning objectives, so students can practice with hands-on skills, thus deepening their understanding of the theories. The affordability of lab setup is crucial since the budget for purchasing the equipment is limited; hence open-source software is adopted instead of vendor-specific commercial software.

The lab materials are compatible with the existing hardware. In addition, the design should be portable to other IT educators. The lab design consists of hardware and software requirements. The labs cover the main topics recommended by IEEE/ACM IT computing curricula such as foundations of networking, routing and switching, physical layer, network management, application, and security areas (IEEE/ACM, 2008).

Affordability: Using Open Source Software and Applications

Open source software is also known as free software, but "free" refers to "freedom and liberty" instead of "cost or price" (Open Source Initiative, 2006). Open source refers to the source code being openly available to the public. The Open Source Initiative specifies that open source software must be distributed under a license that guarantees the right to read, redistribute, modify, and use the software freely (Open Source Initiative, 2006). Open source software is collaboratively built by the developers and shared by the users and developers. Those users and developers may be geographically scattered, work independently, and do not know each other. The code is posted online and the community will test the program, find bugs and improve the program. The community collaboration, contribution, and circulation are the reasons that make OSS evolve rapidly. OSS has developed rapidly, and it is estimated that, within next few years, OSS will take fifty percent of the software market (Lawson & Stackpole, 2006). IBM, SUN, HP, Google, and other big corporations each year spend tremendous amount of money to support the research on OSS (DeKoenigsberg, 2008; Wheeler, 2007).

Initially, the creator of the Linux kernel, Linus Torvalds believed that people shall not sell the open source software (Torvalds & Diamond, 2001); "you can use the operating system for free, as long as you don't sell it, and if you make any changes or improvements you must make them available to everybody in source code" (Torvalds & Diamond, 2001, p. 94). However, today OSS can be released under a variety of different licenses, and four of them are common to educational applications: the GNU public license, the Lesser GNU Public License, the Berkeley Software Distribution License, and the Mozilla Public License.

SourceForge, one of the primary hosting sites for OSS, lists more than 127,000 registered Open Source projects as of 2009 and above 17,000 downloads everyday world wide (SourceForge, 2009). A TechRepublic Research survey shows that 46% of IT professionals support GNU/Linux in their organizations (Wheeler, 2007). Apache, an Open Source software web server, has become the number one web server since 1996.

Linux, the biggest rival to Windows Operating Systems, is one of the most popular open source software. Linux has different distributions; however, the underlying source code can be freely modified and redistributed by anyone under the GNU General Public License (GPL). A lot of open source packages can run on Linux, and most of them are copylefted software. Due to the high availability, outstanding performance, and low cost, Linux has been widely used for many mission-critical systems (Wheeler, 2007). For example, Google has used more than 6000 Linux servers for its search service (Google Press Center, 2000). Linux also serves as a very good tool to learn TCP/IP networking and system administration in the classroom. Students can implement interesting projects on top of it such as building a web server, mail server, and database server.

The advantages of OSS are free of license fee, flexibility to customize the source code, community support, and continuous improvement of the code. Open source software as a viable technological solution has been used in the classroom from high schools to universities over decades. The president of ACM, Dave Patterson, made several suggestions for retaining CS students and reinvigorating the CS curriculum, and one of the suggestions is called "Courses that I would love to take #1: Join the open source movement" (Patterson, 2006).

Over the years, there are lots of open source learning systems being used and developed in primary and secondary schools, as well as higher education institutions. The K through 12 Linux Terminal Server Project (K12LTSP), distributed under the GNU GPL, consists of an open source Linux terminal server which can be connected to lots of low-powered client computers (Pfaffenberger, 2000). Students can login to the server remotely through the terminals to do the computing. Moodle, a learning system, which incorporates the pedagogical features into the system, has been used in more than two hundred countries and by over 200,000 registered users. Moodle, licensed under GPL, allows instructors to customize online courses and deliver the courses to thousands of students. Faber (2002) advocates that students and educators can create forums with ongoing open source projects to share syllabi and teaching materials and access resources. The open-ended feature of open source provides students with motivational and challenging projects; on the other hand, the collaborative nature of the open source process invites input from both the academia and industry.

In addition to these open source learning systems, OSS has been used in the classroom widely as a teaching and learning tool. Many simulation learning tools have been created with OSS, so students can apply the theories they have learned to the level in the simulation environment.

According to Faber (2002), students often find open source based projects interesting, challenging, and engaging. Open source labs instigate students to be inquisitive and try to solve problems with different approaches. The open lab also provides an environment that allows students to practice technical related skills and non-technical skills such as working as a team, reading and writing technical reports.

Torvalds (Torvalds & Diamond, 2001), the creator of the Linux kernel states that "pretty much all of the modern science and technology is founded on very similar ideas to open source" (p. 246). Faber (2002) argues that "open source software takes the simulations one step further into real life, as an educational process, it fits well and is consistent with other practice-based, or activity-based methods of education" (p. 37).

According to Helps (2006), an open source learning lab can create a collaborative, problem-based learning environment that allows students to build a constructive learning community. By working on open source based projects, students can learn collaboration skills. And collaborative learning can help students create ideas and identify different approaches to solving problems (Linn & Burbuels, 1993). Because open source based projects will be handed to the group of new students over the semester, the students have to learn how to write the documentation.

According to Berglund and Priestley (2001), failures are commonly anticipated in open source based projects. Such a learning environment will be different from the traditional classroom setting tremendously which evaluates students only based on exams and quizzes. The open source laboratory will turn the students into innovative problem solvers and investigators instead of passive receivers. To solve problems, students have to know how to locate, interpret, and evaluate information. Learning will be much more enjoyable as a group in a "hands-on" learning environment.

As Faber (2002) states, "open source is more accurately the technological implementation of John Dewey's principles of activity-based education. Open source educational model can build relationships between corporate and academia interests in ways that can benefit both contexts" (p. 33). With such an open source lab, the learning experience is pushed to the environment of real life; it is better than "learning by doing simulation" (Faber, 2002).

However, there have been some debates as to whether using OSS in the organization truly decrease the cost (Wheeler, 2007). For example,

Table 1. Comparison of initial costs of setting up the networking lab with open source software vs. proprietary solutions (Wheeler, 2007)

	Proprietary Software	Open Source Software
Operating systems	Windows XP costs $1510 (25 clients)	Red Hat Linux costs Free or with minimum fee such as $30 (standard), $80 (deluxe), $180 professional
Web Server	Windows Web Server with Windows XP costs $1500	Included
Email Server	Windows mail server costs $1300 (10 clients)	included
Database Server	Windows SQL server costs $2100 (10 CALs)	included
Router	Each low-end Cisco Router costs more than $1500	Free through installing open source software, however, with limited function

Microsoft and Sun claim that their products have the lower total cost of ownership (TCO) than OSS's. Their study shows that there is more technical support and end-user training cost involved with the OSS system. However, it is hard to say that this type of vendor-sponsored survey is not trying to be more favorable to the vendor instead of being fair to other competitors. In addition, the users of proprietary software actually don't own the software and thus don't have the rights of ownership. Table 1 compares the initial costs of setting up the networking lab with open source software vs. proprietary solutions.

There are also some criticisms from commercial software developers that the open source software lacks centralized control and quality (Wheeler, 2007). Additionally, there is a steep learning curve associated with the use of OSS. Furthermore, open source software raises some ethical, economic, and social issues, such as what impact, if any, open source software would have on the national and international intellectual property laws.

Mobility: Installing Linux on USB Flash Drive

USB Linux installation enables students to install a portable Linux operating system on a flash drive or USB key no larger than the thumb. This portable Linux operating system can then be run from any computer that can boot from a flash device, allowing students to bring the operating system with them. It is their own personal operating system that they can carry in their pocket. This mobility allows students to conduct the labs either as an individual or as a group at anytime and anywhere.

One can follow the following steps to install and boot Linux on a flash drive. Download the "*liveusb*" file; extract the file to the laptop. Launch '*liveusb-creator.exe*'. Use the existing Live CD; set the target device to point to the USB flash drive. Click "*create Live USB*" to begin the creation process. Once the process is finished, restart the laptop and set the BIOS to reboot from the USB flash drive. Log in to Linux as "*root*". Connect the USB flash drive to a USB port on a laptop. The USB flash drive is detected and represented as "dev/sda1".

Mount the USB flash drive:

```
#mkdir /mnt/usbflash
#mount /dev/sda1 /mnt/usbflash
```

Check the contents of the USB pen drive:

```
#ls -l /mnt/usbflash
```

Unmount the USB drive:

```
#umount /mnt/usbflash
```

To mount a USB pen drive every time it boots up:

```
# vi /etc/fstab
#/dev/sda1 /mnt/usbdrive vfat
rw;user;noauto 0 0
```

Mobile Open Source Labs

A sequence of six labs is developed. Students can do the labs individually or as a group at anytime and anywhere. Each lab is described as follows:

Lab 1: Set up a Local Area Network

In this lab, students will connect two or more laptops running Linux on the USB pen drive together; they will configure IP addresses with the *ifconfig* command, and set up the routing table using the *route* command; then verify the connectivity with the *ping* command. Through experiments, the students also need to understand the difference between the broadcast domain and the collision domain.

a. *ifconfig* command usage for this part of the lab will be to configure network interface devices. This includes adding the IP address and the netmask for the host.
b. Boot each host machine and log on with your user-name.
c. Open a new terminal on each host and log on as root.

```
$su-
#password
```

d. Check network interface settings using *ifconfig*

```
#ifconfig -a
```

e. Verify that eth0 exists on each host by checking the output from above;
f. Set the ethernet interface at each host using *ifconfig*

This will assign an ip address and network mask to the host. The following is a sample:

```
#ifconfig eth0 192.168.100.1 netmask
255.255.255.240
```

g. Verify communication among the machines by using the *ping* command.

On each host *ping* the other hosts. For example:

```
#ping 192.168.100.2
```

h. On each host run *traceroute* to the other hosts. For example:

```
#traceroute 192.168.1.2
```

i. If the network script is not on, add it using *chkconfig*. This command adds the network script to the current run level.

```
# chkconfig --add  network
```

Lab 2: Set up a Client Server Network: Web Server, DHCP Server, FTP Server

In this lab, students will build upon Lab 1, setting up the DHCP server and Apache web server. The servers will be assigned address statically and the clients will receive the address from the DHCP server dynamically. Test connectivity between the clients and servers. Service packages required for this lab include: *httpd, mysql, mysql-server, php, php-devel, php-devel, dhcpd, vsftpd.* For example, the students follow the following instructions to configure the DHCP server on the laptop.

Box 1.

```
[root@group1 ~] # vi /etc/dhcpd.conf
ddns-update-style                 none;
default-lease-time                 259200;
max-lease-time                         518400;
option routers                         192.168.100.14
option broadcast-address         192.168.100.15
option domain-name-servers         192.168.100.3
subnet 192.168.100.0 netmask 255.255.255.240 {
        range 192.168.100.4 192.168.100.14;
        option subnet-mask 255.255.255.240;
        option nis-domain   "myclass";
        option domain-name "myclass";
host group1-dhcp{
        hardware ethernet         00:13:45:AF:CE:40;
        fixed-address                192.168.100.2;
        hardware ethernet        00:34:24:AD:13:20;
        fixed-address                192.168.100.3;
        }
}
```

a. Configure the DHCP script to reflect the network's information and the range of addresses (see Box 1)
b. Copy the configuration file to /etc/
c. Start the DHCP server

```
[root@group1~]# /etc/init.d/dhcpd
restart
```

d. Remove the ip address from the host's Ethernet interface (a host not running a service)
e. On the DHCP clients, use the following configuration to get the host to contact the DHCP server for a new IP address.

```
root@group1 ~] vi /etc/sysconfig/
network-scripts/ifcfg-eth0
DEVICE=eth0
BOOTPROTO=dhcp
ONBOOT=yes
```

f. Start the DHCP server

```
[root@group1 ~]# /etc/init.d/ network
restart
```

g. Check the dhcp ports (ports # 67, 68) that have been opened.

```
root@group1 ~]# netstat -tlunp
```

h. Check the clients IP leasing information.

```
[root@grop1 ~]# cat /var/lib/dhcp/
dhclient-eth0.leases
```

Lab 3: Using Wireshark and Tcpdump to Analyze the Packets

The Open Source Software used in this lab is Wireshark available at http://www.wireshark.com and Tcpdump available at http://www.tcpdump.

Figure 1. TCP three-way handshake and TCP/IP model

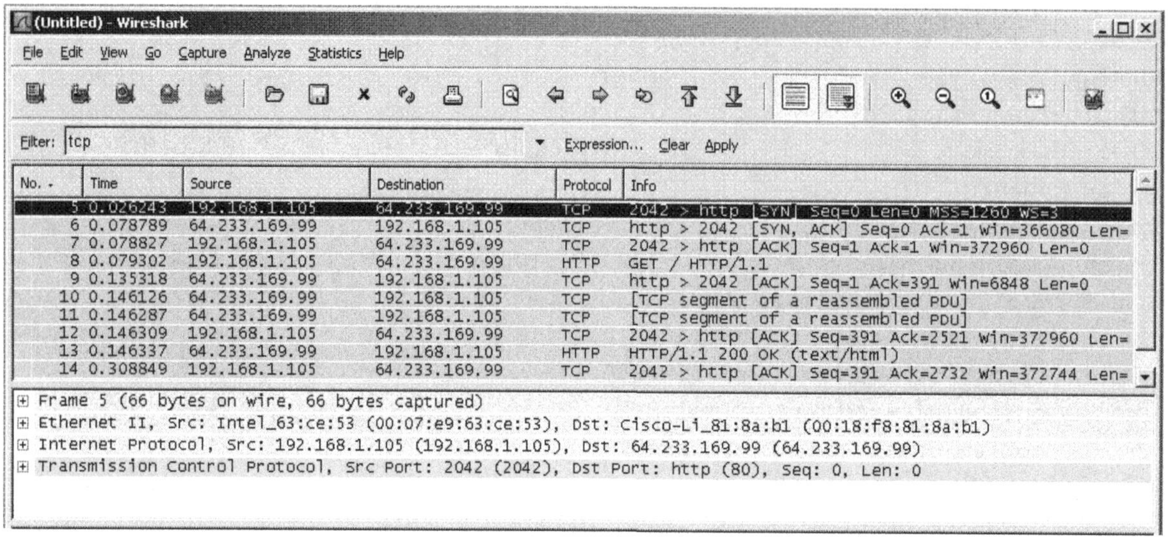

org. Wireshark is a network protocol analyzer for Linux and Windows. Lab 3 is built on Lab 2; students use Wireshark and Tcpdump to capture the packets on the wire and analyze the packets between the clients and server. Students will understand the process of the three-way handshake and TCP/IP 4 layers model. Figure 1 shows the screenshot of Lab 3.

Lab 4: Configure Dynamical Routing Protocol—RIP, OSPF, BGP

The Open Source Software used in this lab is Zebra available at http://www.zebra.org (Zebra, 2009). Linux supports various open source routing software (i.e., Zebra, Routed, Gated), which enables users to set up dynamic routing protocols. Zebra is open source TCP/IP routing software that is similar to Cisco's Internetworking Operating System (IOS). It supports RIP, OSPF, and BGP routing protocols.

In this lab, students will connect different networks and configure a dynamic routing protocol such as RIP, OSPF and BGP. Figure 2 shows the screenshot of Lab 4.

Lab 5: Study and Prevent Denial of Service (DoS) Attacks with IPS, IDS

The Open Source Software: Datapool, Snort available at http://www.packetstormsecurity.org/DoS/index2.html

http://www.snort.org

Datapool is a powerful DoS tool that includes 106 DoS attacks. Snort is a free lightweight network intrusion detection system for UNIX and Windows.

In this lab, students will launch DoS attacks and then learn how to mitigate the attacks through IDS and IPS. Figure 3 shows that the DoS SYN flood attack has started.

Lab 6: Configure a VoIP Service

The Open Source Software used in this lab is Asterisk available at http://www.asterisk.org. Asterisk is an open source telephony switching and private branch exchange service. It allows peer-to-peer calling using a variety of voice over internet protocols.

Figure 2. An example of the students' lab on RIP

```
Password:
ripd> en
ripd# config t
ripd(config)# router rip
ripd(config-router)# network 192.168.2.0/24
ripd(config-router)# network 192.168.3.0/24
ripd(config-router)# exit
ripd(config)# exit
ripd# copy run start
Configuration saved to /etc/ripd.conf
ripd#
```

In this lab, students will set up a VoIP service and build a dial plan. Figure 4 shows the asterisk daemon has been installed and started.

Figure 3. DoS SYN flood attack

```
-----------------------------.-------------------
Option                        |Setting
-----------------------------^-------------------
estination Host:              192.168.1.2
ource IP:                     13.31.16.15
ort Range:                    80-80
ogging:                       OFF
can Only:                     OFF
ine Speed:                    Modem
ontinuous Attack:             ON
Don't stop till they drop":   OFF
ait for online host:          OFF
 of simultaneous attacks:     1
ttacks in initial list:       synful

tarting portstan...
92.168.1.2 resolved to 192.168.1.2
inux host detected...
 TCP port(s) were found open:
0/http
aunching 1 attack(s) at 192.168.1.2 on port: 80
unning SYN flooder (synful)...
aunching 1 attack(s) at 192.168.1.2 on port: 80
unning SYN flooder (synful)...
aunching 1 attack(s) at 192.168.1.2 on port: 80
unning SYN flooder (synful)...
aunching 1 attack(s) at 192.168.1.2 on port: 80
unning SYN flooder (synful)...
aunching 1 attack(s) at 192.168.1.2 on port: 80
unning SYN flooder (synful)...
```

EVALUATION

To measure students' perceptions and attitudes about the labs, an anonymous questionnaire were developed and handed out to the students. The goal of the questionnaire is to assess from the students' viewpoint whether the labs spark the students' interest in taking more networking classes and whether the labs improve their understanding of the concepts, etc. The questionnaire consists of questions in four categories: relevance between the lecture and the labs, the format and contents of labs, relevance between the labs and industry skills, and overall impact of the labs on the students. Each category contains four positive statements about the course. Questions are presented using a five-point Likert scale. The students will be asked to rate the statements based on a scale from 1 (strongly disagree), 2 (disagree), 3 (neither agree nor disagree), 4 (agree), to 5 (strongly agree).

Figures 5, 6, 7 & 8 show the result of the survey. Most of the students finished the class with high confidence and interest in further studies in the area. The students found the labs interesting, challenging, and engaging. They also felt the labs provided flexibility that allowed them to practice technical related skills at anytime and anywhere. However, the survey also shows that the format and instruction of the labs need to be improved for clarity.

Figure 4. Asterisk daemon

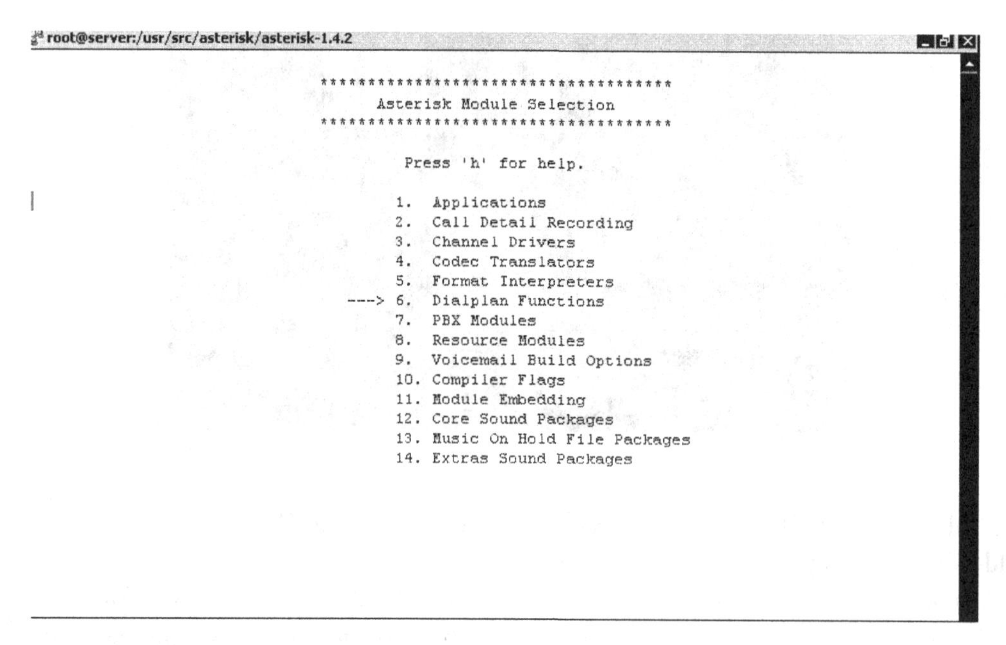

CONCLUSION

With the development of wireless and other mobile technologies, including the mobile computer, cellular phone, GPS, in the past decade, educational practitioners have had the opportunity to develop a ubiquitous learning environment. This chapter provides a detailed account of developing a mobile network laboratory with a set of open source software that allows students to conduct the labs either as an individual or as a group at anytime and anywhere. This chapter also investigates the broader impact of Open Source Software (OSS) on computer education. Then it describes a sequence of six open source labs that successively engage students in learning computer networking. This study is very meaningful and significant since almost all of the universities and colleges

Figure 5. Evaluation result of category 1

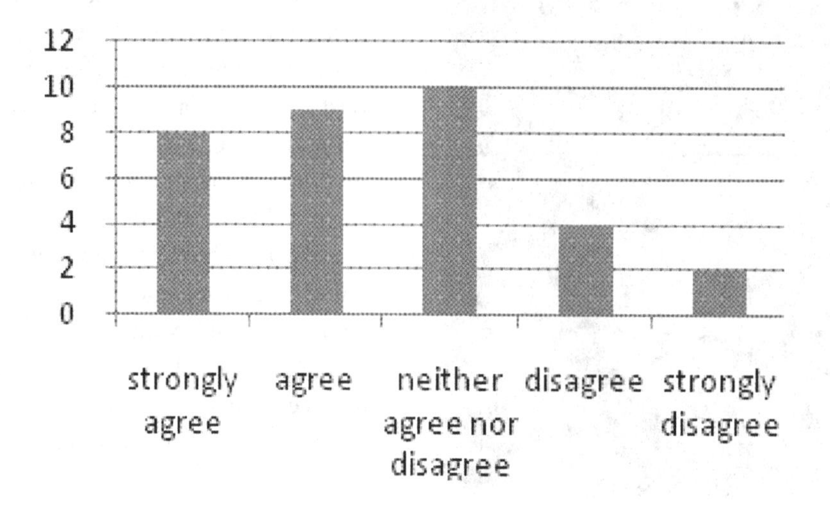

Figure 6. Evaluation result of category 2

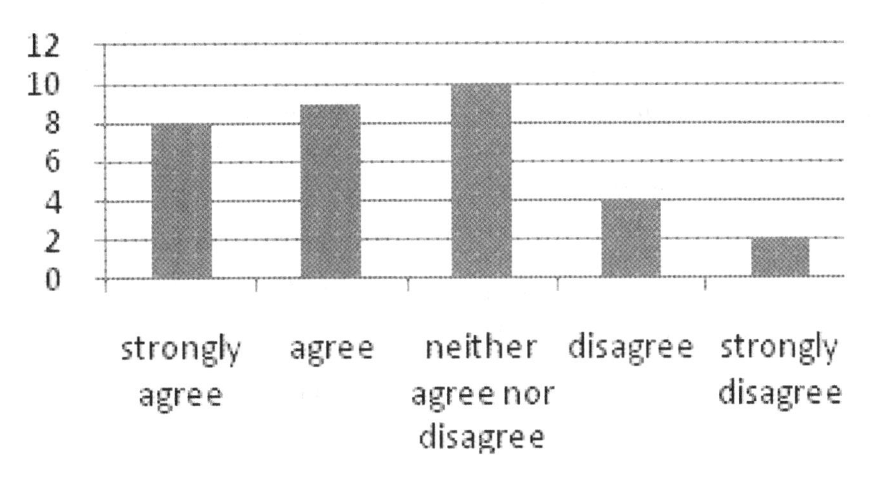

Figure 7. Evaluation result of category 3

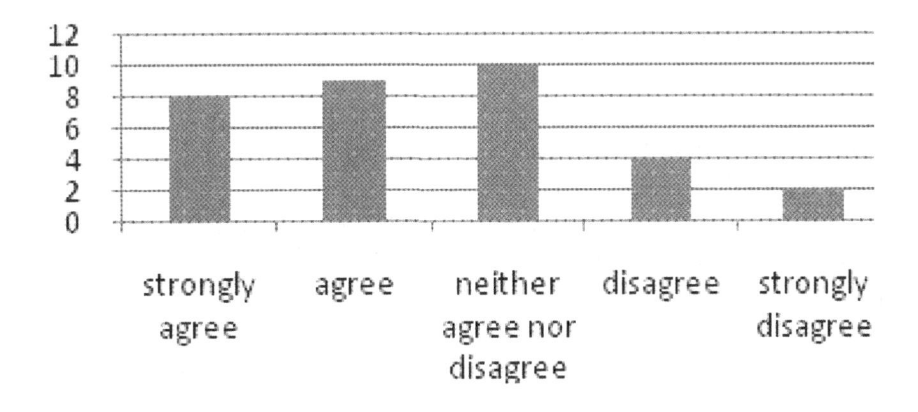

Figure 8. Evaluation result of category 4

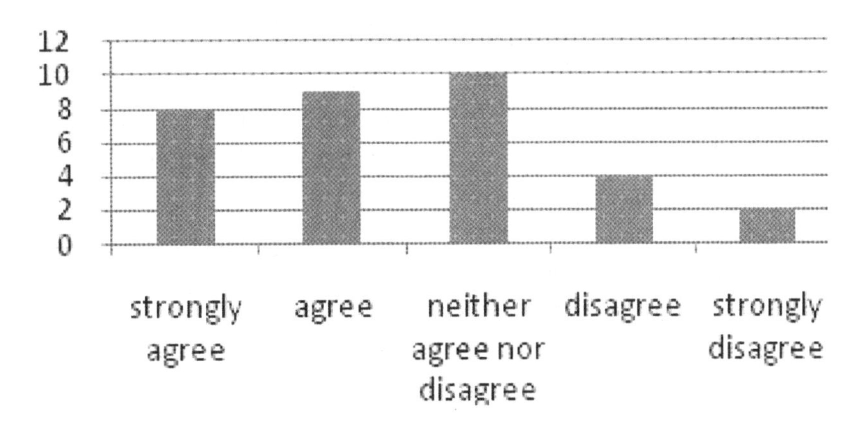

in the U.S. offer networking classes, either at the undergraduate level or graduate level, which provide potential for this research. The research will contribute to the long term improvement of IT workforce qualification in the United States. Specifically, the research has the following significance:

The Linux operating system is installed on a flash drive or USB thumb driver that can be run from any computer, allowing students to bring the operating system with them. This mobility allows students to conduct labs either as an individual or as a group at anytime and anywhere.

All of the tools in the lab are based on OSS. OSS-based projects can make the classroom a more exciting and realistic learning place (Patterson, 2006). In addition, it can provide free or low-cost technology in the classroom (Nelson & NG, 2000). Because the design of the lab is easily replicated, this will benefit other IT educators who face the same problem as the authors who want to incorporate the labs into teaching and meet the ABET accreditation requirements at the same time. The methodology and process to produce the curriculum may be used as a reference to generate similar materials for other computer related disciplines.

REFERENCES

Berglund, E., & Priestley, M. (2001). Open-source documentation: In search of user-driven, just-in-time writing. In *ACM Special Interest Group for Design of Communication Proceedings of the 19th Annual International Conference on Computer documentation* (pp. 132-141).

Casado, M., & Mckeown, N. (2005). The virtual network system. In *The Proceedings of the 36th SIGCSE. Technical Symposium on Computer Science Education (SIGCSE'05)* (pp. 76-80).

Corter, J. E., Nickerson, J. V., Esche, S. K., Chassapis, C., Im, S., & Ma, J. (2007). Constructing reality: A study of remote, hands-on and simulated laboratories. *ACM Transactions on Computer-Human Interaction, 2*(14), 7–27. doi:10.1145/1275511.1275513

DeKoenigsberg, G. (2008). *How successful open source projects work, and how and why to introduce students to the open source world?* In 21st Conference on Software Engineering Education and Training (pp. 274-277).

Denning, P. (2003). Great principles of computing. *Communications of the ACM, 11*(46), 15–20. doi:10.1145/948383.948400

Dewey, J. (1985). *Democracy and education.* Carbondale, IL: Southern Illinois University Press.

Faber, B. (2002). *Educational models and open source: Resisting the proprietary university.* In SIGDOC'02 (pp. 31-39) Toronto, Ontario, Canada.

Gerdes, J., & Tilley, S. (2007). *A conceptual overview of the virtual networking laboratory.* In SIGITE'07 (pp. 75-82).

Google Press Center. (2000). Retrieved July 15, 2009, from **Error! Hyperlink reference not valid.** press/ pressrel/ pressrelease51.html

Goyal, R., Lai, S., Jain, R., & Durresi, A. (1998). *Laboratories for data communications and computer networks.* In 1998 FIE Conference (pp. 1113-1119).

Guo, J., Xiang, W., & Wang, S. (2007). Reinforce networking theory with OPNET simulation. *Journal of Information Technology Education, 6*, 191–198.

Helps, C. R. (2006). *Instructional design theory provides insights into evolving information technology technical curricula.* In SIGITE'06 (pp. 129-135). Minneapolis, Minnesota.

Hnatyshin, V., & Lobo, A. (2008). Undergraduate data communications and networking projects using OPNET software. In *Proceedings of the 39th SIGCSE Technical Symposium on Computer Science Education* (pp. 241-245).

Honey, P., & Mumford, A. (1984). Questions and answers on learning styles questionnaire. *Industrial and Commercial Training, 24*(7), 1–10.

Hughes, L. (1989). Low-cost networks and gateways for teaching data communications. In *SIGCSE '89: Proceedings of the twentieth SIGCSE Technical Symposium on Computer Science Education* (pp. 6–11).

IEEE/ACM. (2008). *IT 2008, curriculum guidelines for undergraduate degree programs in information technology* (Final Draft). Retrieved January 16, 2009, from http://www.acm.org/education/curricula/IT2008%20Curriculum.pdf

Kneale, B., Horta, A. Y., & Box, L. (2004). Velnet: Virtual environment for learning networking. *Proceedings of the Sixth Conference on Australasian Computing Education, 30,* (pp. 161-169).

Kolb, D. (1985). *Learning style inventory.* Boston, MA: McBer and Company.

Lawson, E. A., & Stackpole, W. (2006). Does a virtual networking laboratory result in similar student achievement and satisfaction? In *SIGITE '06: Proceedings of the 7th Conference on Information Technology Education* (pp. 105–114).

Linn, M. C., & Burbules, N. C. (1993). Construction of knowledge and group learning . In Tobin, K. (Ed.), *The practice of constructivism in science education* (pp. 91–119). Hillsdale, NJ: Lawrence Erlbaum Associates.

McAndrew, A. (2008). Teaching cryptography with open-source software. *Special Interest Group Computer Science Education, 40*(1), 325–330.

Meneely, A., Williams, L., & Gehringer, E. F. (2008). A repository of education-friendly open-source projects . In *ACM ITiCSE '08* (pp. 7–12). ROSE.

Minch, R., & Tabor, S. (2003). Networking education for the new economy. *Journal of Information Technology Education, 2*, 191–217.

Morgan, C., Erlinger, M., Davoli, R., & Goldweber, M. (2007). *Environments for a networking laboratory*. In 20th Annual Conference of the National Advisory Committee on Computing Qualification (pp. 53-59).

National Governors Association. (2002). *A governor's guide to creating a 21st-century workforce.* Retrieved January 16, 2009, from http://www.nga.org/ cda/ files/ AM02WORKFORCE.pdf

Nelson, D., & Ng, Y. M. (2000). Teaching computer networking using open source software. In *ITiCSE' 00: Proceedings of the 5th Annual SIGCSE/SIGCUE ITiCSE Conference on Innovation and Technology in Computer Science Education* (pp. 13-16).

Open Source Initiative. (2006). *Frequently asked questions.* Retrieved February 16, 2009, from http://www.opensource.org/docs/osd

Patterson, D. A. (2006). Computer science education in the 21th century. *Communications of the ACM, 29*(3), 27–30. doi:10.1145/1118178.1118212

Pfaffenberger, B. (2000). Linux in higher education: Open source, open minds, social justice. *Linux Journal.* Retrieved March 21, 2009, from http://linuxjournal.com/ article.php? sid=5071

SourceForge. (2009). *SourceForge.* Retrieved May 6, 2009, from http://sourceforge.net

Torvalds, L., & Diamond, D. (2001). *Just for fun: The story of an accidental revolution.* New York, NY: Harper Business.

Wheeler, D. (2007). *Why open source software / free software (OSS/FS, FLOSS, or FOSS)? Look at the numbers!* Retrieved May 5, 2009, from http://www.dwheeler.com/oss_fs_why.html

Wong, K., Wolf, T., & Gorinsky, S. (2007). Teaching experiences with a virtual network laboratory . In *Special Interest Group Computer Science Education, 2007* (pp. 481–485). Turner. J.

Wright, J., Carpin, S., Cerpa, A., & Gavilan, G. (2007). *An open source teaching and learning facility for computer science and engineering education* (pp. 368–373). In FECS.

Zebra. (2009). *GNU Zebra—routing software*. Retrieved June 7, 2009, from http://www.zebra.org

ADDITIONAL READING

Abler, R., Contis, D., Grizzard, J., & Owen, H. (2006). George tech "hands on" network security laboratory. In *IEEE Transactions on Education* (pp. 212-225).

Brustoloni, J. (2006). Laboratory experiments for network security instruction. [JERIC]. *Journal of Educational Resources in Computing*, *6*(4), 5–13. doi:10.1145/1248453.1248458

Chen, L., & Lin, C. (2007). Combining theory with practice in information security education. In *Proceedings of the 11th Colloquium for Information Systems Security Education* (pp. 167-171).

Comer, D. (2004). *Hands-on networking with internet technologies* (3rd ed.). Upper Saddle River, NJ: Prentice Hall.

Corbesero, S. (2006). Rapid and inexpensive lab deployment using open source software. *Journal of Computing Sciences in Colleges*, *22*(2), 228–234.

Kemp, J., Morrison, G., & Ross, S. (1996). *Designing effective instruction*. Upper Saddle River, NJ: Prentice Hall.

Kurose, J., Liebeherr, J., Ostermann, S., & Ott-Boisseau, T. (2002). Curriculum designs and education challenges. In *ACM SIGCOMM Workshop on Computer Networking* (pp. 5-13).

Lunt, B., & Ekstrom, J. (2008). *The IT model curriculum: A status update* (pp. 1–4). SIGITE.

Nakagawa, Y., Suda, H., Ukigai, M., & Miida, Y. (2003). An innovative hands-on laboratory for teaching a networking course. In *33rd ASEE/IEEE Frontier in Education Conference* (pp. 14-20).

Naps, T., RoBling, G., Almstrum, V., Dann, W., Fleischer, R., & Hundhausen, D. (2002). Exploring the role of visualization and engagement in computer science education. *ACM SIGCSE Bulletin, 35*(2), 131-152.

Richards, B. (2001). Rtp: A transport layer implementation project. In *Proceedings of the Sixth Annual CCSC Northeastern Conference on the Journal of Computing in Small Colleges* (pp. 134-141).

Rogers, M. (2000). Working Linux into the CS curriculum. In *Proceedings of the Seventh Annual CCSC Midwestern Conference* (pp. 85-91).

Sarkar, N. I., & Al-Qirim, N. A. Y. (2005). Teaching TCP/IP networking using hands-on laboratory experience. *The second International Conference on Innovations in Information Technology (IIT'05), 49*(2), 285-291.

Schmidt, C., Higgins, J., & Gliser, L. (2007). Implementation of an interactive web application using Foss. In *A participant-Oriented Evaluation Study. Consortium for Computing Sciences in Colleges* (pp. 22-28).

Seel, N. M., & Dijkstra, S. (2004). *Curriculum, plans, and processes in instructional design: International perspectives*. Mahwah, NJ: Erlbaum.

Wolf, M. J., Bowyer, K., Gotterbarn, D., & Miller, K. (2002). Open source software: Intellectual challenges to the status quo. In *ACM SIGCSE'02* (pp. 317-318).

KEY TERMS AND DEFINITIONS

ACM: Association for Computing Machinery (ACM) is the world's largest educational and scientific computing society, and it delivers resources that advance computing as a science and a profession.

Apache Web Server: Apache web server runs on the Linux operating system. It is developed and maintained as open source software. It is the first web server to have over 100 million web sites.

DHCP: Dynamic Host Control Protocol (DHCP) is a network application used by the clients to obtain IP addresses from the server dynamically.

DoS: Denial of Service (DoS) attack exploits TCP/IP three-way handshake. The attack uses spoofing IP addresses to send SYN flooding packets, and the server cannot respond to the requests from the users.

IDS: Intrusion Detection System (IDS) is a network device that passively detects network malicious activities.

IEEE: Institute of Electrical and Electronics Engineers (IEEE) is a non-profit organization. IEEE is the world's leading professional association for the advancement of technology.

IPS: Intrusion Prevention System (IPS) is a network security device that actively monitors network malicious and suspicious activities, and try to block these activities. When such activities are detected, packets will be dropped.

OSS: Open source software (OSS) is also known as free software, but "free" refer to "freedom and liberty" instead of "cost or price". Open source refers to the source code being openly available to the public.

TCP/IP: Transmission Control Protocol/Internet Protocol (TCP/IP) is a suite of communication protocols used for the Internet.

VoIP: Voice over Internet Protocol (VoIP) allows a voice-enabled router to carry voice traffic, such as telephone calls and faxes, over an Internet Protocol (IP) network.

Chapter 8
An Extendible Simulation Game to Promote Team Spirit on Mobile Computing Devices

Vincent Tam
The University of Hong Kong, Hong Kong

Zexian Liao
The University of Hong Kong, Hong Kong

C.H. Leung
The University of Hong Kong, Hong Kong

Lawrence Yeung
The University of Hong Kong, Hong Kong

A.C.M. Kwan
The University of Hong Kong, Hong Kong

ABSTRACT

Game based learning (GBL) has significantly reshaped the latest educational or training technologies through engaging players in many carefully planned learning activities in real-world applications including the training of tactic planning in a simulated combat environment, or resource management for any large organization. Successful examples range from the commercially available simulation games like SimBusiness, being adopted in some business schools in North America, to Quest Atlantis as a 3D multi-user learning game platform to children aged 9 – 15 in meaningful inquiry tasks. In a previous Teaching Development Project, the authors of this chapter developed an interesting simulation game platform for a virtual university campus containing game rooms with different missions for students to fulfill so as to enhance the learners' experience after classes on ubiquitous computing devices. All the missions are focused on engaging players to exercise their logical thinking or problem-solving skills in relevant areas of Information and Communication Technologies (ICT). To promote team spirit, each group of 3 to 4 students will work together to complete all the missions within the virtual campus. The authors used the open-source Nebula Version 2.0 toolkit to build a prototype of our simulation game containing various game rooms inside a virtual campus that can be accessed through any Windows based

DOI: 10.4018/978-1-60960-613-8.ch008

desktop or pocket PCs. To allow the real-time chatting facility among team members, the Yet Another Telephony Engine (YATE) server is employed in the interactive game platform. A preliminary evaluation was conducted to analyze the effectiveness of the simulation game on motivating and/or enhancing the learners' experience in relevant disciplines. All in all, there are many interesting directions for further investigations including the integration of our simulation game with existing learning management system or increasing players' involvement in proposing new missions.

1. INTRODUCTION

Game based learning (GBL) (Prensky, 2002) has significantly reshaped the latest educational or training technologies through engaging players in many carefully planned learning activities in real-world applications including the training of tactic planning in a simulated combat environment, or resource management for any large organization. In addition to such specific training, there are numerous successful examples of commercially developed simulation games including SimCity (SimCity, 2007) or Sims™ 2 Open for Business (SimBusiness, 2007), originally developed for fun and later seriously used by the business schools of various universities in North America to motivate or sustain the students' learning interests in their specialized fields. In many cases, it was shown that the appropriate use of simulation games not only avoids the indispensable costs of human lives or money lost in the hostile combat or investment field, but also effectively motivates and/or raises the learners' interests that may have positive impacts on their actual performance attained in handling the real-world situation, thus reaffirming the important value of simulation in training or education in general. As clearly described in Wikipedia (2007), "Simulation is used in many contexts, including the modeling of natural systems or human systems in order to gain insight into their functioning. Other contexts include simulation of technology for performance optimization, safety engineering, testing, training and education". In many such applications, simulation can be used to demonstrate the eventual effects of alternative conditions and course of actions.

In general, the use of interactive simulation tools can help students better understand the underlying working principles of many biological or physical systems, and therefore motivates the learners' interests for further investigation that may lead to enhanced performance in the concerned subject. For instance, a computer-aided simulation game named "The Incredible Machine" (Ward & Carroll, 1998) was found to enhance the mechanical reasoning capabilities of transition year students, including 17 females and 27 males with their age between 15 and 16 years, to a certain extent in an exploratory study conducted in a vocational school in Ireland. The study strictly followed the pre-test/post-test control group design, with the students' mechanical reasoning skills evaluated by the Differential Aptitude Test for Guidance. There were significant differences in favor of male students in the experimental group with training using "The Incredible Machine" simulation game as shown in the t-test analysis of their pre-test and post-test raw scores. Furthermore, there are various successful applications of simulation tools or games in other sectors of education (Kent NGfL, 2007; SCS, 2007; SuperKids, 2007).

In Hong Kong, where commercial and financial activities are overwhelmingly active in recent years with a lack of long-term development infrastructure for Engineering related disciplines, many tertiary educators in Engineering are faced with the core problems of motivating the students' learning interests and more importantly helping students realize the real values of the professional training in their own specific field. Therefore, in a Teaching Development Project, we proposed to develop an interesting simulation game inside a virtual university campus containing game rooms with different missions for students to fulfill on ubiquitous devices so as to enhance their learning experience after classes. All the missions are focused on engaging players to exercise their logical thinking or problem-solving skills relevant to specific disciplines of Engineering in a virtual environment without the limitation of many physical factors. For example, the Electronic and Communications Engineering students need not to worry about damaging several digital circuit boards before successfully building a working digital device such as a digital computer or network modem. To promote the spirit of teamwork (Ward & Carroll, 1998), each team of 3 to 4 students will work together to complete all the missions within the virtual campus. The team that obtains the highest score after completing all the missions is the winner. Since our simulation game can be accessed through mobile computing devices such as pocket PCs, the teams can continue their missions in the game anytime and anywhere. To demonstrate the feasibility of our proposal, we used the Nebula Version 2.0 toolkit to build a prototype of our simulation game containing various game rooms inside a virtual campus that can be accessed through Windows based pocket PCs. When the prototype is completed, a detailed evaluation will be conducted to analyze the effectiveness of our simulation game on motivating and/or enhancing the learners' experience in relevant disciplines of Engineering. After all, there are many interesting directions for further investigations. The integration of our simulation game with existing e-learning systems or powerful search engines is worth exploring. Besides, any feasible mechanism to increase players' involvement in proposing new missions or acting as mentors to guide other teams after the team has finished its own missions should be thoroughly studied.

The main objective of our study is to carefully investigate and review the possible uses of the game based learning (GBL) approach, particularly through simulation games, to promote learners' experience and also team spirit on ubiquitous computing devices. The remaining part of this chapter is organized as follows. Relevant studies on simulation games and open-source educational software will be reviewed with their evaluation results thoroughly considered in Section 2. Our proposed game design and system architecture of extendible and interactive simulation game platform developed for ubiquitous computing devices will be explained in detail in Section 3. The preliminary evaluation results of our proposal will be carefully analyzed and reviewed in Section 4. Lastly, Section 5 concludes our work with a detailed summary and many interesting directions for future investigation.

2. RELATED WORKS

In this section, we will consider some related works including the use of interactive computer simulation games for learning, the Nebula toolkit that we used to develop the extendible and interactive simulation game platform for our Teaching Development Project, the Session Initialization Protocol (SIP) server to manage user locations and provide real-time responses to handle exchanges of messages between the client machines and also the Yet Another Telephony Engine (YATE) to provide efficient and flexible routing for instant data and voice messaging (using the Voice over Internet Protocol (VoIP)) between the players.

2.1 Simulation Games for Learning

Among the theories of learning, constructivism (Moll, 1990; Piaget, 1963; Wadsworth, 1979) proposed that learners need to construct their own understanding of new ideas, possibly developed in the process of active interaction with the environment, by individuals who must assimilate and accommodate experiences into existing schemata. Piaget (Piaget, 1963; Wadsworth, 1979) accorded a special role to cognitive conflict in which an individual is confronted with different points of view, the individual must reflect on his/her own sets of beliefs, compare one's idea with those of others. As a consequence, the individual restructures and refines his/her own schemata. Vygotsky (Moll, 1990; Vygotsky, 1978) proposed that cognitive change involves internalisation and transformation of what was created. Essentially, Vygotsky focused on what an individual could achieve with help, through interaction with others. While working within this zone of proximal development, the learner can actively construct knowledge based on his/her personal experiences. After all, both theories highlighted the importance of "interaction" with the environment or other people, flexibly provided in many computer simulation games nowadays, for learners to assimilate experiences into existing schemata in order to construct new knowledge.

According to Wikipedia (2007), a simulation game is a game that contains a mixture of skill, chance, and strategy to simulate an aspect of reality, such as a stock exchange. In the industry of computer games, simulation games simply represent a wide super-genre covering many successful titles including the MS Flight Simulator (FSim, 2007), SimCity (SimCity, 2007), Civilization, and the Sims (Sims, 2007). Among these computer simulation games, there were many successful applications like the Sims or Sims™ 2 Open for Business (SimBusiness, 2007) that were originally developed for fun and later seriously adopted by different business schools or academic institutes for training students in their specialized fields. In many cases, it was shown that the appropriate use of simulation games not only avoids the indispensable costs of human lives or money lost in the real-world combat or investment field, but also effectively motivates and promotes the learners' interests, thus likely producing positive impacts on the actual performance attained when handling the real-world situation. Besides, there have been various institutions that tried to develop their own computer simulation games for training or research. For instance, an exploratory investigation (Ward & Carroll, 1998) about the effectiveness of a computer simulation game named "The Incredible Machine" for the mechanical reasoning skills was conducted in a vocational school in Ireland. In a control experiment with groups of 3 or 4 students working in a collaborative environment, it was revealed that significant differences biasing towards male participants were discovered in the t-test analysis of pre-test and post-test raw scores, demonstrating the effectiveness of computer simulation games on the male participants' mechanical reasoning abilities. Furthermore, the students' feedbacks collected in this study also revealed different patterns of mouse usage, reactions to the game, interaction styles, and strategies used by the groups.

2.2 The Nebula Toolkit

The Nebula Device 2, namely the Nebula2 toolkit (Nebula, 2007) is one of the leading open source 3D game and visualization engines used in many commercial games and professional visualization applications. It offers a complete layer of abstraction from the underlying host system for the needs of real-time 3D games, thus providing a more realistic environment for most computer simulation games. Essentially, Nebula2 can be considered as an operating system for games with the following key features:

- an object model which integrates scripting, persistency, and safe referencing through smart pointers;
- wrapper classes for file-system access;
- wrapper classes for multi-threading;
- wrapper classes for networking and inter-process communication across various host machines;
- an accurate time source (milliseconds or better);
- a collection of mathematical classes (vector, matrix, line, bounding box, etc.);

- a flexible scripting subsystem with support for several scripting languages (including Tcl, Lua and Python);
- a 3D graphics subsystem with support for the Microsoft Direct3D 9 High Level Shading Language (HLSL) shaders;
- a 3D audio subsystem on top of DirectSound™ with support for static and streamed sound;
- an input subsystem on top of DirectInput™;
- a graphical user interface (GUI) subsystem for creating 2D user interfaces;
- a complete resource (including textures, meshes, shaders, fonts, animation data, etc.) management subsystem, with support for loading resources in a background thread;
- virtual file-systems (file archives) for faster loading of configuration files or game scenes;
- a flexible scene graph subsystem;
- a statically linked and stripped down version of Tcl called MicroTcl that has 36 core Tcl commands and does not require any additional runtime files.

According to Wikipedia (2010), the High Level Shading Language (HLSL) is a proprietary shading language developed by the Microsoft™ for use with the Microsoft™ Direct3D API, as analogous to the OpenGL Shading Language (GLSL) that is based on the C programming language and used with the OpenGL standard (OpenGL, 2010).

Basically, HLSL programs come in three available forms, vertex shaders, geometry shaders, and pixel (or fragment) shaders. A vertex shader is executed for each vertex and is mainly used to transform the vertex from object space to view space, generating texture coordinates and calculating lighting coefficients such as the vertex's tangent and normal vectors. When a group of vertices (normally 3 to form a triangle) come through the vertex shader, their output position is then interpolated to form pixels within its area. This process is known as rasterisation. Each of these pixels comes through the pixel shader, whereby the resultant screen colour is calculated. Besides, an application using a Direct3D10 interface and Direct3D10 hardware may also specify a geometry shader. This shader takes as its input the three vertices of a triangle and uses this data to generate additional triangles, with each of them being sent to the rasteriser for calculating their resulting screen colors after considering the shading effects to make the underlying objects or scenes look more realistic.

2.3 The SIP and YATE Servers for Instant Messaging and Voice Communication

SIP servers are essential network elements that enable players in our interactive simulation game platform to exchange messages, register user location, and seamlessly move between networks. In general, the SIP standard defines three basic types of server functionalities, including the proxy, redirect and registrar servers. These standard functionalities can be used according to the needs of individual applications. In addition, since SIP defines various event types, including the Winfo and Register, it can also act as an event handling server to handle various SIP events. Nevertheless, the server logic of nowadays applications has become increasingly complex. SIP servers often need to deal with varying network topologies (such as the public Internet networks or broadband residential networks), complex routing policies, and security and SIP extensions. On top of it, SIP servers will need to handle high message/transaction rates and yield real-time performance and scalability, high throughput, and low delay for sophisticated network applications. In our interactive simulation game platform, we employed the Open SIP Server (OpenSIPS) as a mature and open-source implementation of the SIP proxy server providing many application-level functionalities such as the login and gaming sessions for each team to connect to the underlying Nebula2 engine. Together with the Yet Another Telephony Engine (YATE) application, the SIP server can help to build up clustered and IP-based instant messaging and voice communication systems, using the Voice over Internet Protocol (VoIP), among the various teams of registered players in our extendible and interactive simulation game platform.

Besides the SIP server, the open-source YATE (Yate, 2010) application is employed as a powerful telephony engine to provide the online chatting facility in our interactive simulation game. Being focused on the VoIP and PSTN, the main strength of the YATE server is its ease of extendibility. Voice, video, data and instant messaging can all be unified under the flexible routing engine of YATE with the communication efficiency maximized while the infrastructure costs for organizations minimized. This open-source software is mainly written in C++ and also with various supports in commonly used scripting languages such as PHP, Python and Perl. For instance, both PHP and Perl libraries were developed and made available to facilitate the development of external functionalities for integration into the original YATE application.

3. THE DETAILED SYSTEM ARCHITECTURE OF OUR SIMULATION GAME

Based on the Nebula2 engine, the Open SIP server and the Yet Another Telephony Engine (YATE), and also their publicly available open-source libraries for the supported features as clearly described in Section 2, we carefully designed the system architecture of our extendible and interactive simulation game to enhance the learners' experience on mobile computing devices.

Figure 1 depicts the overall system architecture of our extendible and interactive simulation game platform. Through wired or wireless connections to the designated WiFi router, pre-registered players on various client machines can initiate their SIP requests to log into our interactive simulation game platform. Their incoming SIP requests will be processed and then forwarded by the front-end Open SIP server (OpenSIPS) as our SIP proxy server. In case the incoming requests are related to login, logout or various gaming actions, they will be automatically forwarded to another game server installed with the Nebula2 engine for user authentication or performing the requested

Figure 1. The overall system architecture of our interactive simulation game

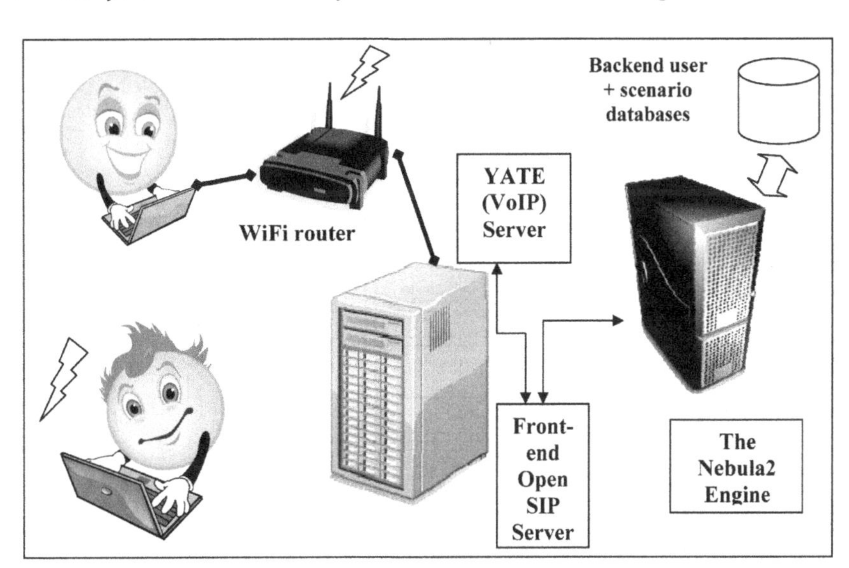

operations to the relevant game scenario after checking the back-end user database or updating the involved game scenario databases. After a successful login for any registered member of a team, a new VoIP session will also be created under the YATE system also installed on the front-end server PC for future voice communication between the team members. Accordingly, if the future incoming request concerns instant messaging or voice communication, the SIP server will then forward the request to the YATE system for further processing. Subsequently, when any member of the same team successfully logs into our simulation game platform, the member will be automatically added into the existing Nebula game session to synchronize their actions or decisions inside the stored game rooms for fulfilling their common mission inside the virtual campus, and also appropriately added into the same VoIP session to enable online chatting among their team inside the simulation game.

It should be noted that after careful and thorough consideration, it is our intent to separate both the SIP server and YATE system installed on the front-end server from the Nebula2 engine installed on another game server machine. Intrinsically, the front-end server machine can act as a data communication server while the backend machine can focus on working solely as a game server to implement possibly complicated game logics and also handle a potentially large number of asynchronous requests for game-related actions from various team players. Obviously, this clear separation of concerns between the communication protocols and game logics will lead to better load balancing, thus more responsive and reliable performance. More importantly, this careful design option may open up the possibility of temporarily buffering (or queuing) the incoming requests in the front-end SIP server for a relatively short period of time while the backend Nebula2 game engine is being rebooted for adding new game rooms into our extendible simulation game platform, or some unpredictable network disruption problems. Overall speaking, this will make our simulation game system more competent in case of any system or network failure.

The software architecture of our refined Nebula2 engine is depicted in Figure 2 below. Essentially, the whole software system of our

Figure 2. The software architecture of the Nebula2 engine used in our simulation game

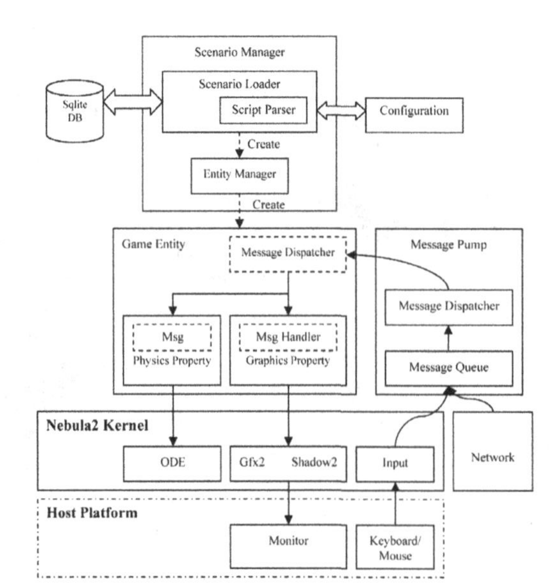

interactive simulation game is composed of two major parts: the game engine which deals with high-level game logic, and the Nebula2 kernel which handles low-level functions such as graphics, keyboard/mouse inputs, etc. The game engine has three components: game entity, scenario loader and message pump. Each game entity represents an object in the game. By attaching different properties to a game entity, the game entity can behave differently. For example, with a physics property, the game entity can collide with other entities whereas without it, the game entity could simply go through. When events arrive, a message dispatcher of a game entity is responsible for passing them to an appropriate property handler. Game entities in a particular scenario are managed by an entity manager created by the scenario loader. When switching the scenario, all the entities in this scenario will be read from the SQLite database by the scenario loader, created by the entity manager, and ultimately stored back to the SQLite database by the entity manager for future retrievals. The game engine is event driven. Events can come from the Nebula2 kernel, such as keyboard/mouse input, or from the network, such as actions of other players. The message pump subsystem dispatches all these messages to the appropriate entity handlers in order to keep the game running. In addition to such high-level game logic as handled by the game engine, all the low-level input functions are implemented by the Nebula2 kernel. For specific details about the Nebula2 kernel, refer to Nebula (2007).

The Nebula2 engine used in our interactive simulation game platform consists of several game rooms that emphasize relevant subjects in Information and Communication Technology (ICT) that should be of interests to Engineering students in general. The relevant subjects in ICT we chose for the game rooms include Computer Architecture, Network Programming and User Interface Design. Clearly, given the flexibility of our system architecture, other relevant subjects can easily be integrated into our simulation game platform by defining new game rooms and missions. Inside the simulation game, each group of 4 students will enter into a different game room, one after another, to perform some designated tasks and then earn a score after accomplishing the tasks. For example, in the game room for Computer Architecture, students will be asked to pick up the correct components out of a diverse range of computer devices lying on the floor, and then configure such components to build a working desktop computer. During the game, the nominated leader of each team will act as a coordinator as well as the ultimate decision-maker to perform the actual actions such as the steps to compose a desktop computer in the simulation game room for Computer Architecture. Each member of the team, possibly located in different places inside or outside our physical campus and being connected by the WiFi networks, can discuss the next action or mission to be performed via the online chat-room facility provided on the mobile devices. After all, the scores obtained by going through various game rooms or missions will be accumulated for each team, and can be checked via the web interface. And the team with the highest score will be selected as the winner of the simulation game.

4. THE PROTOTYPE IMPLEMENTATION & EVALUATION

We used the Nebula2 toolkit (Nebula, 2007) and its supporting technologies such as the SQLite, MicroTcl scripts, and several C++ programs to build a prototype of our interactive game containing several mini-game rooms inside a virtual campus that can be accessed through Windows based desktop or pocket PCs. The use of mini-games like robot-shooting or duct/path connection are aimed to introduce some fun and also the flexibility in the design of each game room that can be an individual game component by itself, or combining with other mini-game rooms in an adventure to focus on some specific subject/topic.

In our prototype implementation, we built an adventure of 3 mini-game rooms inside the virtual campus, that try to emphasize relevant subjects in Information and Communication Technology (ICT) for Engineering students in general. The relevant subjects in ICT we chose for the mini-game rooms include Basic Organization of Computers, the (more advanced) Structure of Microcomputers, and lastly Application Programming on Computers. Clearly, given the flexibility of our system architecture, relevant subjects in ICT such as Network Programming and User Interface Design, or other different topics can easily be integrated into the virtual campus of our simulation game. It is worth noting that the Nebula2 toolkit is open-source software implemented in C++, which is available for free downloading. In fact, the latest version of Nebula3 is also available under the Google Code for free downloading as well. On top of it, the Fedora system we used for prototype implementation is also a free and open-source Linux based operating system. All in all, all the software tools used for our prototype implementation are free and open-source software without any need for proprietary software licenses.

To play the simulation game, each team of 2 to 3 students should "individually" log in the game server via a networked PC and then select to enter into a game room of a virtual campus to perform some designated task. Among the team members, the players should nominate one student as the team leader to submit the final decision to each question/task. The team members are allowed to use the text-based voting mechanism or voice using the peer-to-peer Session Initiation Protocol (SIP) to communicate with each other before making any decision so as to encourage teamwork. After completing a set of questions/tasks, each team will be given a score, based on their time and accuracy, which will be added to their overall score. The team with the highest overall score is the winner. Figure 3 shows the user interface of the "virtual campus" for players to navigate among the game rooms available on our simulation game platform.

As shown in Figure 3, the overall scores of all the participating teams can be checked by clicking on the "*Marks*" button inside the virtual campus from time to time to encourage construc-

Figure 3. The user interface of the virtual campus displayed in our simulation game

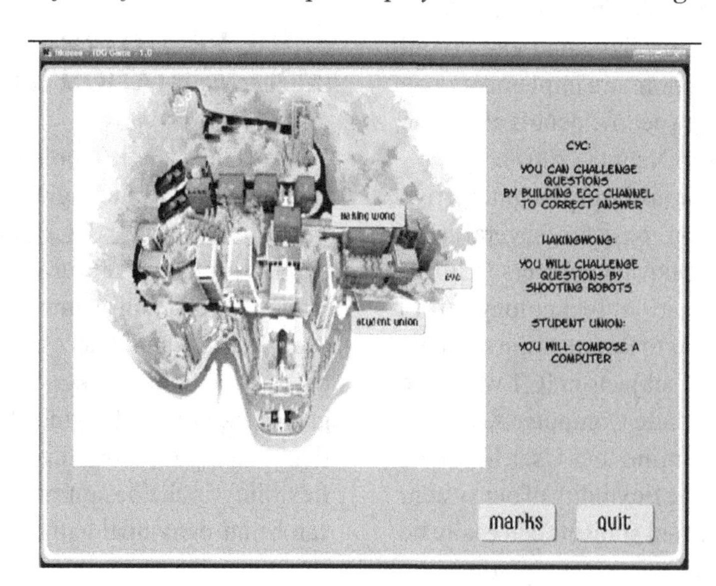

Figure 4. The user interface of the mini-game on the basic organization of computers displayed in our simulation game

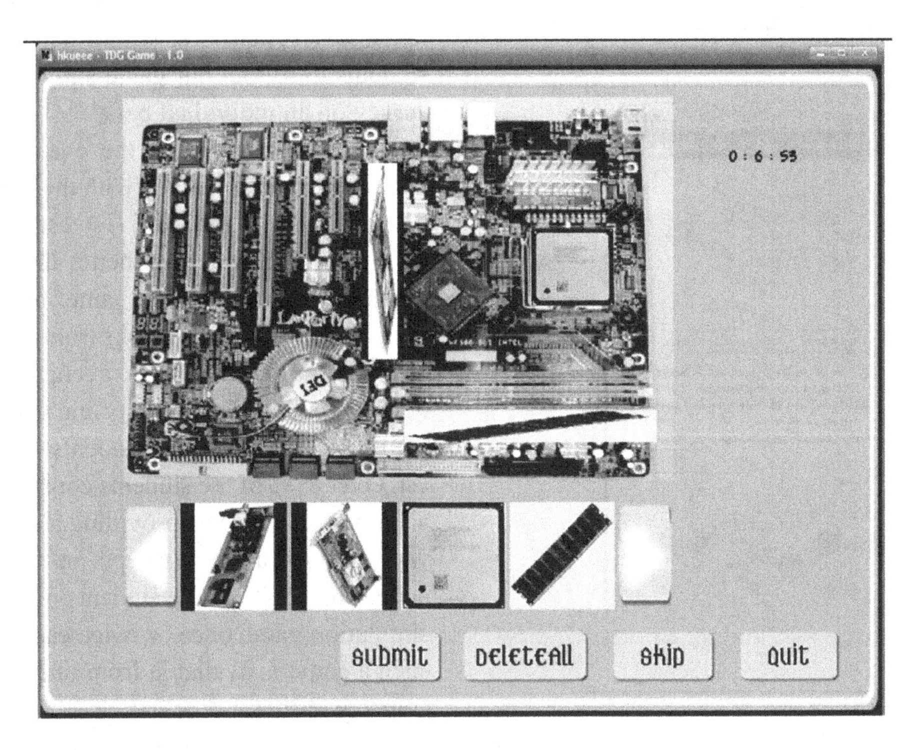

tive competition among the teams. For example, in the mini-game room for Basic Organization of Computers as shown in Figure 4, each team will be asked to select the right components out of a wide range of computer devices including the processor, memory or network card, and also put such components into the correct slots to build a working computer. Each team leader can act as a coordinator to collect opinions from his/her team members via voice communication over the SIP server, and then make necessary decisions to obtain a computer configuration for online submission. Our Nebula2 game engine will proceed to check if it contains all the essential components such as the processor and RAM for a working computer. Accordingly, it will send a message to congratulate the team for their good work. Otherwise, an error message will be displayed to allow the whole team to revise their configuration for later submissions.

Besides, to promote our simulation game through "trials" to be played on ubiquitious devices anytime and anywhere, we ported our TDG game using embedded Visual Basic to run on (standalone) pocket PCs. This will allow us to study the performance attained by different teams with and without the "trial practice" of our TDG game, and more importantly the possible impact(s) of revisions on mobile devices over the retention period of the knowledge/skills covered in our TDG game. Figure 5 shows the similar interface of our TDG game ported onto the Dell™ Axim X41v pocket PC installed with an Intel(R) processor of 624 Mhz, and the Windows Mobile 5.0 operating system.

After the above prototypes were completed with extensive testing, a thorough evaluation was conducted to analyze the effectiveness of our simulation game on motivating and/or enhancing the learners' experience in relevant Engineering disciplines. In particular, we have carefully pre-

Figure 5. The user interface of the pocket PC version of our simulation game

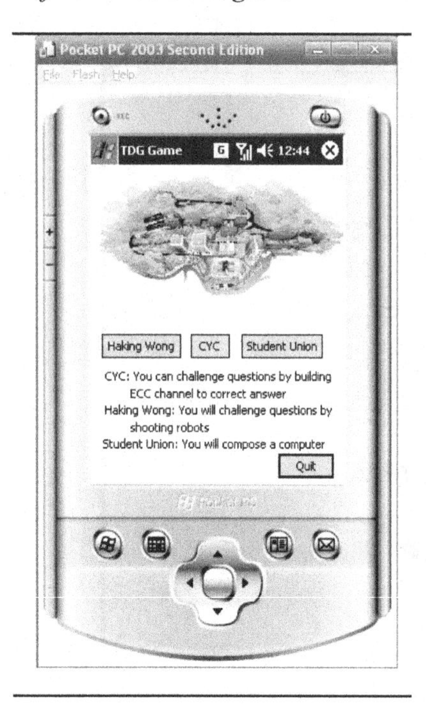

pared a questionnaire and invited individual students from different classes/programmes in the Department of Electrical and Electronic Engineering (EEE) to try out our pocket PC version, which is mobile and can be tested anywhere, and then complete the questionnaire for our detailed analysis.

Essentially, after the (standalone) pocket PC version of our simulation game was built in early January 2007, due to the limited time and resources we had for the project, the pocket PC version was distributed to various (around 8) EEE students, as randomly selected from different classes/programmes, to try out the game with their feedbacks collected through the above-mentioned questionnaire to evaluate the potential benefits of using the simulation game to stimulate and also engage the students' learning interests in different EEE or possibly other courses. It should be noted that due to the "limited number of mobile devices" (only 1 pocket PC) available for this teaching development project, it was impossible to perform the system evaluation on a class or larger scale. Therefore, we had to randomly select students to try out our game and then conduct a survey on an individual basis.

Table 1 summarizes the evaluation results obtained from our survey with the average score displayed out of a scale of 1 (lowest) ~ 5 (highest), the higher the score, the better the impression/impact of our simulation game, for each question. Basically, the feedbacks from these students are fairly positive, with the average score as 3.5 out of a scale of 1 ~ 5 supporting that simulation games can help to initiate students' learning interest. Over 87% of the students consider the use of a simulation game as providing a more relaxing way to learn about concepts in specific areas, and around 50% of the students plan to play the simulation game once or twice each week when such a game is available from our EEE or other course websites.

Besides the formal survey, our team also communicated informally with other EEE students after classes to collect their views and feedbacks on the possible use of simulation games in promoting their learning interests. Most students expressed interests in using our simulation game for learning/revision in the future.

5. CONCLUSION

Game based learning (GBL) has significantly reshaped the latest educational or training technologies through engaging players in many carefully planned learning activities in real-world applications including the training of tactic planning in a simulated combat environment, or resource management for any large organization. With many successful applications developed in the past and to be continuously developed in the future, interactive simulation games will surely form an important part of the GBL approach that aims to enhance or reshape the learners' expe-

Table 1. The evaluation results of our survey about the effectiveness of our simulation game to motivate and/or enhance the learners' interests

Questions	Avg. Score
A. Do you think the simulation mini-game on pocket PC is easy-to-play ? **1.** Strongly disagree **2.** Disagree **3.** Neural **4.** Agree **5.** Strongly agree	**3.9**
B. Do you think the use of simulation game in our EEE or other courses can help to initiate the students' interest in learning? **1.** Strongly disagree **2.** Disagree **3.** Neural **4.** Agree **5.** Strongly agree	**3.5**
C. As compared to the conventional revision exercise, can the use of simulation game in our EEE or other courses simulate and also engage more students' interest in learning? **1.** Strongly disagree **2.** Disagree **3.** Neural **4.** Agree **5.** Strongly agree	**3.4**
D. Do you consider the use of simulation game in our EEE or other courses can provide a more funny and relaxing way to learn basic concepts in specific areas? **1.** Strongly disagree **2.** Disagree **3.** Neural **4.** Agree **5.** Strongly agree	**4.0**
E. Do you consider the use of simulation game in our EEE or other courses can increase the effectiveness of learning from student's perspective? **1.** Strongly disagree **2.** Disagree **3.** Neural **4.** Agree **5.** Strongly agree	**3.4**
F. Do you consider the use of simulation game in our EEE or other courses can provide a more relaxing way to revise about basic concepts after classes? **1.** Strongly disagree **2.** Disagree **3.** Neural **4.** Agree **5.** Strongly agree	**3.8**
G. If such a simulation game is available freely in our EEE or other course website, how likely will you use/play such a game for learning/revision? **1.** Definitely will not use **2.** Will not use **3.** Difficult to decide **4.** Will use **5.** Definitely use	**2.9**
H. If such a simulation game is available freely in our EEE or other course website, how often will you use/play such a game for revision/learning? **1.** Difficult to decide **2.** 0 **3.** once a week **4.** twice a week **5.** 3 times or more a week	**2.0**

riences. In a previous Teaching Development project, we proposed to develop an extendible and interactive simulation game inside a virtual university campus containing game rooms with various missions for students to fulfill on desktop PCs or ubiquitous devices so as to enhance the learners' experience after classes. All the missions are focused on engaging players to exercise their logical thinking or problem-solving skills relevant to specific engineering disciplines. Through wireless mobile devices such as pocket PCs, the teams can continue their missions in the game anytime and anywhere. To demonstrate the feasibility of our proposal, we used the open-source Nebula Version 2.0 toolkit, the Open SIP server, the Yet Another Telephony Engine (YATE) application and their supporting libraries to build a working prototype of our simulation game platform containing various game rooms inside a virtual campus that can be accessed through both Windows based desktop and pocket PCs. When the prototype was completed, a detailed evaluation was conducted through a carefully designed survey to analyze the effectiveness of our simulation game on motivating and/or enhancing the learners' experience in relevant engineering disciplines. From the survey results, some initial and encouraging feedbacks are collected.

There are several interesting directions for future exploration. First, the integration of our simulation game with existing e-learning systems or powerful search engines is worth investigating. Moreover, any feasible mechanism to increase the players' involvement in proposing new missions or acting as mentors to guide other teams after the team has finished its own missions should be thoroughly studied.

REFERENCES

FSim. (2007). *The MS Flight Simulator website*. Retrieved June 5, 2007, from http://www.fsinsider.com/Pages/default.aspx

KentNGfL. (2007). *Simulations – Google Search Kent NGfL website*. Retrieved June 15, 2007, from http://www.kented.org.uk/ngfl/software/simulations/index.htm

Moll, L. C. (1990). *Vygotsky and education: Instructional implications and applications of sociohistorical psychology*. New York, NY: Cambridge University Press.

Nebula. (2007). *The Nebula device*. Retrieved June 13, 2007, from http://nebuladevice.cubik.org

Open, G. L. (2010). *The OpenGL – the industry standard for high performance graphics*. Retrieved March 13, 2010, from http://www.opengl.org

Piaget, J. (1963). *The psychology of intelligence*. Paterson, NJ: Littlefield, Adams.

Prensky, M. (2002). *Digital game-based learning*. New York, NY: The twitchspeed.com. Retrieved June 13, 2007, from http://www.twitchspeed.com/site/news.html

SCS. (2007). *Simulation: Transactions of the Society for Modeling and Simulation International website*. Retrieved June 3, 2007, from http://www.scs.org/pubs/simulation/simulation.html

SimBusiness. (2007). *Aspyr - The Sims™ 2 Open for Business website*. Retrieved June 15, 2007, from http://www.aspyr.com/product/info/3

SimCity. (2007). *SimCity societies*. Retrieved June 15, 2007, from http://simcity.ea.com

Sims. (2007). *The Sims official site*. Retrieved: June 15, 2007, from http://thesims.ea.com

SuperKids. (2007). *SuperKids software review of RockSim – model rocket design and simulation software*. Retrieved June 5, 2007, from http://www.superkids.com/aweb/pages/reviews/science/06/rocksim/merge.shtml

Vygotsky, L. S. (1978). *Mind in society*. Cambridge, MA: Harvard University Press.

Wadsworth, B. J. (1979). *Piaget's theory of cognitive development*. New York, NY: Longman.

Ward, J., & Carroll, P. (1998). *Can the use of a computer simulation game enhance mechanical reasoning ability: An exploratory study.* (Working Paper Series MCE-0998), (pp. 1-20). Retrieved June 10, 2007, from http://citeseer.ist.psu.edu/431172.html

Wikipedia. (2007). *Simulation*. Retrieved June 1, 2007, from http://en.wikipedia.org/wiki/Simulation

Wikipedia. (2010). *High Level Shader Language*. Retrieved March 1, 2010, from http://en.wikipedia.org/wiki/High_Level_Shader_Language

Yate. (2010). *Yate – the next-generation telephony engine home page*. Retrieved June 1, 2010, from http://yate.null.ro

KEY TERMS AND DEFINITIONS

Interactive: Interacting with a human user, often in a conversational way, to obtain data or commands and to give immediate results or updated information.

Learning: The act or process of acquiring knowledge or skills.

Simulation: The representation of the behavior or characteristics of one system through the use of another system, especially a computer program designed for the purpose.

Chapter 9
An Augmented Reality Library for Mobile Phones and its Application for Recycling

M. Carmen Juan
Universitat Politècnica de València, Spain

David Furió
Universitat Politècnica de València, Spain

Leila Alem
CSIRO ICT Centre, Australia

Peta Ashworth
CSIRO, Australia

Miguelon Giménez
Universitat Politècnica de València, Spain

ABSTRACT

This chapter presents the Augmented Reality (AR) library developed for mobile phones. The authors explain the used tools, how they have ported ARToolKit for running on the mobile phone Nokia N95 with 8 GB, and the functionalities they have added. The authors present the mobile phone game they have developed for learning how to recycle using this AR library, ARGreenet. Forty-five children with a mean(SD) age of 11.02(2.33) years played the ARGreenet as well as a 'Team' version of the ARGreenet. The children answered questionnaires both before and after using each game. Several aspects were examined including the level of engagement, fun and ease of use. Analysis of the results showed significant differences between the ARGreenet and the Team ARGreenet with higher means for the Team ARGreenet. A majority of children (59.1%) expressed a preference for the Team ARGreenet.

DOI: 10.4018/978-1-60960-613-8.ch009

INTRODUCTION

In this chapter, we present the Augmented Reality (AR) library we have developed for mobile phones. We explain its hardware and software features. We pay special attention to the portability of ARToolKit for running on the mobile phone Nokia N95 with 8 GB, and the added functionalities. We have developed a game for learning how to recycle using this library. The goal in this game is that people learn about recycling. To achieve this, players have to place as many rubbish items in the right recycling bin as possible and to answer correctly as many questions about recycling as possible. The target audience is children. No special restrictions are considered. The game is played inside, in a room with no special furniture. In fact, in our experiments, the game was played in class rooms. In this chapter, we present in detail this first AR game that is for one player only, the ARGreenet. We also present a study comparing the ARGreenet with a team version of the ARGreenet, Team ARGreenet. In the Team ARGreenet, three players played the ARGreenet. As there are three different levels in the ARGreenet, each player played in one level and the other two players helped him or her during the game, advising where to place correctly the rubbish items and also helped to answer the questions correctly.

The term AR is used to describe systems that blend computer generated virtual objects with real environments. There is a commonly accepted definition for AR. It was given by Ronald Azuma (Azuma, 1997). Azuma's definition states that AR combines real and virtual images, has real time interaction, and it is registered in 3D. The AR ultimate goal is to create a system in which the real world and the virtual objects included in it cannot be distinguished, achieving a total fusion of both environments. There are several types of libraries that can be used for the development of AR systems. One of these types is based on markers. Markers are white squares with a black border inside of which are symbols or letter/s. Normally, these libraries require a camera to capture the real world (USB or FireWire) and identify the marker. The ARGreenet uses markers that are detected by the mobile phone and over them are shown rubbish items and recycling bins.

BACKGROUND

In this section, we focus on available tools for developing applications for PCs and also for mobile devices. There are several tools for the development of AR for PCs, but there are not many for mobile devices. Related to PCs, the most well-known library is ARToolKit. ARToolKit is a vision tracking library that enables the easy development of a wide range of AR applications. It was developed at the University of Washington by Kato & Billinghurst (1999). ARToolKit is made available freely for non-commercial use under the GNU General Public License. Commercial licenses for professional implementation of ARToolKit are also available. Commercial licenses are administered by ARToolworks, Inc., Seattle, WA, USA. The required elements for an AR application developed using ARToolKit are: a USB or FireWire camera, and a marker. ARToolKit uses computer vision techniques to obtain the position and orientation of the camera with respect to a marker. Virtual elements are drawn on these markers. ARToolKit includes code for image acquisition, the detection of markers in the video stream, the calculation of the three-dimensional position and orientation of the markers, the identification of markers, the computation of the position and orientation of the virtual objects relative to the real world, and the rendering of these virtual objects in the corresponding video frame. Other tools have incorporated ARToolKit as plugin, for example:

- DART (Designer's Augmented Reality ToolKit) (MacIntyre et al., 2003). It is based on Macromedia Director.

- AMIRE (Authoring Mixed Reality) (Abawi et al., 2004). It includes different localization systems. ARToolKit was included in its first prototype. It includes 3D graphic libraries such as OpenSceneGraph.
- OsgART (Looser, 2007). It also includes OpenSceneGraph.

Other PC tools are the following:

- Dwarf (Distributed Wearable Augmented Reality Framework) (Bauer et al., 2001). It allows the development of distributed AR applications.
- ImageTclAR (Owen et al., 2003). It is based on ImageTCL, a library for the process of images for Tcl/Tk.
- ARTag (Cawood & Fiala, 2007). It has a library of 2002 markers.
- BazAR (Pilet, 2008). This library is based on feature point detection and matching. It does not use the typical markers.

Related to mobile devices, simple tracking techniques have been studied for offering AR for mobile devices. For example, Möhring et al. (2004) developed a tracking technique based on color-coded 3D markers. Rohs & Gfeller (2004) presented the Visual Code system for smartphones. Another technique that is used is the tracking of natural features. For example, Paelke et al. (2004) presented AR-Soccer in which the player kicks a virtual ball with his/her real foot into the virtual goal. The game tracks the foot. Before our work, ARToolKit was already ported to mobile devices. In 2003, Wagner & Schmalstieg ported ARToolKit to Windows CE. In 2005, Henrysson et al. ported ARToolKit to the Symbian platform. Later, in 2007, Wagner & Schmalstieg presented the ARToolKitPlus library for use on mobile devices (e.g. PDA's). ARToolKitPlus is freely available under the GPL open source license. Studierstube Tracker (http://studierstube.icg.tu-graz.ac.at/handheld_ar/stbtracker.php) is a complete redevelopment that is not related to ARToolKit on a source code basis. Studierstube Tracker runs on: Windows XP; Windows CE & Windows Mobile; Symbian, Linux; MacOS; iPhone. To our knowledge, the latest framework presented by Wagner and colleagues is Studierstube ES (Wagner & Schmalstieg, 2009) which was written from scratch. Studierstube ES does not use code written for other platforms. Studierstube ES operates cross-platform (Windows, Windows CE and Symbian). ARToolKitPlus, Studierstube Tracker and Studierstube ES are all being used for the development of AR applications for mobile devices. For example, ARToolKitPlus was used for developing the Invisible Train (Wagner et al. 2005). The Invisible Train is a multi-player game, in which players steer virtual trains on a real wooden miniature railroad track. One example application developed using Studierstube ES is Cows vs. Aliens (Mulloni, 2007). Cows vs. Aliens is a competitive multi-player game for two groups. The markers are placed on the walls. Each real marker represents a virtual location for the game. The cows can graze in a group of several virtual locations that are connected. There is also a stable for each team (another virtual location). During the game, an alien invasion takes place. UFOs are present on some locations. They are ready to shoot any cows that run in those locations. The goal of the game is to save as many cows as possible; cows are safe when they are at the stable. The players' goal is to collect as many cows as possible and take them safely to their team stable. The team that saves the most cows wins the game. The game uses the game console, Gizmondo.

As mentioned, there are several available libraries for PCs, but there are not many for mobile devices and moreover, to our knowledge, none with the features of our AR library for mobile phones. For developing our AR library, we studied two possibilities. The first one was to port the well-known ARToolKit to mobile phones and later to incorporate additional functionalities. This portability has already been achieved successfully

(Henrysson et al. 2005; Wagner & Schmalstieg, 2007). Therefore, we were sure it was possible. The second one was to use ARToolKitPlus and incorporate to it additional functionalities. The result of both developments should be similar. We decided to choose the first possibility because of our earlier experiences in modifying ARToolKit.

AUGMENTED REALITY LIBRARY FOR MOBILE PHONES

For developing the AR library for mobile phones, we have to consider the hardware and the software features. With regard to the hardware, the only required device is a mobile phone. This research used the Nokia N95 with 8GB. The most outstanding features of this mobile phone for AR are: Large 2.8" QVGA (240 x 320 pixels); TFT display with ambient light detector and up to 16 million colours; Carl Zeiss Optics camera with 5 Megapixels; shoot in DVD-like quality video with up to 30 frames per second; CPU Clock Rate of 332 MHz with vector floating point; memory of 128 MB + 8 GB; video resolution of 640x480. Finally, we would like to highlight the incorporation of 3D graphic HW accelerator that makes possible that AR applications run at an adequate frame rate. The case is different for the Nokia N96 that does not incorporate the 3D graphic accelerator and running AR applications is nearly impossible.

With regard to the software, we have to distinguish firstly between the required development environment and the additional software for programming for the selected mobile phone in C++, secondly the library for the AR facilities, and thirdly the added functionalities.

First, related to the required development environment and the additional software, we have used Microsoft Visual Studio 2005 as the development environment. Developing and running PC applications is quite easy using this development environment. But for running an application simulating its running in the selected mobile phone, an emulator of the mobile phone is required. For including this type of tool, the S60 Platform SDK (3rd Edition) was used. For programming in C++ for Symbian OS, it also requires the installation of Carbide.vs 3.0.1. A more detailed explanation of these tools is outlined below:

- **Microsoft Visual Studio**: Microsoft Visual Studio (MSV) is an IDE developed by Microsoft for Windows computers. The version we have used is MSV 2005, because it supports Carbide.vs. Another version that supports Carbide.vs is MSV 2003. "Service Pack 1" must be installed in order to run Carbide.vs.
- **Symbian OS**: Symbian OS is an OS designed for mobile devices and smartphones, with associated libraries, user interfaces, frameworks and reference implementations of common tools, developed by Symbian Ltd.
- **Carbide.vs**: Developed by Nokia, Carbide.vs is a tool set used for developing applications in Symbian C++ using the MSV 2003 or 2005 IDEs and the corresponding SDKs. Carbide.vs is intended for developers with skills in MSV who want to build C++ applications for Symbian platforms, like 60 or 80 Nokia series, as well as User Interface Quartz. Carbide.vs provides an easy adaptation to Symbian C++ with tutorials, help and automated functions that integrate with MSV. It also contains functionalities for automating a lot of Symbian OS specific development tasks.
- **S60 3rd Edition Feature Pack 1 and S60 3rd Edition Feature Pack 2**: Developed by Nokia, these software development kits (SDKs) allow application development for S60 platform devices using C++. The SDKs are based on the third edition, supporting Feature Pack 1 and Symbian OS 9.2 (Nokia N95 8GB), and Feature Pack 2

and Symbian OS 9.3 (Nokia N96), respectively. The changes between the first SDK and the second are minimal. Therefore, we can use the same classes and functions when implementing the applications for the two mobile phones. It contains all the main functionalities (documentation, application programming interfaces (APIs), emulator) needed for developing applications and has the following features:

- An emulator to test and debug the programs.
- Tools and environments for installing the application on the emulator or the device.
- Documentation about Symbian and S60 platform, and example applications.
- Registration of the product if wanting to use it for more than fourteen days. It is free.

The emulator is one of the most important tools because it allows developers to test and debug applications on the computer before installing them on the phone. It provides a phone graphic interface with its basic functionalities. The emulator simulates accurately the program operations like it was on a real device. Despite this, it is recommended to have a device to test the application on, because sometimes the emulator and the mobile phone may differ in something like memory, execution speed, or others. It is important to observe whether it is possible to debug applications on the phone.

Second, for AR facilities, ARToolKit 2.65 was ported onto a mobile phone running on Symbian OS and Series60. Related to the adaptation of ARToolKit to Nokia N95 with 8 GB, we have taken the minimum necessary functions from the original ARToolKit in order to detect markers and created a class capable of setting up a tracker and tracking markers. We have also modified the configuration file to make it compatible with the Nokia N95 with 8 GB processor and graphic driver. The steps to set up the tracker are:

- Load the camera parameters. We used the file provided with the default settings.
- Update the tracker parameters regarding the image size to analyze.
- Initialize camera parameters.
- Get the projection matrix to be used with OpenGL-ES in order to draw objects.
- Load the patterns to look for.

Our modification used 8-bit ARGB images; e.g the same format obtained from the camera, therefore, no transformation was required.

A very important aspect in a marker-based AR application is the access to the camera. One of the technical challenges in this project was to access the camera of the phone. First, we configured the camera in accordance with our needs (image resolution and format, turning on and off the camera, zoom). Second, it is important to emphasize that the camera accesses restricted resources. The access to these resources is done by activating the 'user environment' capability. If it is not activated, there will be no way to access the camera.

The camera captures data from the environment and returns it in a preselected format. In our case, we choose an 8-bit channel RGB image. The SDKs provide a CFbsBitmap class to store the obtained data. CFbsBitmap has a set of functions for working with the image data and a pointer to the first position of it; therefore, the data can be stored on an array. The data was saved as follows: first byte R, first byte G, first byte B, second byte R, second byte G, second byte G, and so on.

Related to the drawing of the camera images, we used the OpenGL-ES API to draw the image stored in the CFbsBitmap class on the screen, this is provided by the SDKs. OpenGL-ES is an OpenGL subset for embedded systems like phones. The first step is to create a graphic context by defining features such as display attributes (e.g. color depth or z-buffer). The second step is to test if the mobile phone supports the graphic context; if it does not, changes to the context have to be made. Once this step is completed, every object drawn with OpenGL-ES can be displayed on the screen. If we do not want to play the game in fits and starts, the drawing process loop must be repeated at least 24 times per second. The drawing process is as follows.

- Enable OpenGL-ES capabilities (texture, depth, culling, and so on).
- Load projection and model view matrices.
- Save image data on a class or structure.
- Create a texture class. Here we generate the texture and specify its parameters (texture coordinates, filters, and so on).
- Establish the coordinates for drawing the image in the desired position.
- Activate the texture.
- Draw the image.
- Disable OpenGL-ES capabilities.

Third, for additional functionalities, we have added the management of XML files, 3D objects, fonts, and PNG images. XML files allow the inclusion/removal of information without changing the game. Our XML files follow the rules defined in "XML 1.0 Specification" produced by W3C. The two SDKs employed provide a set of classes for working with XML files. In our case, the XML files store the settings of the game like the questions and recycling object configurations, the images used for the objects, the user records, and general settings of the game. Related to 3D objects, we used the library named Model_3DS for loading the 3D Studio Max models. Model_3DS was created by Fairfax and ported to Symbian by Alie Tan. The load and draw on the device screen is possible using a set of library functions. The 3D models with textures can be created with 3D Studio Max or other applications. Related to fonts, we used GLFont that was developed by Fish. The following files compose GLFont:

- **GLfont.exe:** Windows application that creates a texture containing characters of a specific font. It also generates the coordinates automatically for using with OpenGL/OpenGL-ES.
- **GLfont2.h:** Header of the GLFont API version 2.0. It has a set of functions to read textures generated by the executable file and show them on the screen.
- **GLfont2.cpp:** Implementation of the GLFont API version 2.0.
- **GLfont.html:** Documentation, license, and terms of usage.

Related to the management of PNG images, the PNG format allows the use of transparency on OpenGL-ES textures. PNG images in Symbian C++ are loaded through an asynchronous class or active object (similar to C threads but not equal). In our case, the rubbish items and the recycling bins are 2D images stored in the PNG format.

ARGREENET

In this section, we present the ARGreenet. In the ARGreenet, the player has to pick up rubbish items that appear over the objects' marker and place them in the correct recycling bin. The markers are distributed in the room in which the game takes place. The recycling bins appear over four different markers, with the following letters in their interior: A, B, C and G. There are three different levels within the game. In the first level, only two recycling bins and two rubbish items randomly selected among six possibilities for each type of rubbish item appear. In the second and third levels, more recycling bins and more rubbish items appear, specifically, three and four recycling bins and four and six rubbish items, respectively.

During the game, once the object marker is focused and the rubbish item is drawn, the player has to push a button to pick it up. Later, he or she has to place the rubbish item in the right recycling bin. The process is similar to picking up rubbish items. The player has to look for the bin marker where the rubbish item has to be placed. When the bin marker is focused, he or she has to press the same button as before to release the rubbish item in the recycling bin. When the players correctly place a rubbish item, they are rewarded by the game showing two hands applauding over the recycling bin. If participants wrongly place a rubbish item, the game shows a red cross over the recycling bin. The game applies the usual rules for games. A player gains or loses points for correctly/incorrectly recycling or leaving the rubbish items outside the recycling bins. If the rubbish item has been correctly placed, then the player gains points and on the contrary the player loses points if incorrect. If the player is unsure about the correct recycling bin for a rubbish item, he or she has the possibility to place the rubbish item outside all recycling bins. In this case, the game subtracts points, but less than incorrectly placing the rubbish item. The game goes to the next level when the player has achieved a certain number of points for each level. The game also has an estimated time for each level; if the player finishes before this time, he or she gains five points for each second left. The game also includes several questions in each level that are randomly selected in each turn. These questions are also related to recycling. The questions offer three possible answers of which only one is correct. The player has to choose among these three options. Once he or she answers, the game indicates if the answer is correct or not. If not, the correct solution is shown. Again, the player gains points if he answers correctly or loses points if he answers incorrectly. The colours of the recycling bins and the types of rubbish items that have to be placed in each recycling bin are the following:

- **Yellow Bin:** Plastics (bottles, bags, etc.), and light containers like soft drinks or tins.
- **Blue Bin:** Papers, cardboards, magazines, etc.
- **Gray Bin:** Metals and electronic components.
- **Brown Bin:** Organic material, like food leftovers.
- **Red Bin:** Batteries
- **Green Bin:** Any kind of glass: From glass jars to wine bottles.

We would like to highlight that these colours are normally used for these types of recycling bins in Spain. But, these colours could have different meaning in other countries.

The main menu of the game offers the following options: "New Game", "Settings", "Scores", "Help", and "Exit". If the selection is "New Game", another screen appears requiring the user's name. After this introduction, the game loads the necessary data and starts. During the game, it is possible to pause, exit to the main menu or exit the game. If the selection is "Settings", the player can change the volume of the application, etc. If the selection is "Scores", the player can see his records. If the selection is "Help", the player can learn how to play the game and how to recycle. If the selection is "Exit", the game is over. Figures 1 and 2 show two images of the ARGreenet. Figure 1 shows the objects' marker and super imposed over it appears a rubbish item. Figure 2 shows a recycling bin's marker and over it appears the green recycling bin.

RESULTS

Experiment

The research experiment involved forty-five students from the University of Valencia Summer School program. The students were asked to play both the ARGreenet and the Team ARGreenet. All participants experienced both games but in a different order, with one group of students playing the ARGreenet first (n=21), and the second group of students playing the Team ARGreenet first (n=24).

The students first completed an entry questionnaire. They were given time to familiarize themselves with the ARGreenet game with the

Figure 1. Objects' marker with a rubbish item

Figure 2. Recycling bin's marker with the green recycling bin

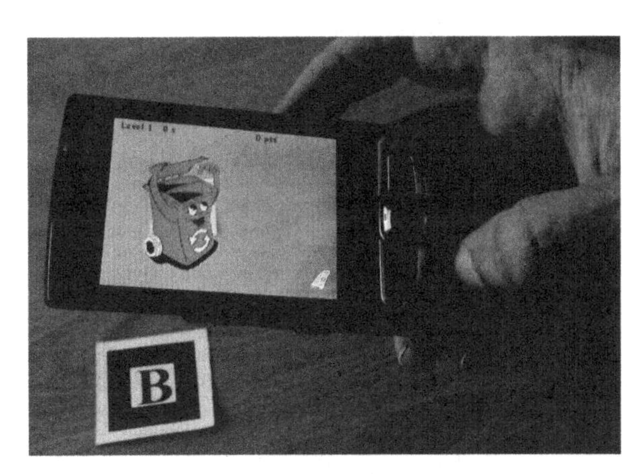

rubbish items and the recycling bins presented on a page. The students then played the two games in a predefined order, upon completion of each game the students were asked to complete a post questionnaire. Once the two games were played, the students were asked to complete an exit questionnaire.

Measures

Quantitative data was collected using questionnaires. Because the target age group was young (<15years), the questionnaire was kept short. All questions were measured on a 7 point Likert scale where in most cases 1 = none and 7 = a great deal. In the case where the meaning of 1-7 was different, the meaning was referred to in the related question.

In addition to basic demographic data including age and gender, there were a number of questions to investigate individuals' experiences with mobile phones and the phone being used in the trial, followed by a question about the students' levels of experience with gaming devices. Further questions were asked to ascertain levels of participants' knowledge of recycling, their attitudes towards recycling and the environment, current recycling behaviours and their perceived willingness to change the behaviours. The questionnaires are presented in Tables 1-3, with different aspects identified by different white/grey back ground colours.

ANALYSIS OF RESULTS

The sample was comprised of forty-five participants with a mean(SD) age of 11.02(2.33) years. Within the sample group, there were more males (53.3%) than females represented.

From the entry questionnaire, the participants reported to have relatively little mobile phone experience, mean(SD)=4.2(1.7), and considered themselves to be novices with Nokia N95, mean(SD)=1.09(0.60). The participants reported to have some experience with gaming, with the group that played the ARGreenet first, mean(SD)=5.19(1.86), reporting more experience than those playing the Team ARGreenet first, mean(SD)=4.42(1.53).

Table 4 presents the rest of the questions included in the entry questionnaire. From the data, we can deduce that overall the participants initially stated they had a moderately high knowledge of recycling. Furthermore most of them reported positive beliefs toward recycling and strongly

Table 1. Entry questionnaire

Question ID	Questions
E1	***Mobile phone experience*** How much experience do you have using mobile phones?
E2	Please indicate your level of expertise with the Nokia N95 phone: (1-Novice, 7-Expert)
E3	***Gaming experience*** How much experience do you have in playing games on a PC or mobile phone?
E4	***Knowledge of recycling*** How much do you know about what can be recycled and how to recycle?
E5	How much do you know about the effect of recycling on your environmental footprint?
E6	Please indicate your level of expertise in what recycling is and why recycling is important: (1-Novice, 7-Expert)
E7	***Beliefs about the environment/Attitudes*** How strongly you agree with the following statement: People should be recycling more in order to reduce their environmental footprint. (1-strongly disagree, 7-strongly agree)
E8	***Behaviours*** How accurate is the following statement? I recycle my garbage and separate the cans, the bottles, newspapers, etc.
E9	Please rate how good you are at recycling: (1-Not good, 7-Very good)
E10	***Intended behavior/Motivation to change*** I am willing to take new actions to improve my recycling behavior. (1-would not accept, 7-would accept).

Table 2. Post questionnaire

Question ID	Questions
P1	***Engagement and fun*** I enjoyed playing this game.
P2	This game was fun.
P3	***Easy to use*** Please indicate if the game has been easy to play (1-not easy, 7-very easy)
P4	***Perceived value*** I think playing this game could help me better recycle.
P5	I would be willing to play this game again because it has some value to me.
P6	***Attitudes*** How strongly do you agree with the following statement?: People should be recycling more in order to reduce their environmental footprint. (1-strongly disagree, 7-strongly agree)
P7	***Intended behavior/Motivation to change*** I am willing to take new actions to improve my recycling behaviour. (1-would not accept, 7-would accept)
P8	***Intention to change*** As a result of playing this game, I will talk to my friends and family members about recycling.
P9	As a result of playing this game, I will think more about recycling and its effect on the environment.
P10	As a result of playing this game, I will make changes to my current behavior.

Table 3. Final questionnaire

Question ID	Questions
F1	***Perceived learning about recycling*** Please indicate the number that most closely describes how much you think you have learned as a result of playing these games: How did you learn about what can be recycled and how to recycle? (1-nothing, 7-very much)
F2	Please indicate your level of expertise about the rubbish you can recycle as a result of playing these games (1-Novice, 7-Expert)
F3	***Preference*** Which game did you like the most? ARGreenet: Team ARGreenet:
F4	Why? Any comment that you like to add
F5	Any comment that you like to add

Table 4. Means(SDs) for the entry questionnaire. Participants' self reported knowledge, attitudes and behaviours to recycling

Question	E4	E5	E6	E7	E8	E9	E10
All	5.40(1.21)	5.40(1.21)	4.89(1.35)	6.91(0.29)	6.11(1.28)	5.38(1.19)	6.13(0.92)
ARGreenet	5.62(1.56)	5.81(1.29)	5.29(1.59)	6.90(0.30)	6.43(0.98)	5.90(1.26)	6.43(0.93)
Team AR-Greenet	5.21(0.78)	5.04(1.04)	4.54(1.02)	6.92(0.28)	5.83(1.46)	4.92(0.93)	5.88(0.85)

agreed that "People should be recycling more in order to reduce their environmental footprint". The participants also reported a willingness to do more, although the majority reported that they were already strong recyclers.

Paired t-tests were applied to the scores given to all questions of the post questionnaire filled out after playing each game. These analyses are shown in Table 5. None of the statistical paired t-tests applied to the results except two showed significant differences between the two games. The significance level was set to 0.05 in all tests.

In order to determine whether using either of the games first has any effect on the scores for the second game, the sample was divided into two groups: the participants who used the ARGreenet first and the participants who used the Team ARGreenet first. One-way ANOVA analyses were applied to the scores for all questions. These analyses are shown in Tables 6 and 7. Sixteen of the twenty statistical ANOVA tests applied to the results showed significant differences between the two games. In both cases, for question 6 the mean(SD) was the same, 7(0). From the data, we can deduce that the order of playing did significantly affect the scores for the second game. For the ARGreenet, the means(SDs) were higher when it was first played, but for the Team ARGreenet, the means(SDs) were higher when it was played second. Based on these results, we analyzed the scores given by the participants based on the game they played first. In total, nineteen participants played either the ARGreenet game first or the Team ARGreenet game first. These results are shown in Table 8. Seven of the ten statistical ANOVA tests applied to the results showed significant differences between the two games, and the means(SDs) for the Team ARGreenet were higher in all questions, except for question 6 that was the same 7(0).

*Table 5. Means(SDs) of the ARGreenet and the Team ARGreenet, and paired t-tests of post-test scores. d.f. 44, '**' indicates significant difference, '+' indicates greater mean*

Question	ARGreenet Post-test	Team ARGreenet Post-test	t	p
P1	5.96(1.11)	6.24(1.07)+	-3.292**	0.002**
P2	6.13(1.08)+	5.87(1.31)	2.383**	0.022**
P3	6.33(0.95)	6.38(1.01)+	-0.467	0.643
P4	6.24(0.96)	6.33(0.88)+	-1.071	0.290
P5	6.00(1.37)+	5.93(1.45)	0.903	0.372
P6	7.00(0.00)+	6.98(0.15)	1.000	0.323
P7	6.40(1.21)	6.51(0.99)+	-1.219	0.229
P8	5.40(1.71)	5.42(1.88)+	-0.240	0.811
P9	6.00(1.09)+	5.82(1.09)	1.835	0.073
P10	5.87(1.58)+	5.62(1.63)	1.914	0.062

*Table 6. Means(SDs) of the ARGreenet used first and second, and one-way ANOVA analysis of post-test scores. d.f. 1, 43, '**' indicates significant difference, '+' indicates greater mean*

	ARGreenet – 1st	ARGreenet -2nd	F	p
P1	6.48(0.60)+	5.50(1.25)	10.614**	0.002**
P2	6.52(0.75)+	5.79(1.22)	5.711**	0.021**
P3	6.71(0.90)+	6.00(0.88)	7.167**	0.011**
P4	6.71(0.72)+	5.83(0.96)	11.821**	0.001**
P5	6.90(0.30)+	5.21(1.44)	27.849**	0.000**
P6	7.00(0.00)=	7.00(0.00)	=	=
P7	6.95(0.22)+	5.92(1.50)	9.787**	0.003**
P8	6.33(1.24)+	4.58(1.67)	15.607**	0.000**
P9	6.71(0.46)+	5.38(1.10)	27.071**	0.000**
P10	6.71(0.56)+	5.13(1.80)	15.034**	0.000**

*Table 7. Means(SDs) of the Team ARGreenet used first and second, and one-way ANOVA analysis of post-test scores. d.f. 1, 43, '**' indicates significant difference, '+' indicates greater mean*

	Team ARGreenet – 1st	Team ARGreenet -2nd	F	p
P1	5.96(1.27)	6.57(0.68)+	3.927	0.054
P2	5.50(1.41)	6.29(1.06)+	4.354**	0.043**
P3	6.25(1.07)	6.52(0.93)+	0.826	0.369
P4	6.08(0.97)	6.62(0.67)+	4.490**	0.040**
P5	5.17(1.61)	6.81(0.40)+	20.774**	0.000**
P6	7.00(0.00)+	6.95(0.22)	1.147	0.290
P7	6.29(1.23)	6.76(0.54)+	2.612	0.113
P8	4.63(1.95)	6.33(1.32)+	11.493**	0.002**
P9	5.29(1.00)	6.43(0.87)+	16.338**	0.000**
P10	4.96(1.83)	6.38(0.92)+	10.379**	0.002**

*Table 8. Means(SDs) of the ARGreenet and the Team ARGreenet only using the data from the first use, and one-way ANOVA analysis of post-test scores. d.f. 1, 43, '**' indicates significant difference, '+' indicates greater mean*

Question	ARGreenet Post-test	Team ARGreenet Post-test	F	p
P1	5.96(1.27)	6.48(0.60)+	2.922	0.095
P2	5.50(1.41)	6.52(0.75)+	8.819**	0.005**
P3	6.25(1.07)	6.71(0.90)+	2.426	0.127
P4	6.08(0.97)	6.71(0.72)+	5.969**	0.019**
P5	5.17(1.61)	6.90(0.30)+	23.795**	0.000**
P6	7.00(0.00)	7.00(0.00)=	==	==
P7	6.29(1.23)	6.95(0.22)+	5.855**	0.020**
P8	4.63(1.95)	6.33(1.24)+	11.882**	0.001**
P9	5.29(1.00)	6.71(0.46)+	35.776**	0.000**
P10	4.96(1.83)	6.71(0.56)+	17.839**	0.000**

A paired t-tests analysis was conducted in order to explore the extent to which the participants' attitudes towards recycling as captured by the question "People should be recycling more in order to reduce their environmental footprint" was influenced by playing the two recycling games. As the order of use is important, we only used the data for the first use. The results did not show significant differences between ARGreenet, t(20)=-1.451, p=0.162; and Team ARGreenet, t(23)=-1.446, p=0.162. Therefore, the participants' attitudes have not been influenced by the games. However, our explanation for this result is that the initial mean scores were very high (6.92 for the Team version and 6.91 for the basic version) and the mean scores after the game were 7 in both cases. With such initial high scores, it is challenging if not impossible to obtain significantly higher scores after playing the two games. A t-test analysis was conducted to check if the participants' intentions to change behaviours were altered after playing the games. The intention to change was captured by the following question "I am willing to take new actions to improve my recycling behavior". The results for the ARGreenet, t(20)=-2.586, p=0.018, showed significant differences; and for the Team ARGreenet, t(23)=-1.926, p=0.067, did not show significant differences. However, the means after playing the games were higher than the initial scores and therefore, playing the games induced the players to give a higher score in this question.

Checking if the participants' perception for learning about recycling has been influenced by playing the games, the scores for E6 "Indicate your level of expertise in what recycling is and why recycling is important" and the question answered after playing both games, F2 "Indicate your level of expertise about the rubbish you can recycle as a result of playing these games", were compared using paired t-tests. The results showed significant differences when ARGreenet is first used, t(20)=-3.022, p=0.006; and did not show significant differences when Team ARGreenet is first used, t(23)=-1.701, p=0.102). We also compared, using paired t-tests, the scores for E4 and the question answered after playing both games, F1 "How did you learn about what can be recycled and how to recycle?". The results showed significant differences when the ARGreenet is first used, t(20)=-2.584, p=0.018; but failed to show significant differences when the Team ARGreenet is first used, t(23)=0.123, p=0.903. Therefore, playing the ARGreenet influenced the players' feel that they have learnt about what can be recycled and how to recycle. For the Team ARGreenet, the initial mean (5.21) and the mean after playing the Team ARGreenet (5.17) were nearly the same. Therefore, playing the Team ARGreenet did not influence the players' feeling that they have learnt about what can be recycled and how to recycle.

When asked the question F3 "Which game did you like the most?", most participants (59.1%) preferred the Team ARGreenet.

FUTURE RESEARCH DIRECTIONS

The game and the trial can be improved in several ways. Related to the Team ARGreenet version, a real collaborative game could be used instead of using the Team ARGreenet. We have developed a collaborative and a competitive version of ARGreenet. With respect to the ARCollaborativeGreenet, the game is the same as ARGreenet, but players can play in pairs. Each player plays with his or her partner. There is no competition, just collaboration. Each player while playing can see on their Nokia screen, the current performance of their team. In addition to this, at the end of the play they can see their team performance in relation to other teams' performance. With respect to the ARCompetitiveGreenet, the game is the same as ARCollaborative, but groups of players compete against other groups. There is both competition and collaboration. Similar to the ARCollaborative system, users of the ARCompetitiveGreenet can see, while playing the game, the performance of their team. At the end of the game they can compare their team performance in relation to other teams' performance. In another study, twenty-eight children played with the ARGreenet, ARCollaborativeGreenet, and ARCompetitiveGreenet. We are still analyzing the data of that study. After we finish that analysis, a comparison with the results presented in this chapter will be possible.

Related to the game, first, the images that appear in the ARGreenet are 2D images and it is likely that a 3D version would improve the quality of the game. Although the inclusion of 3D objects is possible in the developed AR library, due to the time restriction, we only included 2D images. Second, the game could also incorporate more rubbish types apart from those already included and more questions relating to these types of rubbish items.

Related to the trial, first, as the order of use was important, a different design could be considered, using a between-subjects design in which participants play either only the ARGreenet or only the Team ARGreenet. Second, other ways of checking if children learn from the game could be studied instead of using questionnaires.

Related to the SDKs, it would be interesting if a better way of loading PNG images were supported by the SDKs for these mobiles. Having active objects for loading many PNG images causes some trouble like never finishing the load. Another possibility could be the use of BMP images with OpenGL-ES, but a mask image is required for achieving transparency. The advantage of this method is that the images are loaded in a synchronous way, unlike PNG images. The disadvantage is the consumption of more memory.

Related to our adaptation of ARToolKit to mobile phones, it would be interesting to compare the performance of our adaptation with respect to ARToolKitPlus and Studierstube ES.

CONCLUSION

We have presented a new mobile phone AR library and the steps we have followed for its development. This information could help to perform similar development. We have presented the game for learning how to recycle that we have developed using the library, ARGreenet. We have also included a study involving forty-five children that played the ARGreenet and a 'Team' version of the AR-

Greenet. Children experienced more fun playing the Team ARGreenet. They considered the Team ARGreenet as easy to use as the ARGreenet. We can expect this result because the two games are the same, the only difference is that in the Team ARGreenet two or three children are playing at the same time with the same game. The results showed significant differences between the ARGreenet and the Team ARGreenet for children's perceived value, intended behaviour, motivation to change, and intention to change. The Team ARGreenet offered higher means in all these cases. Therefore, it is possible to deduce that the Team ARGreenet had a greater influence on these aspects than the ARGreenet. Moreover, a majority of children (59.1%) expressed a preference for the Team ARGreenet when they were asked about which game they liked the most.

Finally, not many years ago, AR was mainly used on PCs. However, with the increased capacity of today's mobile phones and the technological advancements being made, there are increased opportunities to develop new and more powerful applications that were not possible five years ago. This explains the current interest in AR mobile applications and the increasing number of them that are appearing, especially for the iPhone. It is almost certain that in the near future the technology of mobile devices will go further and almost all of the applications that now run on PCs will run on mobile devices. Nowadays, phones are incorporating more and more interesting utilities like the tactile screen, accelerometer, GPS, better graphics and camera that help in the development of better applications, in general, and in this case, for AR. The AR community is aware of these advantages and for sure it will develop new and interesting applications for many areas.

ACKNOWLEDGMENT

We would like to thank the following:

This work was partially funded by: the APRENDRA project (TIN2009-14319-C02-01) and the GreenHunt project (PCI2006-A7-0676).

The Summer School of the Technical University of Valencia for its collaboration and for allowing us to include the games as activities.

Simone Carr-Cornish for her assistance in analyzing the data of the experiment.

REFERENCES

Abawi, D., Dörner, R., Haller, M., & Zauner, J. (2004). *Efficient mixed reality application development.* In First European Conference on Visual Media Production (pp. 289-294).

Azuma, R. T. (1997). A survey of augmented reality. *Presence (Cambridge, Mass.)*, *6*(4), 355–385.

Bauer, M., Bruegge, B., Klinker, G., MacWilliams, A., Reicher, T., & Rib, S. ... Wagner, M. (2001). *Design of a component-based augmented reality framework.* In The Second IEEE and ACM International Symposium on Augmented Reality (pp. 45).

Cawood, S., & Fiala, M. (2007). *Augmented reality: A practical guide*. Raleigh, NC: Pragmatic Bookshelf.

Henrysson, A., Billinghurst, M., & Ollila, M. (2005). *Face to face collaborative AR on mobile phones.* In International Symposium on Augmented and Mixed Reality (pp. 80-89).

Kato, H., & Billinghurst, M. (1999). *Marker tracking and HMD calibration for a video-based augmented reality.* In Second IEEE and ACM International Workshop on Augmented Reality (pp. 85–94).

Looser, J. (2007). *AR magic lenses: Addressing the challenge of focus and context in augmented reality*. Doctoral Dissertation, University of Canterbury, New Zealand.

MacIntyre, B., Gandy, M., Bolter, J., Dow, S., & Hannigan, B. (2003). *DART: The Designer's Augmented Reality Toolkit*. In The Second International Symposium on Mixed and Augmented Reality (pp. 329-339).

Möhring, M., Lessig, C., & Bimber, O. (2004). *Video see-through AR on consumer cell phones*. In IEEE and ACM International Symposium on Augmented and Mixed Reality (pp. 252-253).

Mulloni, A. (2007). *A collaborative and location-aware application based on augmented reality for mobile devices*, Master Thesis, Facoltà di Scienze Matematiche Fisiche e Naturali, Università degli Studi di Udine.

Owen, C., Tang, A., & Xiao, F. (2003). *ImageTclAR: A blended script and compiled code development system for augmented reality*. In The International Workshop on Software Technology for Augmented Reality Systems (pp. *23-28*).

Paelke, V., Reimann, C., & Stichling, D. (2004). *Foot-based mobile interaction with games*. In ACM SIGCHI International Conference on Advances in computer entertainment technology (pp. 321-324).

Pilet, J. (2008). *Augmented reality for non-rigid surfaces*. Doctoral Dissertation, Ecole Polytechnique federale de Lausanne, France.

Rohs, M., & Gfeller. B. (2004). Using camera-equipped mobile phones for interacting with real-world objects. *Advances in Pervasive Computing*, 265-271.

Wagner, D., Pintaric, T., Ledermann, F., & Schmalstieg, D. (2005). *Towards massively multi-user augmented reality on handheld devices*. In Third International Conference on Pervasive Computing (pp. 208-219).

Wagner, D., & Schmalstieg, D. (2003). *First steps towards handheld augmented reality*. In the 7th International Conference on Wearable Computers (pp. 127-135).

Wagner, D., & Schmalstieg, D. (2007). *ARToolKitPlus for pose tracking on mobile devices*. In 12th Computer Vision Winter Workshop (pp. 139-146).

Wagner, D., & Schmalstieg, D. (2009). Making augmented reality practical on mobile phones, part 1. *IEEE Computer Graphics and Applications*, *29*(3), 12–15. doi:10.1109/MCG.2009.46

ADDITIONAL READING

Azuma, R., Baillot, Y., Behringer, R., Feiner, S., Julier, S., & MacIntyre, B. (2001). Recent advances in augmented reality. *IEEE Computer Graphics and Applications*, *21*, 34–37. doi:10.1109/38.963459

Bimber, O., & Raskar, R. (2005). *Spatial augmented reality: Merging real and virtual worlds*. Wellesley, MA: A. K. Peters, Ltd.

Bucolo, S., Billinghurst, M., & Sickinger, D. (2005). *Mobile maze: A comparison of camera based mobile game human interfaces*. In 7th International Conference on Human Computer Interaction with Mobile Devices & Services (pp. 329-330).

Cheok, A. D., Goh, K. H., Liu, W., Farbiz, F., Fong, S. W., & Teo, S. L. (2004). Human Pacman: A mobile, wide-area entertainment system based on physical, social, and ubiquitous computing. *Personal and Ubiquitous Computing*, *8*(2), 71–81. doi:10.1007/s00779-004-0267-x

DiVerdi, S., & Höllerer, T. (2007). GroundCam: A tracking modality for mobile mixed reality. *IEEE Virtual Reality*, *3*, 10–14.

Feiner, S., MacIntyre, B., Höllerer, T., & Webster, A. (1997). *A touring machine: Prototyping 3D mobile augmented reality systems for exploring the urban environment. In First International Symposium on Wearable Computers* (pp. 74-81).

Föckler, P., Zeidler, T., Brombach, B., Bruns, E., & Bimber, O. (2005). *PhoneGuide: Museum guidance supported by on-device object recognition on mobile phones.* In International Conference on Mobile and Ubiquitous Computing (pp. 3-10).

Hakkarainen, M., & Woodward, C. (2005). *Symball-camera driven table tennis for mobile phones.* In *ACM SIGCHI International Conference on Advances in Computer Entertainment Technology* (pp. 321-324).

Höllerer, T., Feiner, S., Terauchi, T., Rashid, G., & Hallaway, D. (1999). Exploring MARS: Developing indoor and outdoor user interfaces to a mobile augmented reality system. *Computers & Graphics, 23*(6), 779–785. doi:10.1016/S0097-8493(99)00103-X

Rekimoto, J., & Ayatsuka, Y. (2000). Designing augmented reality environments with visual tags. In *Designing Augmented Reality Environments* (pp. 1–10). CyberCode. doi:10.1145/354666.354667

Schmalstieg, D., & Wagner, D. (2007). *Experiences with Handheld Augmented Reality*, The Sixth IEEE and ACM International Symposium on Mixed and Augmented Reality (pp. 1-13).

Wagner, D. (2007). *Handheld augmented reality.* Doctoral dissertation, Graz University of Technology, Austria.

Wagner, D., Reitmayr, G., Mulloni, A., Drummond, T., & Schmalstieg, D. (2008). *Pose tracking from natural features on mobile phones.* In 7th IEEE and ACM International Symposium on Mixed and Augmented Reality (pp. 125-134).

Wagner, D., & Schmalstieg, D. (2009). *History and future of tracking for mobile phone augmented reality.* In International Symposium on Ubiquitous Virtual Reality (pp. 7-10).

KEY TERMS AND DEFINITIONS

API (Application Programming Interface): It is a set of functions (an interface) offered by a program for being used by other software.

Augmented Reality: It enriches the real world with virtual elements that are superimposed over it.

IDE (Integrated Development Environment): It is a software application that provides a graphical interface, a source code editor, a compiler, build automation tools and a debugger.

Plug-in: It is a computer program that extends the capabilities of the program in which it is incorporated.

SDK (Software Development Kit): It is a set of development tools that allow the development of applications.

Symbian C++: It is a programming language used for developing programs for Symbian OS.

Symbian OS: It is an operating system for mobile devices.

S60: It is a platform for mobile devices that use Symbian OS.

S60 3rd Edition: It is an SDK for programming in Symbian C++ for devices that use Symbian OS and the S60 platform.

Chapter 10
Using a Process-Aware Information System to Support Collaboration in Mobile Learning Management Systems

Roberto Perez-Rodriguez
University of Vigo, Spain

Manuel Caeiro-Rodriguez
University of Vigo, Spain

Luis Anido-Rifon
University of Vigo, Spain

ABSTRACT

This chapter deals with the support of collaboration in mobile Learning Management Systems. The authors propose a collaborative game, to be taken place in an enhanced reality environment, as an example of collaboration. Several alternatives to support this scenario are analyzed, and the proposed architecture to integrate process-based collaboration in mobile Learning Management Systems is discussed in detail. Finally, an implementation of this scenario using open-source technologies is detailed.

INTRODUCTION

The interest in Information and Communication Technology (ICT) to support learning activities both in industry and academia has had a dramatic increase in recent years. Companies and universities are using Learning Management Systems (LMSs) to deliver courses to employees and students respectively.

Mobile learning presents some particularities because of two factors: the most important one is the mobility, the other one is the hardware itself. Modern mobile devices are provided with a digital camera, video player, microphone, GPS, internet connectivity, etc. And they can be used in mobile learning scenarios. The usual approach to the design of mobile LMSs is not to treat the mobile

DOI: 10.4018/978-1-60960-613-8.ch010

built-in devices as first-class objects. Therefore, a participant in a mobile learning course may have to take, for example, a picture using the camera app, and later to upload the picture to the LMS. When the built-in camera is treated as a first-class object, participants can take photos that are immediately accessible to their partners in a collaborative graphical environment. In a similar way, the GPS position of a participant may be public for all group members. So, data acquisition and data flow require to be automated as fist-class objects to minimize the cognitive load and error.

LMSs provide functionalities for the control of the learning process such as user management, storage of learning materials, communication facilities, assessment activities, notification, and more. Traditionally, commercial LMSs such as Blackboard (Blackboard Website), and Claroline (Claroline Website) have been used to carry on the learning process. Recently, a great shift towards open-source has taken place both in industry and academia. There are popular web-based Learning Management Systems such as Moodle (Cole, 2005), dotLRN (Santos, Boticario, Raffenne, & Pastor, 2007), Sakai (Farmer & Dolphin, 2005), and OLAT (Fisler & Schneider, 2008) which support online courses in companies and universities. In respect with mobile environments, mobile front-ends to LMSs, such as MLE for Moodle, represent the most outstanding trend nowadays.

These LMSs are used to support learning experiences in accordance with different styles and contexts, including collaborative practices. In the literature, there are a number of works addressing common metaphors for collaboration, such as the one by Sanderson (1997). Figure 1 shows collaboration in the classroom, typically viewed as people sitting around a table, over which objects are placed. This type of collaboration is not of an informal kind, but it is framed and structured to achieve a group goal. Therefore, interactions among participants as well as the information they interchange have to be structured (Hernandez-Leo, Asensio-Perez, & Dimitriadis, 2005). If the collaboration has an explicit purpose, such as to carry on an augmented reality game, typically a roadmap serves as a guide to divide and assign work among all participants.

In the Information Technology field, there are a group of systems that are classified as *process-aware*. Dumas, van der Aalst and Ter Hofstede (2005) classify Process-Aware Information Systems (PAISs). PAISs have been suggested to support collaboration in Learning Technology settings. We have analyzed different types of PAISs, according to their suitability to support

Figure 1. A collaborative educational game in the classroom

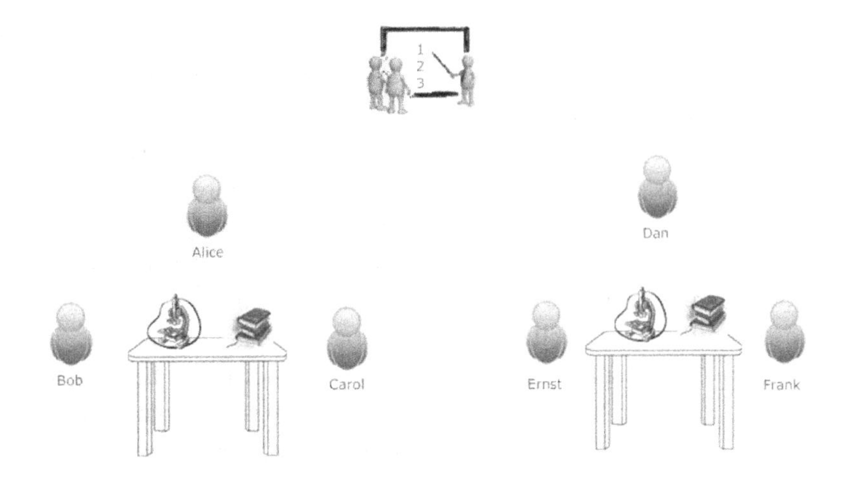

collaborative learning. We will show, for example, that ad-hoc workflow (Voorhoeve & van der Aalst, 1997) presents a lack-of-generality, and it is not optimal to support collaborative games.

This chapter is structured as follows. The chapter begins with an introduction above, which outlines the area under review. This is followed by an introduction to PAISs, and how they can be used to support collaboration. The chapter continues with a discussion on the internal organization of LMSs. This includes a review of Moodle. This is followed by the problem description and the analysis of possible alternatives. The chapter continues with the description of the *platform-independent architecture* and the implementation of the proposed solution. The chapter ends with some conclusions.

PROCESS-AWARE INFORMATION SYSTEMS THAT SUPPORT COLLABORATION

In this section, we analyze collaboration using a metaphor based on educational games in the classroom. We also analyze PAISs from the point of view of their support to collaboration.

Educational Games in the Classroom

In this situation, we can identify several aspects in enabling collaborative games:

1. The classroom, composed by participants (learners and teachers) who share a common space.
2. The table, which is the basic grouping unit.
3. The objects, which are manipulated by people.
4. The roadmap, which informs people on the steps that have to be followed, as well as on the assignment of tasks to people.

The life-cycle of a collaborative educational game in the classroom is typically composed of the following stages: the *design-time* stage, in which the teacher creates the roadmap of the practice, including the number of participants per table; the *instantiation-time* stage, in which the teacher communicates the assignment of people to groups and the game starts; and the *run-time* stage, in which participants collaborate following the instructions in the roadmap, and at the same time the teacher walks around tables and sees the progression of groups.

Now, let's see how the above collaborative practice is typically supported using ICTs: the *classroom* is supported by a Learning Management System:

1. An LMS provides authenticated access to courses, in which participants are enrolled. So, the LMS serves as a virtual meeting point for learners.
2. The *table* is supported by a course inside the LMS.
3. The *objects* are supported by the tools in the LMS, such as wikis, weblogs, document editors, forums, etc.
4. The *roadmap* is typically supported by a textual description in the form of a text file or similar.

The problem with this approach is that the roadmap only provides indications on how to divide work among participants, and on the different steps that have to be followed. In other cases, the roadmap is reified into groupware interfaces (Dillenbourg, 2002). This approach lacks some important features, mainly regarding process automation: (i) it does not automate the formation of groups of participants; (ii) it does not automate the assignment of activities to participants; (iii) it does not automate notifications; (iv) it makes monitoring difficult, because the game steps are not computer-understandable; (v) it does not automate conditional activities or secondary paths;

(vi) participants do not have a to-do list at their disposal, etc.

Process-Aware Information Systems

PAISs have been used extensively from the very beginning of Information Technology. Many industrial processes can be successfully modeled by using Petri-nets and programmable automata (van der Aalst, 1998). From the early 90s, PAISs have been employed to automate business processes. In recent years, human-workflows are being used to model and execute business processes involving many people that have to be coordinated to achieve a shared goal (Andrew, et al., 2005).

Figure 2 shows the two dimensions in the modeling of collaboration: space and time. On the one hand, the space-axis represents the definition of virtual spaces for participants. Following the collaboration metaphor introduced in Section 1, the space-axis contains the definitions of classrooms, groups, and even objects. On the other hand, the time-axis contains the definition of roadmaps, that is, the definition of what has to be done, the order in which it has to be done, and when it has to be done.

Groupware software allows non-framed collaboration. In this group of tools, there are ones such as wikis, shared document edition, etc. More recent applications such as Google Wave fall into this category as well. Groupware tools structure well the space-axis; but they are not suitable to support complex and framed processes, which require a lot of automation in order not to overload the participant's cognitive capabilities (Pinelle, Gutwin, & Greenberg, 2003). The problem is that the roadmap, the description of what has to be done, is not automated. And what has to be done, the roadmap, is time-aware; therefore, the time-axis has to be modeled too. Workflow systems are the other face of the coin in collaboration-enabling software. Workflow systems focus on time-axis of collaboration: task sequencing. In many workflow systems, there is even no common space for team members to share objects freely.

Learning Technologies, from the beginning, provide tools for online publication of contents, as well as communication services to put participants into contact with each other. Nowadays, social networks provide a horizontal structure of virtual spaces really complex, allowing for configuring private, group-of-friends, and world spaces. Some of these tools have an implicit, trivial, time-axis

Figure 2. Two dimensions of collaboration-enabling software

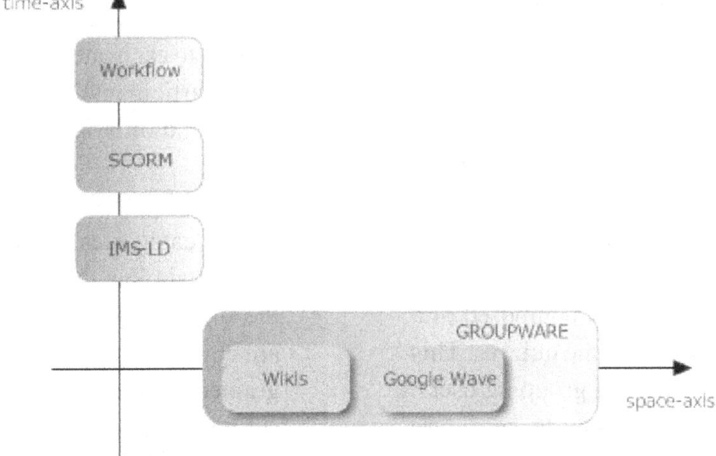

underlying collaboration process. Forums, for example, are of a first-in-first-out kind. Wikis provide a plain space structure based on word entrances.

We have scripting languages such as SCORM (Bohl, Schellhase, Sengler, & Winand, 2002) that focus on the time-axis of collaboration: task sequencing. The theatrical metaphor has been employed many times to describe collaborative processes in the learning field. This metaphor, which has been made explicit in IMS-LD (Van Es & Koper, 2006), the most outstanding Educational Modeling Language (EML) (Botturi, Derntl, Boot, & Figl, 2006) to date, describes collaboration as the play of several actors' script. The problem with this approach is that it does not integrate well with well-established learning tools, such as wikis and forums. So, learning tools are non configurable black boxes from the IMS-LD point of view. From the point of view of the provider of the infrastructure, the collaboration process is more like the preparation of a theatrical play, or more explicitly, like the organization of events such as concerts. Space has to be reserved inside tools, and user actions on tools have to be encapsulated and treated as first-class objects in the process definition.

INTERNAL ORGANIZATION IN LEARNING MANAGEMENT SYSTEMS: MOODLE CASE STUDY

The course is the most frequent form of assigning virtual space to learners (Mason, 1998). Learners are enrolled in a course and, from it, they obtain learning resources and may participate in online learning activities supported by some tool such as a wiki or a weblog. In its most basic form, a collaborative course may be just composed of a pdf file, a support forum, and some quizzes. This is a very coarse-grained way of grouping users. Regarding tools, they must allow to configure course-identified access control lists in order to support grouping. In order to model the time-axis, a Learning Management System may structure spaces in sections, which have to be completed in a sequential form, or it may allow for liberating content on a weekly basis (Cole, 2005).

Moodle

Moodle is a very popular open-source LMS, and many universities are basing their e-learning solution on this platform. Among the reasons for choosing Moodle, we highlight the following ones:

1. Moodle is open-source, and it is distributed under the GNU license, which allows for downloading the code, modifying it, and redistributing it.
2. Moodle has a good documentation.
3. Moodle has a good reputation.

Moodle is written in the PHP (Lerdorf, Tatroe, Kaehms, & McGredy, 2002) programming language, which is specially suited to developing web applications. Regarding persistence, Moodle can work with a wide range of database systems, including the popular MySQL.

Analysis of Moodle Following a Separation-of-Concerns Approach

Following our conceptual framework for the modeling of courses, we analyze Moodle in accordance to its support to each one of these concerns: course structure, participants' organization, temporal constraints, and ordering constraints:

- Structurally, Moodle allows for defining course categories, which aggregate courses of a similar theme. Inside the courses, the unique structures that can be created are course sections, which are a more fine-grained structure to aggregate the learning activities and resources (Cole, 2005).

- Regarding the organization of participants, those are enrolled in a course with one of the predefined roles: administrator, teacher, and learner. In Moodle, it is allowed to define roles at a tool-level; for example, a participant may possess a learner role in the Math course, but in the social forum of the Math course the same participant may have an administrator's role. In Moodle, we encounter no facilities to automate formation of groups of participants, as well as other advanced functionality to deal with participants in an advanced way.
- In Moodle, course sections can be organized in a weekly format. No more advanced or fine-grained time-control structures are at our disposal to set temporal constraints.
- Finally, and regarding ordering constraints, there exists neither mechanism to sequence course sections nor to block course sections to certain participants, etc.

Mobile Learning Engine

Mobile Learning Engine (MLE) (Mobile Learning Engine) is a mobile interface for the Moodle Learning Management System (LMS). MLE features a stand-alone Java application (see Figure 3), which is capable of being run in the majority of modern mobile phones. MLE is also a mobile web interface for accessing Moodle in those devices that do not enable Java.

One of the most interesting characteristics of the stand-alone version of MLE is that it integrates a QR-code reader. QR-codes are a type of bidimensional codes, in which information can be stored in both the horizontal and the vertical axis. The information codified into a QR code may contain a word, a URL, etc.

Figure 3. MLE Java mobile app

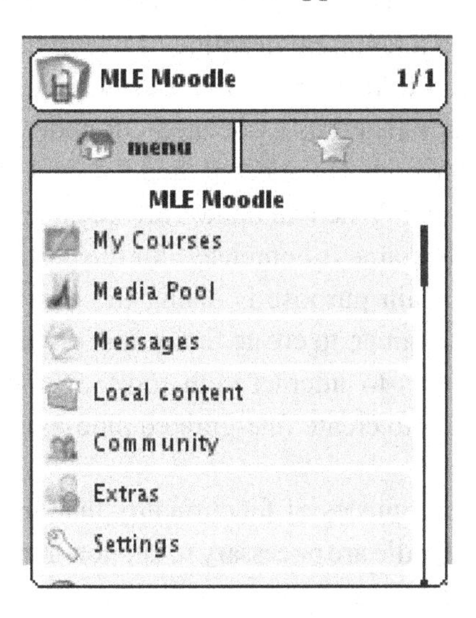

PROBLEM DESCRIPTION

Most LMSs provide a number of groupware tools. These collaboration tools enable participants to interact with each other, as well as with contents. Our aim is to extend LMSs with the capability to support framed collaboration that enables collaborative games. In the literature, these framed collaboration structures supporting educational games are named gameflows. In its most basic form, a gameflow can be defined as follows:

- A gameflow process is composed of a set of activities.
- Participants are enrolled in the process, and activities are assigned to them.
- The activities have to be completed in a certain order.

We illustrate gameflows with an example (see Figure 4). This game is about animal recognition, and makes use of QR codes for automation. The game uses as tools the mobile camera as well as the mobile keyboard. The game is proposed to take place in a room with pictures of animals. Near each picture a QR code is placed. At this stage, every participant is requested to recognize seven animals, given by their names. Besides, it is provided a temporal deadline of twenty minutes for this task. At this point, participants take a photo of an animal and its associated QR code. At this point, the code is validated, and the next animal is proposed. When all photos are taken and validated, the game is completed for the participant.

Our main purpose is, using the course as a space container, to create fine-grained spaces for participants to interact with tools. Our second purpose is to create fine-grained time-axis structures.

Several pieces of functionality that are lacking in Moodle are necessary to support gameflow definition and execution: capability to assign different activities to different participants in the same gameflow; capability to assign different activities to different participants in the same gameflow; capability to mark the end of an activity; capability to sequence activities, etc. Therefore, our problem is to extend Moodle in order to support both the definition and execution of gameflows.

ANALYSIS OF ALTERNATIVES

There are some alternatives for supporting process-based collaborative games in LMSs. All of them imply the use of a PAIS that supports the definition and execution of collaboration structures. Therefore, regarding the type of the selected PAIS, some alternatives can be listed:

1. To develop an ad-hoc PAIS to support collaboration structures in the LMS of choice.
2. To use an Educational Modeling Language.

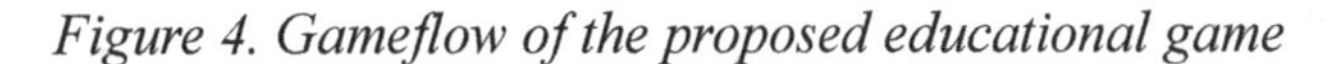

Figure 4. Gameflow of the proposed educational game

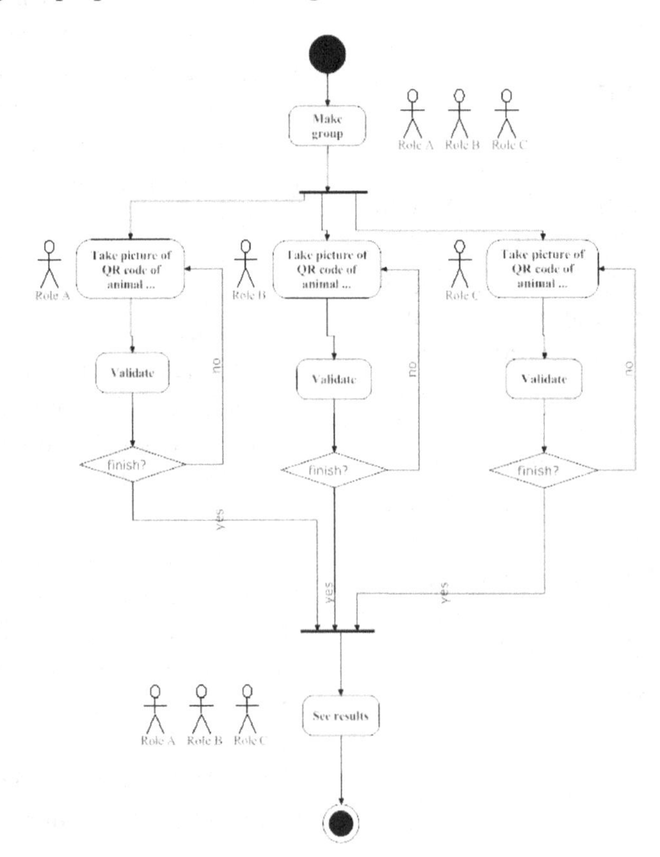

Table 1. Analysis of alternatives

Process metamodel/execution metamodel	Ad-hoc PAIS	EML	Workflow Engine
Ad-hoc PAIS	YES (1)	NO	NO
EML	NO	YES (2)	YES (3)
Workflow Engine	NO	NO	YES (4)

3. To use a generic-purpose workflow engine.

In the next subsections, we discuss the implications of each one of those approaches both for the process metamodel and for the execution metamodel. The process metamodel provides a formalism for defining process models: how the process can be structured, which are the different kinds of activities, which are the routing mechanisms, etc.; whilst the process execution metamodel provides a formalism for defining the run-time behavior associated with the process metamodel (Breton & Bezivin, 2002).

Process Metamodel

In a PAIS, the process metamodel provides the definition and format of data involved in processes.

The first alternative is to develop an ad-hoc process metamodel to represent elements involved in collaboration structures. Systems supporting collaboration can be developed ad-hoc with no interoperability and reusability concerns in mind. Thus, collaborative sequences cannot be treated as first-class objects. Collaborative sequences are codified using a programming language, and it is required to change the base code of the application in order to modify a collaborative sequence.

The second alternative is to use a process metamodel compliant with an EML, such as SCORM or IMS-LD. Thus, the process metamodel of the EML of choice will serve to represent spaces, participants, timing, and ordering concerns in collaboration structures.

The third alternative is to use the process metamodel of a generic-purpose workflow engine. Thus, we can define a wide range of collaboration structures, which depend on the process definition language of choice.

Execution Metamodel

Regarding the execution metamodel, there are some alternatives that are marked as 1,2,3,4 in Table 1:

1. If we use an ad-hoc process metamodel, it implies to use its corresponding ad-hoc execution metamodel.
2. If we use a process metamodel compliant with an EML, then we can use the corresponding EML execution metamodel.
3. We can also use a process metamodel compliant with an EML and the execution metamodel of a generic-purpose workflow engine (Marino, Casallas, Villalobos, & Correal, 2007).
4. The last alternative is to use a workflow process metamodel and its related execution metamodel (Perez-Rodriguez, Caeiro-Rodriguez, & Anido-Rifon, 2009).

Our Choice

Each alternative for supporting gameflows has its own advantages and disadvantages. It follows an analysis of advantages/disadvantages (summarized in Table 2):

Table 2. Analysis of advantages/disadvantages

	Advantages	Disadvantages
(1)	Custom process metamodel	(a) The process metamodel, as well as the execution metamodel, has to be designed and developed from scratch. (b) This solution does not conform to any standard, nor from the workflow field nor from the e-learning field.
(2)	(a) The EML metamodel is specific to the learning domain.	(a) Difficult integration reported in literature. (b) Several non-compatible standards are available.
(3)	(a) To use well-proven, high-performance software as workflow engines. (b) The EML metamodel is specific to the learning domain.	(a) Possible mismatches between the EML process metamodel and workflow execution metamodel.
(4)	(a) To conform to workflow standards. (b) To use well-proven, high-performance software as workflow engines.	(a) The workflow metamodel is not specific to the learning domain.

- The first alternative, to use both an ad-hoc process metamodel and its corresponding execution metamodel, is very inefficient, because both the process metamodel and the execution metamodel have to be designed and developed from scratch. Moreover, the developed solution would not be in accordance with current standards, neither from the e-learning field, nor from the workflow field.
- The second alternative is to use an EML as a process metamodel, with its corresponding execution metamodel. The main advantage in this approach is that EMLs provide metaphors that are particularly well suited to the learning domain. The disadvantages are basically for the current state-of-the-art in the development of EML execution engines. In this way, there are some attempts to integrate EML execution engines in Learning Management Systems (Burgos, Tattersall, Dougiamas, Vogten, & Koper, 2007). In addition, the most outstanding EML to date, IMS-LD, presents some flaws regarding collaborative practices, as those reported by Hernandez-Leo, Asensio-Perez, and Dimitriadis (2004) and by Jurado, Redondo and Ortega (2006).
- The third alternative, to use an EML process metamodel over a workflow execution metamodel, is well covered in the literature (Marino, Casallas, Villalobos, & Correal, 2007). The main drawback in this approach is that some mismatches may occur between the EML execution metamodel and the workflow execution metamodel.
- The fourth alternative has been treated in the literature (Perez-Rodriguez, Caeiro-Rodriguez, & Anido-Rifon, 2009). The main advantage in this approach is that workflow technologies are a very mature field in software engineering, and there are some good workflow engines reliable for this work. Contrary to EML-based alternatives, the modeling metaphors specific to the learning domain are not available here. Taking into account the advantages and disadvantages of this alternative, we consider it to be the most suitable one.

Assessment of Workflow Engines

The Workflow Management Coalition (WfMC) (Workflow Management Coalition) is a consortium concerned with the definition of standards for the interoperability of Workflow Management Systems. The Workflow Reference Model (Hollingsworth, 1995) defines a Workflow Management System (WfMS) as a set of generic components that interact in a defined set of ways. A standardized set of interfaces and data interchange formats are provided. The Workflow Reference Model forms the basis of WfMSs today. XPDL is a language used for describing a process definition, developed by the WfMC. XPDL is not an executable language, but a process design format that serves as a "lingua franca" both for authoring tools and execution engines.

Some of the most outstanding open-source WfMSs are Enhydra Shark, Bonita, and jBPM, according to our literature survey. Some of their characteristics are:

- Enhydra Shark (Enhydra Shark) is an extendable and embeddable Java workflow engine completely based on WfMC specifications. Enhydra Shark uses XPDL as its native process definition format. Enhydra Shark can be used as a standalone application or as a Java library used from our own context. Service wrappers can be built on top of the Enhydra Shark Workflow Engine.
- Bonita (Bonita) is downloadable, as well as Enhydra Shark, under the LGPL license. Bonita is compliant with the WfMC standard. Bonita is based on J2EE, and it is compliant with the Enterprise JavaBeans 2.0 (EJB) specification. Bonita EJBs are deployed into JOnAS, the application server by Bull S.A.S. From our point of view, the EJB API can make the integration process hard.
- jBPM uses Java Process Definition Language (jPDL) for describing processes. jBPM is a very stable and potent Workflow Engine, already tested in stable/production projects, both commercial and open-source ones (Alfresco).

Most of the open-source workflow engines provide very similar functionalities. Our choice of a concrete Workflow Engine is driven by the stable/production projects (Wohed, Russell, ter Hofstede, Andersson, & van der Aalst, 2009). Under these criteria, jBPM is the most accepted open-source workflow engine.

Browser-Based vs. Standalone Apps

There exists a considerable controversy nowadays on the convenience of browser-based mobile services versus installable apps. The following points treat to shed some light on this issue:

- There are services that perform better as installable apps, as those that need fine-grained control of the mobile built-in devices, such as the screen. Other services, such as news readers or online forums, can work well in a mobile browser.
- Installable apps are usually developed in proprietary frameworks and are thus not portable between the different operating systems of different vendors. The web, on the contrary, is cross-platform and achieves portability between different mobile platforms.
- Web-based applications do not need installation, whilst stand-alone applications do. The issue of locating the desired app is trivial in the Web since a unique URL identifies it. Installable apps, on the contrary, depend on different vendor-dependent strategies to download them.
- The main difference is that browser-based apps suffer from the *sandbox* effect; that is, in most cases browser-based apps are blocked from accessing the device due to security (accelerometer, GPS chip, etc.). Mobile Widgets have been proposed as an intermediate solution to this issue.

There are some initiatives aiming to enable Web-based apps to access the local capabilities of the mobile device:

- The OMTP BONDI (OMTP Bondi) activity is focused on enabling a set of JavaScript APIs that can access local capabilities in a secure way.
- Palm webOS (Allen, 2009) appears to treat web apps as first class citizens by providing access to the GPS location, accelerometer, camera, and messaging.
- In the iPhone, the only way to access the mobile built-in devices is to rewrite the Web app using Xcode.
- In Google's Android platform, there is a WebKit-based web browser with sandbox limitations. Notwithstanding, there is a proprietary native SDK based on the Java SDK to write first-class citizen apps.
- The PhoneGap project (PhoneGap) aims at creating a single JavaScript API for use on the Android, Blackberry and iPhone platforms. Hence, PhoneGap injects a JavaScript API into the browser that allows it to access the native functions of the phone: location, camera, accelerometer, and vibration functions. The main limitation of PhoneGap is that it builds native applications, so they have to be installed prior to being used.

As a conclusion on this discussion, we point out that, in our opinion, both mobile Web and stand-alone apps will converge in the near future, so an integrated approach to the development of apps is needed.

We choose the MLE standalone app as a mobile client for Moodle because of its potential to use mobile built-in devices.

PLATFORM-INDEPENDENT ARCHITECTURE

This section explains a generic architecture to support collaborative educational games in mobile Learning Management Systems. The proposed architecture follows the principles of Service-oriented Architecture (SOA). Figure 5 shows an UML diagram of the architecture.

The course inside the LMS is logically divided into three views: authoring, monitoring, and delivering. The authoring component provides the view for creating new process definitions, which are incorporated to the models schema in the gameflow engine. The monitoring component provides the view for following the progress of participants through the collaboration structures. Finally, the delivering component provides the

Figure 5. Platform-independent architecture

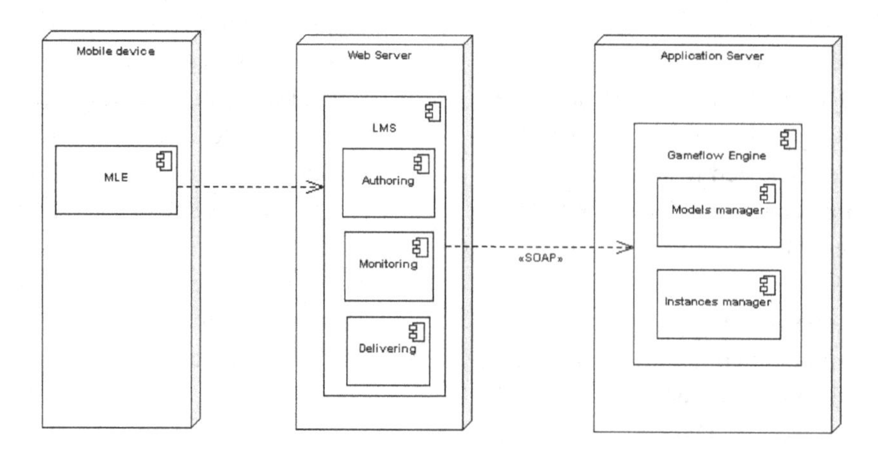

working view for participants, including a to-do list that contains the pending assignments.

The gameflow engine uses the functionality of an embedded workflow engine. The engine stores information on structures, participants, ordering, timing, and the rest of the elements in a gameflow, and makes the state of the system evolve depending on events. The gameflow engine is integrated into the e-learning system through a well defined interface. Inside the gameflow engine, we can distinguish two subcomponents: the gameflow models manager, and the gameflow instances manager.

- The gameflow models manager deals with process definitions, that is, gameflow models. This subcomponent maintains the versions of the models, and it also updates models when required by an authorized user. Communication from the exterior of the gameflow engine is made by making use of the authoring interface, which provides the needed methods for authoring gameflow models.
- The gameflow instances manager is in charge of managing process instances. Communication from the LMS is made by making use of both the information retrieval interface and the events interface. The LMS is able to access the gameflow engine in a passive way using the information retrieval interface, just to get information on structures, participants, ordering, timing, and the rest of the elements. In a similar way, the gameflow engine has to manage the events from the events interface. An event that is external to the gameflow engine, such as the finishing of an activity, may trigger several events that are internal to the execution engine, such as to update the to-do list of all participants in that process. In summary, the interaction between the LMS and the gameflow engine may be passive information retrieval as well as communication of events generated by participants.

Gameflow as a Service

The functionality provided by the gameflow engine is published in a WSDL file. LMSs that want to embed gameflow capabilities can invoke service methods as specified in the WSDL file. The service methods are:

- getPendingTasks() - This service method returns the pending tasks for a participant.
- deployProcessDefinition() - Deploys the process definition of an educational game.
- endOfActivity() - Signals the end of an activity.

Some Final Notes on Architecture

The gameflow engine acts as a wrapper for the out-of-the-box workflow API, encapsulating users' actions as sequences of workflow queries. The mechanism for transferring functional control from the LMS to the gameflow engine is message-passing: the LMS transfers the functional control by invocating a service method and waits until the gameflow engine returns the functional control.

This architectural design is a *platform-independent model* because any workflow engine can be wrapped by a Web service façade for publishing its basic functionalities. In the next section, we explain the low-level details of the integration of a concrete LMS, Moodle, with the workflow engine of choice, jBPM. The exposition of the implementation is a *platform-dependent model*.

IMPLEMENTATION

This section details the implementation of the solution described in the previous section using some technologies of choice, translating thus the *platform-independent model* into a *platform-*

dependent one. Our LMS of choice is Moodle, whilst our PAIS of choice is jBPM.

The platform-dependent model is as follows:

- jBPM provides a workflow interface for making workflow queries.
- The gameflow engine implements its service methods by making workflow queries to jBPM.
- Communication from Moodle to the gameflow engine is performed by making use of SOAP middleware. At the Moodle side we use NuSOAP (NuSOAP), whilst at the gameflow engine side we use Axis (Axis).
- In the mobile device, the MLE app is running.

Querying the Workflow Engine

In this subsection, we detail the procedure for making workflow queries to jBPM, being thus specific to the java platform. *JbpmContext* is a service-provider object, and it is created using a *Factory Method Pattern*, which is a useful creational pattern when the object creation process depends on settings in configuration files. *JbpmContext* provides access to the most common operations on jBPM or workflow queries, representing each *JbpmContext* a transaction.

JbpmConfiguration is a thread safe object and serves as an object factory for *JbpmContext*. All threads can use *JbpmConfiguration* as a factory for *JbpmContext* objects. Hence, *JbpmConfiguration* can be used to create *JbpmContext* (one per user request).

Implementation of the Gameflow Engine

Gameflow Service Methods encapsulate workflow queries that provide certain functionality. The gameflow engine customizes jBPM to be used for collaborative learning purposes.

The getPendingTasks method queries jBPM to retrieve the list of pending tasks for a participant in the gameflow, by querying jBPM using the getTaskList() method.

The endOfActivity method signals the end of a process task to jBPM. This is accomplished in a three step procedure: (i) it retrieves all the tasks assigned to the user; (ii) it finds out the addressed task by its name, using exhaustive comparison; and (iii) it signals the end of the task to jBPM.

Middleware

The communication between Moodle and the gameflow engine is done by using SOAP messages. Using SOAP middleware, we can rely on mature SOAP engines, both for publishing and consuming Web services.

Gameflow service methods are published in a WSDL file, making use of Apache Axis, a SOAP engine that we use to create a SOAP server. Using the Axis framework, we can easily write an interface definition in Java and, using JavaToWSDL, generate a WSDL file that contains the web service interface.

The binding of Web services with gameflow service methods is done in a class that Axis generates automatically, SoapBindingImpl, and serves as a skeleton to fill with the binding code.

Control transfer from Moodle to the gameflow engine is done by a Web service invocation. We use the NuSOAP library.

The Moodle Part

Moodle queries the gameflow engine with the getPendingTasks() method to retrieve the to-do list for a participant. Then, the participant selects an activity from the to-do list and performs it; at that point, Moodle signals the end of the activity to the gameflow engine.

The invocation of gameflow service methods is done at well-known points inside Moodle (see Figure 6):

Figure 6. Sequence of operations

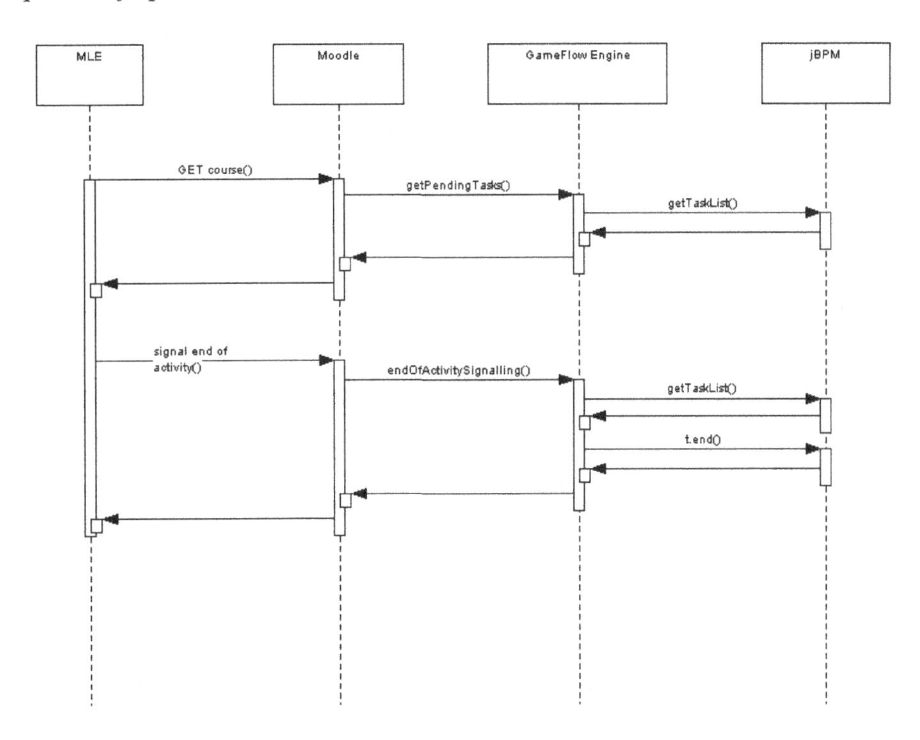

- *Before a user displays the course main page* - The getPendingTasks() service method is invoked to get a list with all the tasks that the user has pending to be done.
- *After a user performs an activity* - The endOfActivity() service method is invoked to signal the end of a task to the gameflow engine.

Using Mobile Apps from MLE

The local applications that must be accessible are declared. In Figure 7, we show how, from the MLE app, a QR Code Reader application is launched, then the participant takes a picture using the mobile camera (launched in this case from the QR app), the picture is decoded, and the code is passed to the MLE app.

Course Authoring and Deployment

The jBPM component provides an authoring tool that is an Eclipse (Eclipse) plugin. The main limitation of using the Eclipse plugin is that a user has to install Eclipse in order to be able to author gameflows. The most user-friendly approach is to offer the authoring tool via the Web. At this point, there are two alternatives: to offer a graphical editing tool, or to offer a form-based editing tool.

It is obvious that to edit processes with a graphical editing tool is more user-friendly. Regrettably, this approach is not feasible with HTML-only pages. It could be possible to offer a graphical tool via the Web either by means of a Java applet or by means of a Flash application.

The other alternative is to offer a form-based editing tool. In this case, the user can author a process by using HTML form controls. The user can define control activities such as *Fork* and *Join*, and the *Start* and *End* activities. In the same way, the user can define transitions between activities.

Figure 7. Sequence diagram to get a QR code

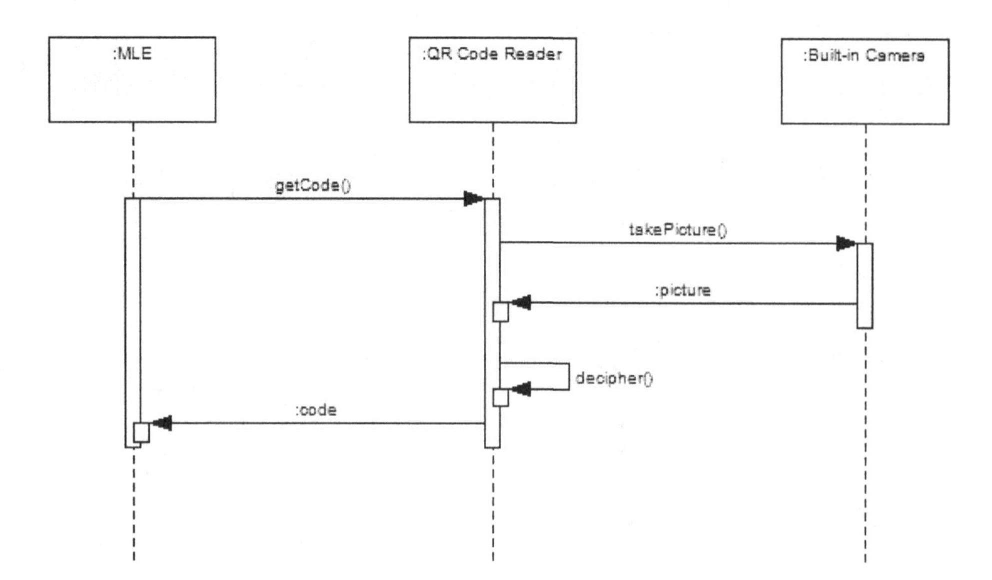

Once the activities and transitions that compose the workflow have been defined, the authoring subsystem creates an XML output with the process definition in the JPDL language ready to be deployed into jBPM.

The importation process is composed of three steps:

- The authoring system creates a valid XML file with the course definition.
- The Moodle system sends the XML file with the course definition to the gameflow service method that deals with deployment.
- When the XML file is received at the gameflow engine, it is deployed into jBPM.

CONCLUSION

As Learning Technologies are being increasingly used both in industry and academia, new technological solutions are required for supporting advanced uses. In this chapter, we have shown how PAISs are the natural candidates for supporting collaborative educational games in mobile Learning Management Systems. Approaches based on ad-hoc processes do not treat gameflows as first-class objects, thus lacking the needed capabilities to create and instantiate gameflows in an easy way. In our approach, collaboration structures are defined with a process definition language and are enacted by a PAIS.

We have explained a procedure to build a gameflow engine by wrapping a generic-purpose workflow engine. This approach is extensible to processes belonging to different domains, which can be translated into a process lingua franca and enacted by a workflow engine.

The proof-of-concept of our approach was built on jBPM and integrated in Moodle. In our opinion, LMSs like Moodle greatly benefit from using a PAIS to control collaborative practices.

Technically, the integration of a PAIS in Moodle was designed following the SOA paradigm. Hence, process capabilities are made available as Web services. This approach can be named gameflow as a service. SOA enables to decouple presentation-related concerns from the business logic, and it is a good approach to integrate gameflow capabilities in mobile devices.

Since we place the gameflow engine as the central component of the system, different LMSs may serve the same collaborative practices using different front-ends, resulting in interoperability advantages.

REFERENCES

Alfresco. (n.d.). Retrieved 2010, from http://www.alfresco.com

Allen, M. (2009). *Palm webOS*. Sebastopol, CA: O'Reilly Media, Inc.

Andrew, P., Conard, J., Woodgate, S., Flanders, J., Hatoun, G., & Hilerio, I. … Willis, J. (2005). *Presenting Windows Workflow Foundation*. Indianapolis, IN: Sams.

Axis. (n.d.). Retrieved 2010, from http://ws.apache.org/axis

Blackboard Website. (n.d.). Retrieved 2010, from http://www.blackboard.com

Bohl, O., Schellhase, J., Sengler, R., & Winand, U. (2002). The sharable content object reference model (SCORM)--a critical review. In *Proceedings of the International Conference on Computers in Education* (pp. 950--951).

Bonita. (n.d.). Retrieved 2010, from http://bonita.objectweb.org

Botturi, L., Derntl, M., Boot, E., & Figl, K. (2006). *A classification framework for educational modeling languages in instructional design*. In 6th IEEE International Conference on Advanced Learning Technologies (ICALT 2006). Kerkrade (The Netherlands).

Breton, E., & Bezivin, J. (2002). Weaving definition and execution aspects of process meta-models. In *Proceedings of the Annual Hawaii International Conference on System Sciences* (pp. 3776-3785).

Burgos, D., Tattersall, C., Dougiamas, M., Vogten, H., & Koper, R. (2007). A first step mapping IMS learning design and Moodle. *J. UCS*, *13*(7), 924–931.

Claroline Website. (n.d.). Retrieved 2010, from http://claroline.net

Cole, J. (2005). *Using Moodle: Teaching with the popular open source course management system*. Sebastopol, CA: O'Reilly Media, Inc.

Dillenbourg, P. (2002). *Over-scripting CSCL: The risks of blending collaborative learning with instructional design*. In Three worlds of CSCL. Can we support CSCL (pp. 61-91).

Dumas, M., van der Aalst, W., & Ter Hofstede, A. (2005). *Process-aware information systems: Bridging people and software through process technology*. Hoboken, NJ: Wiley-Blackwell. doi:10.1002/0471741442

Eclipse. (n.d.). Retrieved 2010, from http://www.eclipse.org

Enhydra Shark. (n.d.). Retrieved 2010, from http://shark.enhydra.org

Farmer, J., & Dolphin, I. (2005). *Sakai: eLearning and more*. In EUNIS 2005-Leadership and Strategy in a Cyber-Infrastructure World.

Fisler, J., & Schneider, F. (2008). *Creating, handling and implementing e-learning courses and content using the open source tools OLAT and eLML at the University of Zurich*. In ISPRS Conference (pp. 3-11).

Hernandez-Leo, D., Asensio-Perez, J. I., & Dimitriadis, Y. (2004). IMS learning design support for the formalization of collaborative learning patterns. In *Proceedings of the IEEE International Conference on Advanced Learning Technologies* (pp. 350-354).

Hernandez-Leo, D., Asensio-Perez, J. I., & Dimitriadis, Y. (2005). Computational representation of collaborative learning flow patterns using IMS Learning Design. *Journal of Educational Technology & Society*, *8*(4), 75–89.

Hollingsworth, D. (1995). *The workflow reference model. WfMC. TC-1003*. Workflow Management Coalition.

Jurado, F., Redondo, M., & Ortega, M. (2006). Specifying collaborative tasks of a CSCL environment with IMS-LD. *Lecture Notes in Computer Science*, *4101*, 311–317. doi:10.1007/11863649_38

Lerdorf, R., Tatroe, K., Kaehms, B., & McGredy, R. (2002). *Programming PHP*. Sebastopol, CA: O'Reilly & Associates, Inc.

Marino, O., Casallas, R., Villalobos, J., & Correal, D. (2007). *E-learning networked. Environments and architectures: A knowledge processing perspective* (pp. 27–59). New York, NY: Springer-Verlag. doi:10.1007/978-1-84628-758-9_2

Mason, R. (1998). Models of online courses. *ALN Magazine*, *2*(2), 1–10.

Mobile Learning Engine. (n.d.). Retrieved 2010, from http://mle.sourceforge.net

NuSOAP. (n.d.). Retrieved 2010, from http://sourceforge.net/projects/nusoap

OMTP Bondi. (n.d.). Retrieved 2010, from http://bondi.omtp.org/1.1

Perez-Rodriguez, R., Caeiro-Rodriguez, M., & Anido-Rifon, L. (2009). Enabling process-based collaboration support in Moodle by using aspectual services. In *Proceedings of 9th IEEE International Conference on Advanced Learning Technologies* (pp. 301-302). Riga, Latvia.

PhoneGap. (n.d.). Retrieved 2010, from http://phonegap.com

Pinelle, D., Gutwin, C., & Greenberg, S. (2003). Task analysis for groupware usability evaluation: Modeling shared-workspace tasks with the mechanics of collaboration. *ACM Transactions on Computer-Human Interaction*, *10*(4), 281–311. doi:10.1145/966930.966932

Sanderson, D. (1997). Virtual communities, design metaphors, and systems to support community groups. *SIGGROUP Bulletin*, *18*(1), 44–46.

Santos, O., Boticario, J., Raffenne, E., & Pastor, R. (2007). Why using dotLRN? UNED use cases. In *Proceedings of FLOSS International Conference* (pp. 88-107).

van der Aalst, W. (1998). The application of Petri nets to workflow management. *Journal of Circuits Systems and Computers*, *8*, 21–66. doi:10.1142/S0218126698000043

Van Es, R., & Koper, R. (2006). Testing the pedagogical expressiveness of IMS LD. *Journal of Educational Technology & Society*, *9*(1), 229–249.

Voorhoeve, M., & van der Aalst, W. (1997). Ad-hoc workflow: Problems and solutions. In *Proceedings of the 8th International Workshop on Database and Expert Systems Applications* (pp. 36 - 40).

Wohed, P., Russell, N., ter Hofstede, A., Andersson, B., & van der Aalst, W. (2009). Patterns-based evaluation of open source BPM systems: The cases of jBPM, OpenWFE, and Enhydra Shark. *Information and Software Technology* .

Workflow Management Coalition. (n.d.). Retrieved 2010, from http://www.wfmc.org

Chapter 11
OpenLaszlo:
Developing Open Rich Internet Applications for Mobile Learning Environments

Chris Moya
Open University of Catalonia, Spain

ABSTRACT

Programming a rich Internet application (RIA) in any Web environment is the goal of Laszlo Systems. The open source software, OpenLaszlo Presentation Server, allows a user to run, on any device, applications that blend to perfection a user-centered design. It facilitates development from the basic levels such as creating forms, menus and other components for a website, up to high-level tasks like focusing on the attention of the user, to easily create, for example, an e-commerce website, a full management back office or a trip booking site, all this using animations comparable to those created with proprietary software.

INTRODUCTION

In this chapter we will introduce, develop, deploy, and execute applications made with OpenLaszlo to be run on mobile devices. We are not going to develop apps for a unique device, so everything developed will run on almost any device equipped with an Internet browser. Although there are ways to run these applications off-line, we will upload them into a web server to make the user experience friendlier and more intuitive. OpenLaszlo is an incredible open source way to make rich applications run with no device restrictions.

The simplicity of its language, its versatility, and the potential of applications generated within this framework make OpenLaszlo a serious po-

DOI: 10.4018/978-1-60960-613-8.ch011

tential competitor in the market of applications for mobile phones.

In this section, we will deal with basic aspects of this language based on XML and JS, from the installation of the software within a Linux distribution, to the aspects of programming in this language, and to the best ways for a programmer to feel comfortable creating high level applications. We will also show how to develop code in OpenLaszlo in an efficient way within an environment in which, as we will see, no application of proprietary software will be necessary.

History

Laszlo Presentation Server was set up in 2001 by a team of 20 developers coming from Apple, Adobe, Macromedia, Allaire, Excite and Go companies. Trying to get involved in the web business, two years later the platform was released as Open Software with the goal of having a free and open framework to create web based RIA applications.

Nowadays, OpenLaszlo has a community with hundreds of developers that are using this software to create incredible web applications and even lending a hand in coding the next releases of OpenLaszlo. When Adobe's Flash emerged, a revolution was started. Before that, designing and developing apps were two different worlds and nobody was expecting this barrier to disappear. Adobe's Flash was released and this scenario changed.

Developers were capable of "seeing" how their work would appear in the final version and, at the same time, designers were introduced to code, making the barrier vanish. After some releases of Adobe's Flash, swf web applications were well designed and coded, making flash content almost essential in any web page that wanted to give users the pleasure of a rich web experience.

Laszlo Systems wanted to take part and give users the chance to make the same content without depending on any plug-in software and, of course, totally free. As a result, we have a huge number of applications created with OpenLaszlo that are at least as good as Flash applications.

At the beginning of OpenLaszlo, the results of our applications were files with the .swf format. We know of the existence of open source players of this type, but let us be honest, we were all using the Flash Player by Adobe, at that time Macromedia. As a matter of fact, it was installed in 90% of computers worldwide as it was the best player of .swf files. Therefore, OpenLaszlo was really an application which depended on another piece of software (Flash).

Ever since OpenLaszlo 4.0, we have been able to obtain compiled codes in DHTML, so all the applications, events and animations that we generate, which were before dependent on the Flash player, now become scripts which will be compiled and executed natively on any browser; that is, we will obtain the same result in an open source manner.

Another strong point of OpenLaszlo executed as DHTML is that, by means of next generation mobiles (iPhone, HTC...), it allows us to execute natively applications which were previously developed on the .swf format. By using OpenLaszlo-DHTML, they will become excellent platforms for the use of our applications.

We are aware of the latest progress of applications in FlashLite by Adobe, but we are not looking at this now, as this solution would correspond to the development of proprietary software.

Let us not forget that now, at the beginning of the year 2010, there are certain aspects of these systems which are still obscure for developers. One of these points is the difference existing in the performance of browsers such as Safari (Apple) or Chrome (Google), and whether they are used on mobile devices or on personal computers. Differences such as the maximum memory size of scripts or the speed of the execution of scripts may have effects on the applications developed for these devices.

We may also find problems, with very specific devices, restricted to the brand, for instance, iPhone, whose browser substitutes events we launch with OpenLaszlo for native events of the browser, causing events of the mouse such as Drag&Drop or Double-click to become events controlled by the hardware and not by the software, as we were used to until now. Also we have to keep in mind that big companies such as Apple or Google nowadays are promoting the use of HTML5 instead of Flash technologies into smartphones. In spite of these inconveniences, the generation of free content and execution on these types of devices may become quite impressive.

Deploying OpenLaszlo

OpenLaszlo provides two ways to deploy applications:

- **SOLO** (Standalone OpenLaszlo Output): The applications are precompiled for any HTTP Web server. It supports data integration through XML over HTTP and drastically simplifies data center requirements and minimizes service costs. When executed on the client, the application contacts other servers directly, without mediation by the OpenLaszlo Server. This "serverless" or Standalone OpenLaszlo Output deployment is called SOLO.
- **OpenLaszlo Server**: Using the OpenLaszlo compiler to "precompile" programs and make the resulting file (in SWF or JavaScript) available on your server. Applications are compiled and cached in a J2EE or Java Servlet Container Environment. This deployment supports applications that require data integration SOAP, XML-RPC or Java-RPC or require persistent connections, among others.

In Figure 1, we see a diagram of the architecture of OpenLaszlo that shows the two ways of working with this technology, as discussed above:

In Figure 2, we see a diagram of the architecture in which OpenLaszlo shows us two ways to work with this technology, as discussed above:

Installation

To proceed with the installation we will have to connect to the OpenLaszlo website (www.openlaszlo.org) and download the OpenLaszlo Server version 4.6.1. The OpenLaszlo Explorer direct link will open a browser in which we will find a large number of examples and guides from a beginner user guide to an advanced user guide. I truly recommend starting viewing and developing the first examples within this web page because the documentation is very accurate.

Since OpenLaszlo Server version 3.3.3, there has been no IDE for Eclipse available, so coding will be a little bit less user friendly because we will have to write code within the a single Open-

Figure 1. Two ways of deploying. ([OpenLaszlo, Inc]. Used with permission)

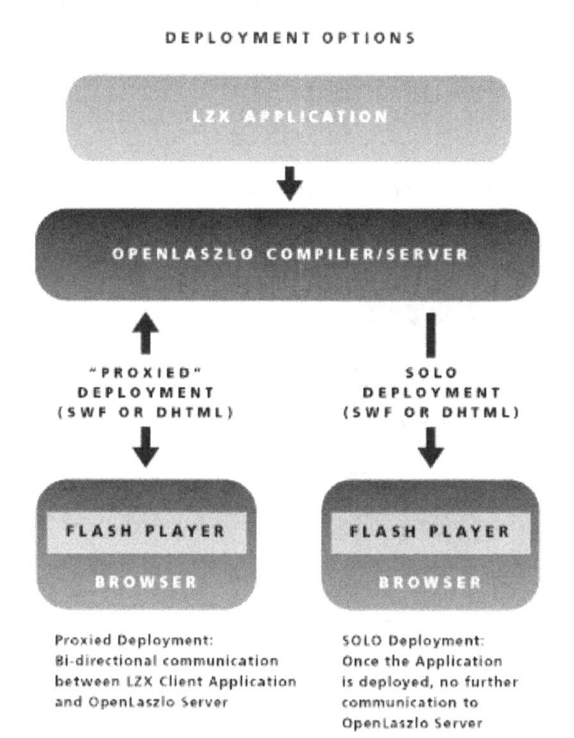

Figure 2. Architecture. ([OpenLaszlo, Inc], Used with permission)

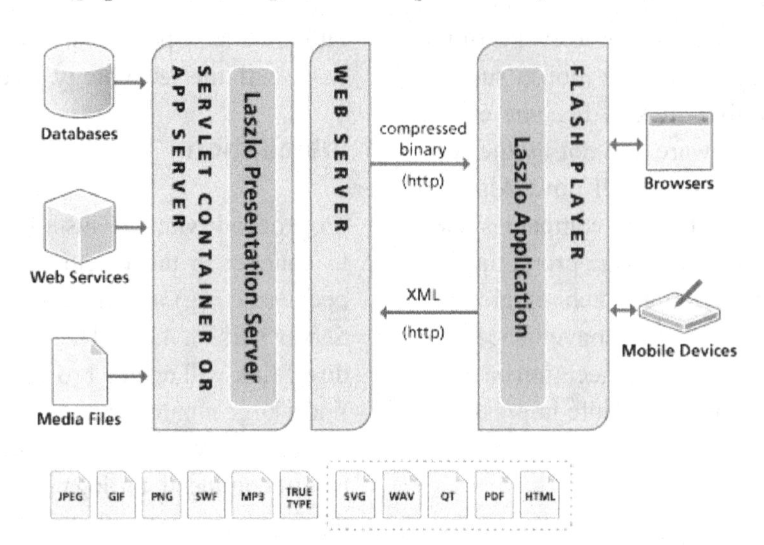

Laszlo Explorer page for non-complex examples or with any text editor for a complex development. Please note that in the official documentation you will find how to configure *vim* editor in order to integrate LZX files correctly.

For an example of coding within the browser using the OpenLaszlo Explorer, try to use any of the examples written and modify them in order to get used to OpenLaszlo syntax (see Figure 3).

As we can see, coding a simple example is easy and highly recommended to get used to the OpenLaszlo way of programming.

In a further case, we will develop a complex example, so using the browser to code will not be a good idea; thus, I recommend using a text editor. As we will see, a structure of folders and resources will be needed to deploy our last example, so structuring our program and distributing classes and resources will be a must (Figure 4).

Figure 3. Example of compiling in browser

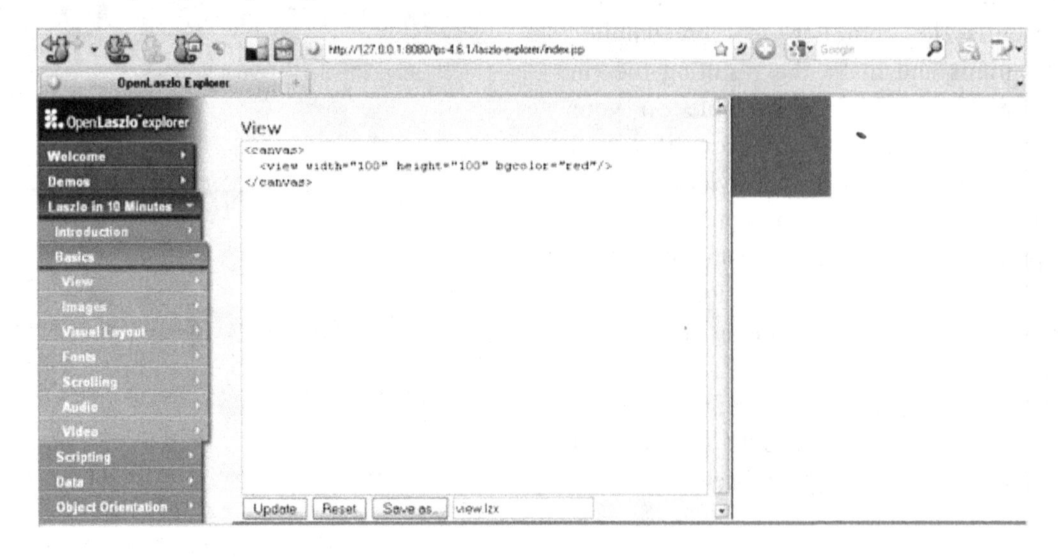

Figure 4. Folder structure

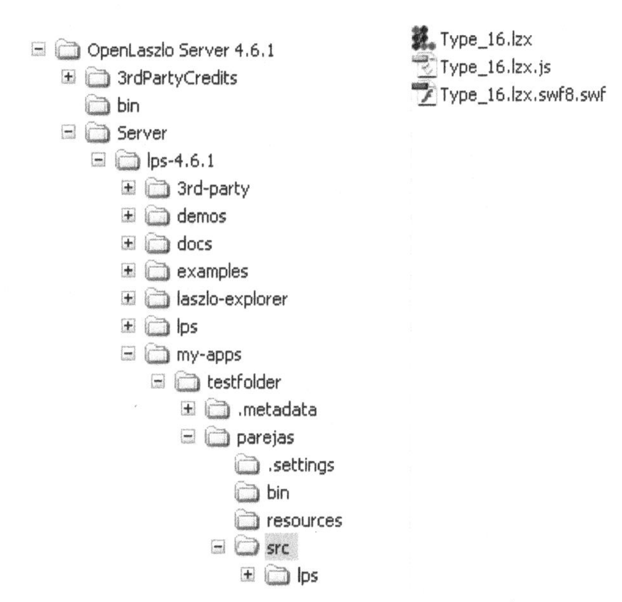

There are two ways we can compile our application: first, we can compile it into a web browser, so the path to it (remember to start tomcat first) will be:

```
http://127.0.0.1:8080/lps-4.6.1/
path_to_your_directory
```

As we can see above, the path to our LZX file is:

```
OpenLaszlo Server 4.6.1/ Server/lps-
4.6.1/ my-apps/memo/src
```

So instead of the URL given when the OpenLaszlo Explorer opens,

```
http://127.0.0.1:8080/lps-4.6.1/lasz-
lo-explorer/index.jsp
```

We will have to write our personal URL, in our case it will be,

```
http://127.0.0.1:8080/lps-4.6.1/my-
apps/testfolder/memo/src/Type_16.lzx.
```

We will look at this issue in more depth in the next sections.

On the other hand, we can compile our application using the standalone compiler, lzc. It is a great way to compile when a big number of applications have to be compiled at once, like using ANT software. Please feel free to visit the official documentation about this kind of compiling.

Basic Aspects (XML, Canvas, js, View, Event Handlers, Run)

All LZX files are written in the XML format, and they will be compiled always as long as they are well-formed. So, the basic structure will be like any XML file but using OpenLaszlo tags. Here we are going to see the basic aspects of principal tags anyone coding in OpenLaszlo must know.

Canvas

It is the container of any tag we will write in our program. It is the main shape where any tag or element will be inserted. It is possible to declare its dimensions in order to create a proper container for our application. This example will give us an application for almost any smart phone's screen.

Its syntax is as follows:

```
<canvas height="480" width="320">
</canvas>
```

View

The *view* tag is the most basic element displayed in an OpenLaszlo application. Anything displayed on screen will be a view. It can be defined with parameters, width, height, background color, opacity, include a resource, etc. This element is strictly hierarchical; a view always has one parent and has multiple views. In order to read more about the usage or parameters of views, please refer to the official documentation.

Its syntax is as follows:

```
<canvas height="480" width="320">
        <view width="100"
height="100" bgcolor="red">
                <view width="50"
height="50" bgcolor="black" />
        </view>
</canvas>
```

Events

These are elements that call to a specific method in order to execute some actions when an event is thrown. They can be any word called at any time when a thread points to it.

```
<view>
        <handler name="onclick">
                // code to be ex-
ecuted when click. In a smart phone,
when view is touched
        </handler>
</view>
```

Attributes

We can define any attribute of instances of classes. An attribute can have any name we want to give to it and use it during run time. We can define three types of attributes, number, boolean, and string. In this example, we will change the "found" property from a 0 value to 1 when a layer is clicked.

Its syntax is as follows:

```
<view name="cell" bgcolor="0xffcc99"
width="80" height="64" clickable =
"true">
        <attribute name="found"
value="0"/>
        <handler name="onclick">
                this.
setAttribute("found","1");
        </handler>
</view>
```

Animations

When compiled in DHTML, JavaScript will give us an application with animations that will behave as a native .swf file. This is one of the most impressive aspects of OpenLaszlo. Let's see an example. This layer will move when clicked from the position 0px to 100px on the x-axis during one second. Note that the duration is set in milliseconds. This is how it works:

```
<canvas height="480" width="320">
    <view width="50" height="50"
bgcolor="black" onclick= "this.anim.
doStart()">
        <animator name="anim"
attribute="x" to="100" duration=
"1000" start="false"/>
    </view>
</canvas>
```

Figure 5. Our first example

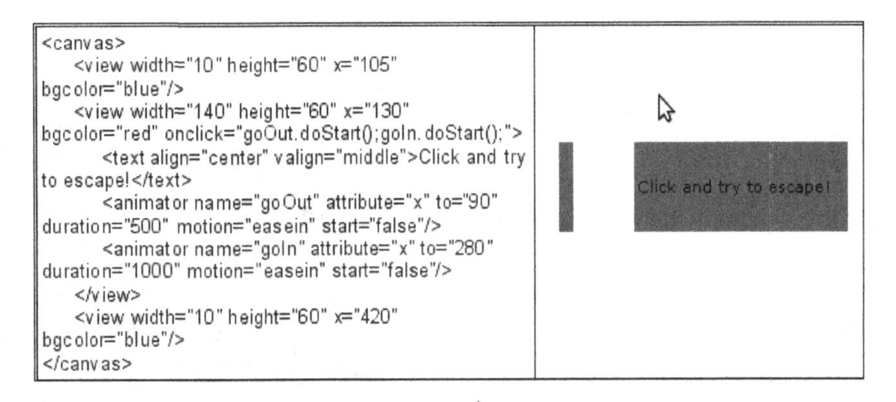

```
<canvas>
    <view width="10" height="60" x="105"
bgcolor="blue"/>
    <view width="140" height="60" x="130"
bgcolor="red" onclick="goOut.doStart();goIn.doStart();">
        <text align="center" valign="middle">Click and try
to escape!</text>
        <animator name="goOut" attribute="x" to="90"
duration="500" motion="easein" start="false"/>
        <animator name="goIn" attribute="x" to="280"
duration="1000" motion="easein" start="false"/>
    </view>
    <view width="10" height="60" x="420"
bgcolor="blue"/>
</canvas>
```


Compiling My First App in DHTML Format

Let's develop our first application to be deployed into a web browser. We will always have to be sure of taking the correct measures of viewport in order to keep an app usable in a device. Not in this first case.

The next example is shown in Figure 5.

OPENLASZLO DHTML: INCLUDE RIA APPS INTO HOSTILE ENVIRONMENTS

Mobile System Apps [Orbit Project]. OpenLaszlo & Java

Sun Microsystems and Laszlo Systems, Inc have joined forces to create a project, Orbit, in which the power of the development of OpenLaszlo may be executed within a JavaTM Platform, Micro Edition (Java ME) application platform. Through this innovation, the result of applications is no longer restricted to the platform and becomes independent from it, with the resulting portability to an endless variety of devices based on the Java technology. Nowadays, there are 1,100 types of mobiles which execute Java natively, with a worldwide distribution of over 3,800 million mobiles. Orbit performs the task of visualizing LZX code, as the Flash player may be for .swf files, in an application that will run under the engine of JavaScript Rhino executed by Java. This will be a very notable issue when released, for now we can only highlight it (see Figure 6).

Figure 6. Sun / Laszlo Project Orbit. ([OpenLaszlo, Inc], Used with permission)

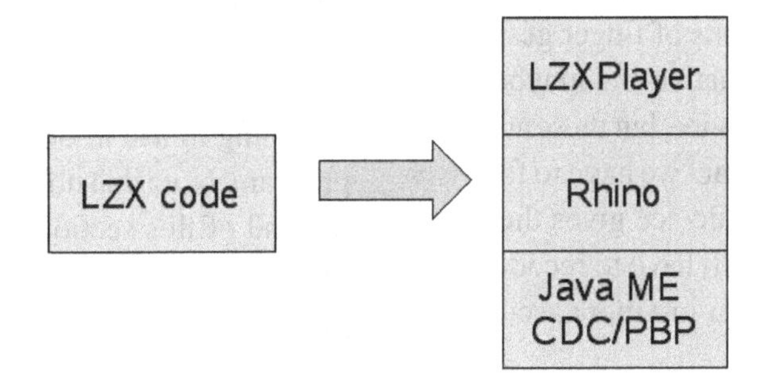

MOBILE WEB APPS

Getting away from Mobile Native's OS Coding, Universal Coding for Any Device

We will dedicate this chapter to the development of RIA, making the most of the possibilities of the browsers implanted in the devices. As we have stated before, they have great possibilities and some disadvantages too. Nevertheless, it is the best way of programming in open source for any mobile. We will explain the different elements that we have for each of the most common browsers or devices, as well as the technical limitations that we will have to face when beginning development.

When trying to develop an open source program for a mobile device, we will have to consider how to implement it if we want to deploy it in as many devices as we can. The Android platform uses a SDK based on Java, a good way to develop, but since the iPhone has over ten million users, we have to choose our target. So, in my opinion, the best way to develop an app is to get as standard as we can, and this is by using the browser. If not, we will have to code in basically two languages, Java and Objective-C, using two different SDKs.

OpenLaszlo, as mentioned earlier, has issued a Dynamic-HTML compiler that will allow our apps to run without the swf player, and this is an incredible solution for accomplishing our purpose. Now, all our apps will run in any smart device with no code property restrictions and without having to put our app into any online store. As Apple is not going to release Flash player for Safari (2009-2010), this will be a great way to face this issue.

There are some limitations of finger gestures that the device will take as native and maybe our solution is blocked by the device, but these are not going to be many. Another fact we have to face is the amount of memory the device gives the app to use and sometimes we will have to recode our app to make it better, faster or just more accurate.

Figure 7. Compressor rate http://compressorrater.thruhere.net

One good way to make js run optimized is to minimize loops, global variables and for-ever-listeners. Once our code is properly done, one way to make the browser faster is to erase any blank lines or comments. The best solution will be to put all our code into one line of text.

In this URL, you will find multiple solutions to recode a JavaScript file (see Figure 7).

Creating an OpenLaszlo App for a Language Learning Course Using the Internet Mobile Browser: Examples for iPhone's Browser, Nexus One's Browser

In this subsection, we will deal with the main steps in the development of an application for the browsers of some devices. Let's start coding!

To start our app, we have to declare the canvas properties and our datapath to an XML file that we are going to use in order to import data into our program. Note that not all the code is included. At the end of this section, there are some QR code links to all the code.

Our aim will be to develop and put into full swing a game based on the famous MEMORY. With this game, we will see how to create animations, user-centered design and accessible content with a few lines of code and using only the OpenLaszlo Framework. The game consists of a number of cells that the user will have to uncover and attempt to recall its pair. However, we would see that animations, alpha channel changes or finger events on the multitouch screen of the smartphone do not involve any extra effort in developing the code. The result may be performed in any browser, without depending on the operating system, as far as a PC, MAC, a notebook or a smartphone. We will just have to adjust the viewport size to the device, which is a big advantage.

```
<?xml version="1.0"
encoding="UTF-8"?>
<canvas height="480" width="340">
<dataset name="exer_XML" src= "../re-
sources/data.xml"/>
```

Note that the XML file points to a folder in an upper level called resources, in which we will also keep our images.

There is an inclusion in which we will declare the cell class that is going to be the main section of the application.

```
<!--  INCLUDES   -->
<include href="../ resources/memo.
lzx"/>
```

We can use the *script* tag to include some JavaScript code. In this case, we will declare some variables for coordinates.

```
<script>
var posYini =15;
var incY = 66; ...
</script>
```

Now we are going to draw some views to set the title and one start button. Note that we are using the *simplelayout* tag which will arrange views vertically. In this section of the code, we can see how to import an image to a view that will contain it, and a *button* tag that has a listener *on click* to execute a method called *Unsort*.

```
<view width="${canvas.width}">
        <simplelayout axis="y" spac-
ing="10"/>
        <text fgcolor="#175DDE">
                <font face="arial"
size="3">
                        <b>Test your
memory!!</b>
                </font>
        </text>
        <view resource="line"/>
        <button id="btn"
text="Begin" onclick= "canvas.
Unsort(TotalCells);"/>
</view>
```

Now we see another view positioned in the y-axis with yproperty. An *animator* tag is used to declare a native method to create animations. As an overview, we can see attributes as duration (expressed in milliseconds), autostart put in false, and the attribute x that will make the view move horizontally from the position 0px to 500px.

Moreover, we see the first declarations of our principal class, cell. We will see more of this in the coming pages.

```
<view id="originLayout"
y="${posYini+incY}">
        <animatorgroup name="linear_
move" process="simultaneous"
start="false" duration="200" >
                <animator
attribute="x" from="0" to="500" />
        </animatorgroup>
        <cell id="C1" Number="1"
```

```
x="${posX1}" y="${posYini}"
clickable="false"/>
        <cell id="C2" Number="2"
x="${posX2}" y="${posYini}"
clickable="false"/> ...
</view>
```

We are also using an animatorgroup named *gameRotationMoveEnd* with a duration of almost one second whose rotation attribute ranges from 0 degree to 360 degrees. This method is launched when all the pairs are guessed and the panel will make an animation consisting of completing a circle movement.

```
<animatorgroup
name="gameRotationMoveEnd"
process="simultaneous" start="false"
duration="950" >
        <animator
attribute="rotation" from="0"
to="360" />
</animatorgroup>
```

Since our LZX file has to be well-formed, do not forget to close the *canvas* tag.

```
</canvas>
```

Now let's see the simple but important XML files in which resources are written, giving us the opportunity to split coding from content. As we can see, there is a *root* tag called EXER which will define all the beginning of the XML structure. This tag has one descendant called PAIRS which will include all the PAIRS of our example. The *PAIR* tag has two attributes, whose values are the names of the images; in this case the file extensions are JPG, but we can import almost any media file to our app.

```
<EXER>
  <PAIRS>
      <PAIR>
          <IMAGE>f0_7_voc_07</IMAGE>
          <MIRROR>f0_7_vob_07</MIR-
ROR>
     </PAIR>...
  </PAIRS>
</EXER>
```

Now let's see the main library called memo.lzx.

As always, we declare the file as XML well-formed and use the *library* tag to declare it as an inclusion that we have referenced before in the main LZX file. We use the *script* tag to declare some variables used in run time.

```
<?xml version="1.0" encoding="UTF-8"
?>
<library>
    <script>
         var couple=1;
         var totalPairs=0;
         var TotalCells=0;
         var firstClick=false;
         var even="";
         var sortElem=new Array(); //
Saves data in correct sorted ...
    </script>
```

Now we can explain the main class *cell*. Declaring the *cell* tag, we can give it some attributes such as background color, width, height and whether this view will be clickable or not. As we saw in previous subsections, we can add unlimited attributes to views or in this case our class called cell.

```
<class name="cell" bgcolor="0xffcc99"
width="80" height="64" clickable =
"true">
    <attribute name="Number"/>
    <attribute name="found" val-
ue="0"/>
```

We will use finger events, but as they are not native in OpenLaszlo, we will use them as if a mouse was being used, declaring an *onMouseUp* method. As known, in an XML document or externally parsed entity, a CDATA section is a section of element content that is marked for the parser to interpret as character only data, not markup. A CDATA section is merely an alternative syntax for expressing character data; there is no semantic difference between character data that is shown as a CDATA section and character data that uses the common syntax in which "<" and "&" would be represented by "<" and "&", respectively.

As seen above, we can modify elements in runtime with the setAttribute method, so if this is not our first click in a cell, we are going to make it visible and make it fully opaque.

```
<handler name="onmouseup">
   <![CDATA[
      ...
         if(firstClick==false){
            this.setAttribute("clic
kable",false);
            this.textOfCell.
setAttribute("visible",true);
            this.textOfCell.
setAttribute("opacity",1); ...
```

A *for* loop will be declared as we do in JavaScript and other languages; in this portion of code, the more interesting thing is to acknowledge the similarity to JavaScript. There are some calls to methods in this code. The most important here is switchOff(firstClickedCell) that will change the opacity of the cell managing an animation. Note Debug.write(pairsCounter); this line will be used for debugging. OpenLaszlo explorer brings a really good debugger in which every code error will be commented and in which we will have the option to evaluate variables and functions in run time.

```
for(var i=1;i<=TotalCells;i++){
   if(unSortElem[firstClickedCell]==
sortElem[i]){
      if(i%2!=0){
         if
(unSortElem[secondClickedCell]! =
sortElem[i+1]){
            if (this.found==0) swi-
tchOff (firstClickedCell);
         }else{
            Debug.write
(pairsCounter); ...
   ]]>
</handler>
```

Method described above, switchOff. See that does an opacity animation and other actions.

```
<method name="switchOff"
args="firstClickedCell">
  this.textOfCell.animate
('opacity',this.opacity = = 1?0:0,
500, false);
  this.setAttribute
("clickable",true);
</method>
```

The next method, which is called when the user makes all pairs correctly, will call the animation described before the gameRotationMoveEnd event and make it run.

```
<method name="endOfGame">
  <![CDATA[
    btn.setAttribute
("text","Begin");
    gameLayout.gameRotationMoveEnd.
doStart();
  ]]>
</method>
```

The following portion of the code is a declaration of a view inside our *cell* class. An event called at the start of run time using an *oninit* statement will import data from the XML file. Notice that an attribute is declared as a datapath, and we use the Xpath language in order to get some information from the XML. First of all, we can know the number of PAIR tags using the parameter length. See how to create a new instance of view class with the OpenLaszlo native method *new lz.view(this,{resource:imageName});*. In this case, we use as parameters of the function the *this* tag that is used to reference the actual view called *textOfCell*, and give it an attribute to include it in a resource (image in this case), with the *imageName* value.

```
  <view name="textOfCell">
    <handler name="oninit">
      <![CDATA[
        mixed=false
        this.
setAttribute("datapath","exer_XML:/
EXER/PAIRS");
        var totalPairs=this.datapath.
xpathQuery('PAIR').length;
        this.
setAttribute("datapath","exer_XML:/
EXER/PAIRS/PAIR["+couple+"]/IMAGE");
        var imageName = this.data-
path.xpathQuery('text()');
        new lz.view(this,{resource:im
ageName});
```

Figure 8. Our app shown in iPhone 3G.

```
        sortElem[count]=imageName;
        count++;
      ]]>
    </handler>
  </view>
</class>
</library>
```

With all this done, we have developed an application as shown in Figure 8.

At this web URL, it will be possible to download all code and resources of the examples shown in Figures 9-11.

CONCLUSION

To conclude this chapter devoted to the use of OpenLaszlo for web browsers, we may say that apps should not depend on the device. Using this framework, we can have an in-depth RIA application design and programming level. This chapter has intended to prove first, that the design and programming do not always need to be seen as two irreconcilable worlds, and second, that the use of standards and open source languages can compete with proprietary software.

ACKNOWLEDGMENT

I would like to thank a number of people who have encouraged me to write this chapter. David Trelles and J.Antonio Recio - both from Open University of Catalonia had helped me along this writing journey. My gratitude goes to Coral Bru, Julio Rico and Kyle Knoblock (MIT) for the patience in the language revision. Not least, perhaps, I should thank my love, Eli Bru, for her wittiness and for giving me the chance to learn lots of new things every day.

Figure 9. QR code: http://www.hostalis.net/mobileLearning/solo_deploy/Type_16.lzx.html

Figure 10. QR code: http://www.hostalis.net/mobileLearning/solo_deploy.zip

Figure 11. QR code: http://www.hostalis.net/mobileLearning/memofiles.zip

REFERENCES

Apple Computer, Inc. (2009). *Mac OSX reference library*. Retrieved February 5, 2010, from http://developer.apple.com/ Mac/ library/ documentation/ Cocoa/ Conceptual/ ObjectiveC/ Introduction/ introObjectiveC.html

Apple Computer, Inc. (2010). *iPhone section*. Retrieved February 5, 2010, from http://www.apple.com/iphone

Denso Wave Incorporated. (2000). *QR code section*. Retrieved March 2, 2010, from http://www.denso-wave.com/ qrcode/ index-e.html

Google, Inc. (2010). *Nexus One section*. Retrieved March 7, 2010, from http://www.google.com/phone

Klein, N., Carlson, M., & McEwen, G. (Eds.). (2008). *Laszlo in action: Rich Web applications with OpenLaszlo*. Greenwich, CT: Manning Publications Co.

Laszlo Systems. (2006-2009). *Official documentation*. Retrieved March 15, 2010, from http://www.openlaszlo.org/ documentation

World Wide Web Consortium. (W3C). (2008). *Extensible Markup Language* (XML) 1.0 (Fifth edition). Retrieved March 19, 2010, from http://www.w3.org/TR/REC-xml

KEY TERMS AND DEFINITIONS

iPhone: Smartphone from Apple that integrates cell phone, iPod, camera, text messaging, e-mail and Web browsing. Data and applications can be sent to the phone wirelessly or via Apple's iTunes software, which is used to organize music, videos, photos, and applications.

JavaScript: JavaScript is the most popular scripting language on the Internet and works in all major browsers, such as Internet Explorer,

Firefox, Chrome, Opera, and Safari. JavaScript's official name is ECMAScript.

Nexus One: Smartphone launched by Google in January 2010. Nexus One uses the Android open source mobile operating system and is manufactured by Taiwan HTC Corporation.

Objective-C: The Objective-C language is a simple computer language designed to enable sophisticated object-oriented programming. Objective-C is defined as a small but powerful set of extensions to the standard ANSI C language. Its additions to C are mostly based on Smalltalk, one of the first object-oriented programming languages. Objective-C is designed to give C full object-oriented programming capabilities and to do so in a simple and straightforward way.

QR Code: QR Code is a kind of 2-D (two-dimensional) symbology developed by Denso Wave (a division of Denso Corporation at the time) and released in 1994 with the primary aim of being a symbol that is easily interpreted by scanner equipment.

SDK: A software development kit (SDK or "devkit") is typically a set of development tools that allows for the creation of applications for a certain software package, software framework, hardware platform, computer system, video game console, operating system, or similar platform.

XML: Extensible Markup Language, abbreviated XML, describes a class of data objects called XML documents and partially describes the behavior of computer programs which process them. XML is an application profile or restricted form of SGML, the Standard Generalized Markup Language [ISO 8879].

XPath: XPath is a language for finding information in an XML document. XPath is a major element in the XSLT standard. Without XPath knowledge, you will not be able to create XSLT documents.

Chapter 12
Open Source LMS and Web 2.0 in Mobile Teaching

Elisa Spadavecchia
Provincial School Authority, Italy

ABSTRACT

Can students learn a foreign language at school meeting their real communicative needs? Is it possible to exploit the potentialities of the 2.0 Web tools and the advantages of the Open Source software to guide students towards effective linguistic competence and autonomy? The chapter describes an experience of using simple Web 2.0 and teaching support online tools for learning English in an Italian secondary school, pointing out the achievements and the drawbacks of the integration of e-learning 2.0 with classroom teaching.

INTRODUCTION

The idea of attending the information technology laboratory regularly during the English as a foreign language class sprang, in a certain sense, from the students themselves. The school where the activities have been carried on, an Italian secondary high school (Liceo Scientifico Quadri in Vicenza, a town near Venice), with a large population of students between 14 and 19 made up of a great number of commuters, many face-to-face activities have been developed to help students in their learning process. This project is situated in a complementary position among the other opportunities offered in the Liceo. Indeed, it is a sort of integration, not an alternative to ordinary teaching practice.

The school is well equipped with such facilities as ICT laboratories, and almost all the students have a computer and an Internet connection available at home. At the beginning of the school year, during some warming up classes carried out in the ICT laboratory, the students asked to attend it regularly during the whole school year. They found

DOI: 10.4018/978-1-60960-613-8.ch012

it much more interesting and involving than the traditional lesson in the ordinary classroom. So, it was decided to exploit the Open Source Learning Management System (LMS) Moodle (https://gibi.liceoquadri.it/moodle) that had been implemented and successfully used for a supplementary course in summer 2008 together with a popular helpdesk carried out on a blog (http://www.sportelloinglese.it) plus a podcast for English as a foreign language (http://www.quadripodcast.it) to integrate, sustain and enrich the learning opportunities offered in the traditional English course. The challenge was to show that the use of ICT was not just a diversion from traditional classroom teaching; on the contrary, an effective use of LMS and Web 2.0 tools could facilitate communication, knowledge sharing, cooperation, and language learning.

After the preparation of the specific courses for different grades, the students started to interact in the ICT lab and could decide freely whether to use the course or not during their studying activities at home. At the beginning, some resistance came from the weakest or the most de-motivated students, but after the first shocking impact of the novelty, dragged by the overwhelming majority, almost everybody enrolled in the course. At the end, only 3 pupils out of the 127 students from the five classes involved in the project have never enrolled, even though they have participated in the online activities with the rest of the class in the ICT lab and have contributed actively to it also through their assessment of the project.

ORGANIZATIONAL ASPECTS

The language activities were developed with a pragmatic approach, starting from the students' needs and suggestions. The technology and resources already available at school were used together with tools and materials freely available on the Internet and some personal materials to enrich on the basis of the learners' needs.

The initial idea was to give the students enough motivation to use the web as an alternative to the textbook to attain specific learning and linguistic objectives. That was a real challenge because students are generally difficult to involve in school learning activities. Normally, after the first curiosity produced by the novelty, a passive attitude prevails, particularly in the classroom. The only interest that really drives them, even though it has much to do with an extrinsic motivation rather than an intrinsic one, is their eagerness to know the marks they get at school. This was exactly what the teacher played on at the beginning of the experience to encourage them to enroll and involve them in the online activities, relying on their curiosity to attain other educational kinds of objectives. If the students wanted to check the web activities their teacher had prepared for them, or to know the marks they got during their learning activities in real time, with the total respect of their own privacy, first of all they had to learn to use the platform and do the activities well. The teacher's hope was that they would do them not only at school because they had to but above all at home. In that way, they could start to consider the positive impact of the use of technologies in their learning, while sitting in front of their home computers downloading music, chatting with friends, playing videogames, and texting messages at the same time.

Another important consideration stemmed from the awareness of how, after the success of the so called Web 2.0 that is characterized by a more and more active role of its users in the production of contents, some authors have started to criticize the distance learning approach based on the exclusive use of Learning Management Systems and foster new types of approach (Cross, 2006). The objective to attain for an effective e-learning is the integration of different kinds of knowledge acquisition, from formal to informal, as it happens in traditional learning. So, it was chosen to integrate the formal experience provided by the online knowledge management and traditional classroom learning activities with others derived from informal e-learning because of the strong motivation that drives towards what has come to be called "e-learning 2.0" (Downes, 2005). E-learning 2.0 requires a new way of considering online learning. It is not a technical question but rather a methodological issue; in other words, it deals with the opportunity of becoming authors in the web as well as readers by means of the creation of blogs and podcasts, photo and document sharing and so on through any kind of social interaction. Thanks to the Web 2.0 tools, the traditional distance learning practice based on the transmission of contents is turned into a more stimulating, appealing interactive process, an aspect of great importance in one's own learning.

THE LEARNING ENVIRONMENT

From the didactic point of view, the integration of formal education as it is provided by LMS and informal learning as it can be found in the cooperative Web 2.0 embraces the principles of Constructivism and Socio-Constructivism. In these theories, cooperative learning is viewed as an interactive process through forms of social integration and negotiation where people learn the one from the other, both formally and informally.

Moodle (http://www.moodle.org) is a system that gives educators a great variety of tools to manage and promote learning. They range from the most traditional teaching methods to the new cooperative approaches based on the constructivist learning theories (Piaget, 1976). Moodle was preferred to other similar Open Source e-learning platforms in the undertaking of the project because it is commonly used in schools and universities all over the world and has features that allow communication, interaction at different levels, and activity tracing. One of its positive characteristics is its ergonomics, in other words, the immediacy and relative simplicity for both the teacher/author and the students in creating and performing the activities provided on the platform. Moreover, with Moodle, a teacher can also monitor the students' actions and connection times.

There are two important factors to consider in the use of this tool from the educational point of view. The former is the observation that school times are often too narrow and concentrated to give enough opportunities to establish an effective pedagogic relationship with the learners. The exploitation of these technological tools allows to extend the learning dialogue to extra-scholastic times, when teenagers are, often alone in their bedrooms, in front of their computers at homework time. When they connect to the Internet to check, for example, if their teacher has uploaded the following day's learning activities they are made aware, at the same time, of the wide range of hidden teaching activities that are performed beyond face-to-face classes. The students and their families often underestimate this submerged work.

The latter is the opportunity of tracing the students' learning process provided by the platform. In Moodle, it is always possible to monitor the users' activities and their connection times. In this way, teachers can focus their attention not only on the final product of their study but also, more importantly, on their learning process as such.

So, thanks to the use of an e-learning platform integrated with the Web 2.0 tools, the notion of the learning environment changes considerably from a traditional classroom situation where there is a strong vertical teacher/learner interaction and the role of the textbook is central. In this case, the Internet becomes the learning place par excellence and the teacher is turned into a facilitator, an assistant on demand who is always available and sensitive to the students' needs. At the same time, the teacher learns to use the technological resources and web tools in a parallel developing process together with his/her own students, thanks to their advice and suggestions.

The position of the learners changes as well. In this learning context, the students use an environment that is generally congenial to them and contribute actively to it also in a cooperative way. Each student can organize his/her own learning time, space, and modalities with great flexibility, activating his/her cognitive style based on his/her own kind of intelligence (Gardner, 1999), different from the other class mates'. For example, during the explanation of a lesson shown on a series of slides uploaded to the platform in the lab, a student can do different things to learn. He/she can focus on the teacher's voice, discuss the topic with him/her, look at the pictures or the key ideas on the slide projected from the classroom overhead projector and, at the same time, go forward or backward to recollect his/her ideas on the slides on his/her own computer screen, download the presentation and edit it by a word processor, copy it down or write notes in his/her textbook or notebook, etc. In other words, a learner is stimulated to react to the stimuli he is getting in an active way on the basis of his own kind of intelligence and learning style.

Horizontal scaffolding, both in the classroom and at home, can be another great advantage for the learners because it cuts down anxiety and helps develop self-confidence.

TEACHING ACTIVITIES

Various kinds of documents were uploaded to the LMS: texts, audio files, songs, video clips, links to websites and presentations, not just video transpositions of papers but materials that try to exploit the theory of multiple intelligences through the new opportunities offered by multimedia (Torresan, 2008). Besides the teaching materials uploaded to the repository of the e-learning platform (https://gibi.liceoquadri.it/moodle), the course was enriched with some posts on the blog (http://www.sportelloinglese.it) and some podcasts (http://www.quadripodcast.it) to allow more informal interaction. Donath (2008) points out that "the task of teachers is to create a multimedia learning environment which means to structure and to organize the learning process" (p. 100).

The students were asked to do the activities provided on the platform individually, in pairs or very small groups of 3-4 students with different roles (coordinator, surfer, secretary, and monitor). The aim of these activities was to favor the development of specific language strategies. Each activity was revised, shared and discussed with the rest of the class and the teacher's guidance. As regards the class-tests, the students used the traditional test only for the checking of their writing skill. The listening and reading abilities were assessed through the use of the school LAN (local area network), while for the oral test, apart from the traditional face-to-face conversation, the classical 'interrogation' between the students and the teacher, videos were shot for the interactive and pair work activities between the students. This activity had a very positive effect on their motivation to do the best they could thanks to the recording effect.

As regards the creation of communicative virtual places, the forum and above all the chat were used. As often happens, the forum soon became a virtual place of formal discussion and sharing of problems and tools for peer and vertical assessment of the activities. The platform provides an e-mail notification of the new posts in the forum. The use of the chat instead, used for quick communication and the making of decisions, met the demands of an informal learning environment where further peer interaction is allowed, something that is particularly appealing and motivating for teenagers, many of whom are passionate and experienced chatters.

By the use of both technologies, the condition for effective e-learning through the integration of the formal and informal dimension was fulfilled.

ASSESSMENT AND FEEDBACK

For the assessment of the projects by the students, two questionnaires were provided using an online questionnaire generator implemented by Prof.

Roberto Trinchero from the FAR (Formazione Aperta in Rete) at the University of Turin. A questionnaire was set to assess the use of the LMS Moodle (http://www.farnt.unito.it/trinchero/qgen/richiama.asp?codice=elspad_coop_2009) and another was planned for the evaluation of the use of the blog and the podcast (http://www.farnt.unito.it/trinchero/qgen/richiama.asp?codice=elspad_sport_podcast_2009). The students were asked to answer a short set of questions, in accordance with the formative plan (POF) of the Liceo as regards the project evaluations, making a choice among four given alternatives. The possible options were very much, enough, little, and nothing at all.

The four questions were the following:

1. How do you assess the validity of the activity with respect to your expectations?
2. How do you evaluate the utility of this initiative on the cultural and teaching level?
3. How do you evaluate the clarity of the expression?
4. How do you evaluate the organization of the activity?

For each answer, the score to be given was 3 points for the evaluation very much, 2 points for the evaluation enough, 1 point for little, and 0 for nothing at all. An open question was also provided at the end of the questionnaire for suggestions and advice on how to improve the service. The average results are shown in the following figures (Figure 1 and Figure 2).

Figure 1. Evaluation of the e-learning platform

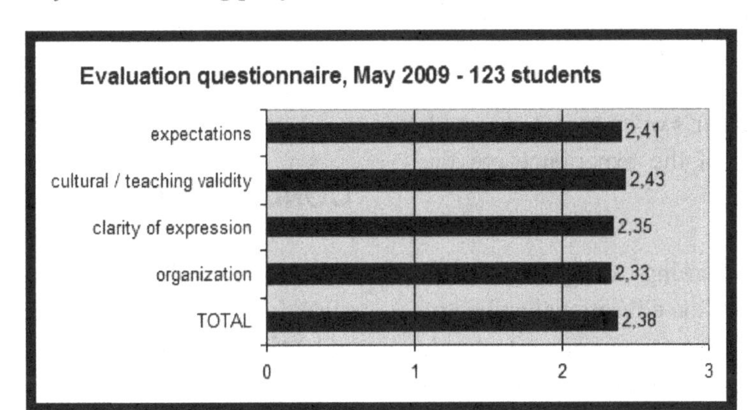

Figure 2. Evaluation of the blog and the podcast

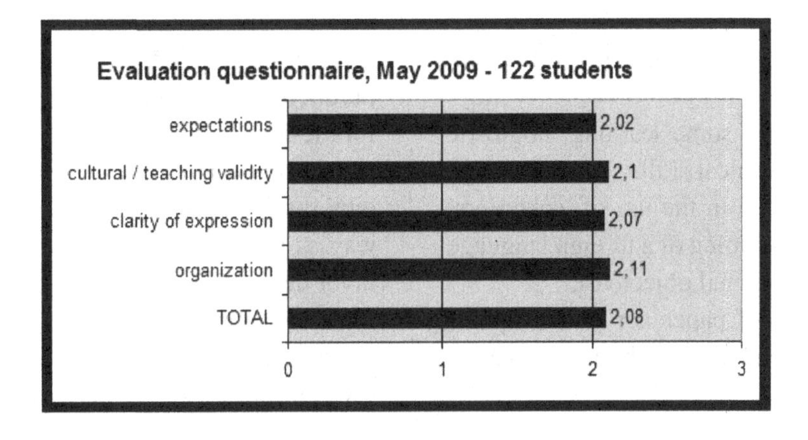

The evaluation of both projects was positive, with various considerations from the students, e.g. about the higher degree of interactivity in the study through the Web 2.0 social tools, the necessity of more links of favorites, the efficacy of studying literature through the exploitation of the LMS, the importance of inserting more colors in the e-learning environment, the establishment of a more positive human relationship between the teacher and learners, the scaffolding between students and e-tutor or among peers for mutual help. Only one student wrote that he preferred "traditional classroom activities where concentration and desire to learn are higher". However, apart from a high level of involvement in the students' use of the technology, curiosity and active participation in the activities, as regards their linguistic competence, there is no precise evidence that the students have learnt the language quicker or better. Comparing learning outcomes obtained with different media can be misleading (Salomon, 2000).

As far as the teacher's opinion is concerned, the positive aspects of the experience are the following:

- efficacy of the learning experience;
- demonstration of the educational and democratic value of the Open Source technology and Web 2.0 tools that allow learning without the necessity of great economic investments or the use of pilot programs;
- validity of cooperative learning in the students' growth;
- higher degree of autonomy in the learning method;
- improvement of some learning, linguistic and/or technological skills;
- more motivation in the use of technology and/or in the learning of a foreign language to attain educational objectives;
- neither waste of paper nor of money for photocopies and the possibility of re-using the same materials adapting them to future similar experiences.

As regards the disadvantages of this kind of activity, the most important points can be listed as follows:

- the necessity of producing teaching materials continuously to keep the students' interest alive;
- a great organizational effort compensated for by a very modest budget, as online supplementary activities are not recognized explicitly by the Italian Ministry of Education;
- difficulty of humanizing the virtual relationship between students and teacher;
- the risk of silence, you have to intervene as soon as possible to arouse the students' interest again;
- students' difficulty in meeting the deadlines.

CONCLUSION

Without pretending to offer a universally valid answer to the problem of the effectiveness of the use of LMS and the Web 2.0 tools in teaching practice, the experience described in this chapter points out that their progressive introduction can represent a further learning opportunity in both teaching and learning. This does not imply that being able to use the new technologies leads automatically to the acquisition of an effective digital competence as the one recommended among the key competences for life-long learning by the European Parliament (EEC, 2006), but the use of ICT can offer some opportunities that can be seized just in the same way as a rich semantic-lexical competence can favor deeper thinking abilities.

The challenge is now many-sided. First of all, it is important to keep the students' interest towards this learning experience alive, perpetuating it in time and adapting it to their educational needs. A more demanding challenge is, on the other hand, the professional recognition of the online work in terms of both quality and quantity by the Italian institutions. It is also important to overcome the e-tutor's sense of loneliness when he/she assumes his/her teaching responsibilities and makes his/her educational choices favor the sharing of teaching practices.

REFERENCES

Cross, J. (2006). *Informal learning, rediscovering the natural pathways that inspire innovation and performance*. San Francisco, CA: Pfeiffer.

Donath, R. (2008). Learning languages in Web.20. *Lend – Lingua e Nuova Didattica, 37*(3), 99-101.

Downes, S. (2005). *E-learning 2.0*. Retrieved March 29, 2010, from http://www.elearnmag.org/ subpage.cfm? section=articles& article=29-1

EEC. (2006). *Recommendation of the European Parliament and the Council of 18 December 2006*. Retrieved March 29, 2010, from http://eur-lex.europa.eu/ LexUriServ/ site/ en/ oj/ 2006/ l_394/ l_39420061230 en00100018.pdf

Gardner, H. (1999). *Intelligences reframed: Multiple intelligences in the 21st century*. New York, NY: Basic Books.

Piaget, J. (1976). *The grasp of consciousness*. Cambridge, MA: Harvard University.

Salomon, G. (2000). *It's not the tool but the educational rationale that counts*. Retrieved March, 16, 2010, from http://www.aace.org/ conf/ edmedia/ 00/ salomonkeynote.htm

Torresan, P. (2008). *Intelligenze e didattica delle lingue*. Bologna, Italy: EMI.

ADDITIONAL READING

Angeli, C., & Valanides, N. (2009). Epistemological and methodological issues for the conceptualization, development, and assessment of ICT-TPCK: Advances in Technological pedagogical content knowledge (TPCK). *Computers & Education, 52*(1), 154–168. doi:10.1016/j.compedu.2008.07.006

Bennett, S., Maton, K., & Kervin, L. (2008). The «digital natives» debate: A critical review of the evidence. *British Journal of Educational Technology, 39*(5), 775–786. doi:10.1111/j.1467-8535.2007.00793.x

Calvani, A. (2009). ICT, Information and Communication Technologies in schools: What rationale? A conceptual frame for a technological policy. *Educational Technology, 49*(4), 33–37.

Dishon, D., & O'Leary, P. W. (1984). *Guidebook for cooperative learning. A technique for creating more effective schools*. Holmes Beach, FL: Learning Publications.

Ferdig, R. E. (2006). Assessing technologies for teaching and learning: Understanding the importance of technological pedagogical content knowledge. *British Journal of Educational Technology, 37*(5), 749–760. doi:10.1111/j.1467-8535.2006.00559.x

Hennessy, S., & Deaney, R. (2007). *Exploring teacher mediation of subject learning with ICT: A multimedia approach. T-Media Project (2005-2007)*. Cambridge, UK: ESRC. Retrieved March 30, 2010, from http://www.educ.cam.ac.uk/ research/ projects/ istl/ T-MEDIA_Fin _Rep_Main.pdf

JISC Development Group. (2006). *Designing spaces for effective learning. A guide to 21st century learning space design*. University of Bristol. Retrieved March 30, 2010, from http://www.jisc.ac.uk/ uploaded_documents/ JISClearningspaces.pdf

Jonassen, D. H. (2005). *Modeling with technology: Mindtools for conceptual change*. Upper Saddle River, NJ: Prentice Hall.

Jonassen, D. H., Howland, J., Marra, R. M., & Crismond, D. P. (2007). *Meaningful learning with technology*. Upper Saddle River, NJ: Pearson.

Kennewell, S. (2005). *Researching the influence of interactive presentation tools on teacher pedagogy*. Paper presented at the British Educational Research Association Annual Conference, University of Glamorgan. Retrieved March 30, 2010, from http://www.leeds.ac.uk/ educol/ documents/ 151717.htm

Koehler, M. J., Mishra, P., & Yahya, K. (2007). Tracing the development of teacher knowledge in a design seminar: Integrating content, pedagogy and technology. *Computers & Education*, *49*(3), 740–762. doi:10.1016/j.compedu.2005.11.012

Margaryan, A., & Littlejohn, A. (2008). Are digital natives a myth or reality? Students' use of technologies for learning. Retrieved March 30, 2010, from http://www.academy.gcal.ac.uk/ anoush/ documents/ DigitalNatives MythOrReality- MargaryanAnd Littlejohn-draft-111208.pdf

Mayer, R. E. (2001). *Multimedia learning*. Cambridge, UK: Cambridge University Press.

Mills, C. J., & Durden, W. G. (1992). Cooperative learning and ability grouping: An issue of choice. *Gifted Child Quarterly*, *36*(1), 11–16. doi:10.1177/001698629203600103

Mishra, P., & Koehler, M. (2009). Too cool for school? No way! Using the TPACK framework: You can have your hot tools and teach with them, too. *Learning and Leading with Technology*, *36*(7), 14–18.

Oblinger, D. (2006). Learning spaces. Washington, DC: EDUCAUSE. Retrieved March 30, 2010, from http://www.educause.edu/ elements/ cdn. asp? id=learning spaces_e-book

Oblinger, D., & Oblinger, J. L. (Eds.). (2005). *Educating the net generation*. Washington, DC: EDUCAUSE.

Oppenheimer, T. (2003). *The flickering mind. The false promise of technology in the classroom, and how learning can be saved*. New York: Random House.

Pedró, F., & OECD-CERI. (2006). The new millennium learner. Challenging our views on ICT and learning. Retrieved March 30, 2010, from http://www.oecd.org/ dataoecd/1/ 1/38358359.pdf

Prensky, M. (1998). Reaching younger workers who think differently. Retrieved March 30, 2010, from http://www.twitchspeed.com/ site/ article.html

Prensky, M. (2001). Digital natives, digital immigrants. *On the Horizon, 9*(5). MCB University Press.

Raymond, E. S. (2000). Homesteading the Nooshpere. Retrieved March 30, 2010, from http://catb.org/ ~esr/ writings/ cathedral-bazaar/ homesteading

Robinson, A. (1990). Cooperation or exploitation? The argument against cooperative learning for talented students. Point-counterpoint-cooperative learning. *Journal for the Education of the Gifted, 14*(3), 9-27 & 31-36.

Ruthven, K. (2004). Teacher representations of the successful use of computer-based tools and resources in secondary-school English, mathematics and science. *Teaching and Teacher Education*, *20*(3), 259–275. doi:10.1016/j.tate.2004.02.002

Spiro, R., Feltovich, P. J., Jacobson, M. J., & Coulson, R. L. (1995). Cognitive flexibility, constructivism and hypertext. Random access instruction for advanced knowledge acquisition . In Steffe, L. P., & Gale, J. (Eds.), *Constructivism in Education* (pp. 85–107). New Jersey: Lawrence Erlbaum Associates.

Tapscott, D. (1999). *Growing up digital: The rise of the net generation*. New York: McGraw-Hill.

Watson, D., & Tinsley, D. (1995). *Integrating information technology into education*. London: Chapman and Hall.

Wetzel, K., Foulger, T. S., & Williams, M. K. (2009). The evolution of the required educational technology course. *Journal of Computing in Teacher Education*, *25*(2), 67–71.

Chapter 13
Mobile Web 2.0:
New Spaces for Learning

Clara Pereira Coutinho
University of Minho, Portugal

ABSTRACT

In this chapter, the author reflects on the emergence of Mobile Web 2.0, a new paradigm for learning in the 21st century, made possible by the combination of a powerful generation of mobile devices with Internet access and the Web 2.0 technologies that allow collaboration, participation, knowledge sharing and construction. The author presents the theoretical framework which sustains learning with mobile devices, and reflects on the potential of Mobile Web 2.0 for the development of informal learning and the construction of personal learning environments. Finally, the chapter presents educational scenarios for the development of mobile learning using Web 2.0 tools, in particular, those made possible using Twitter and m-Flickr.

INTRODUCTION

In the global society of the XXI century, the Internet is not a simple communication technology, but the epicenter of many areas of the social, economic and political activity, becoming, as pointed out by Castells (2004), "as the technological instrument and the organizational means that distribute the power of information, the creation of knowledge and the ability to network in any of the human activity scope" (p. 311). In fact, the evolution of the scientific and technological knowledge have been, in the last decades, structural for new forms of work organization (telework, mobile work, blended work both present and at distance), production and consumption (e-business and e-commerce), communication, new relations with the information and knowledge construction (e-learning, m-learning, b-learning) (Coutinho & Bottentuit Junior, 2009a).

DOI: 10.4018/978-1-60960-613-8.ch013

Mobile technologies did not emerge recently; it is an old concept, but it has been evolving constantly due to the huge possibilities that are being added to the devices. Nowadays, owning a mobile phone is no longer a luxury or fashion; it is a need. Another reason why mobility is very appealing is that these devices allow access to data and information in any given moment or place, becoming powerfully attractive to individuals and organizations (Bottentuit Junior & Coutinho, 2007).

Mobility, portability and ubiquitousness are the three main features of mobile technologies that make them unique for the promotion of effective and quality learning. The school needs to explore alternative classroom methodologies, in order to shorten the gap between the classroom and the real world while improving students' motivation and learning outcomes. We need to open doors to new scenarios, in either formal or non formal education, provided by the mobility offered by a new generation of mobile devices that break time and space boundaries and allow learning to occur in accordance with each individual's cognitive level, as well as any geographical place (Bottentuit Junior & Coutinho, 2008).

The emergence of the versatile, free, and easy to use open-source Web 2.0 applications opens additional reasons for the development of mobile learning. The review of the research by Coutinho & Bottentuit Junior (2009a) shows that Web 2.0 technologies facilitate the adoption of more learner centered pedagogies that allow the students to play a more active role in the educational scenario.

The marriage of the open source Web 2.0 technologies with mobile devices may determine a true revolution in education. Mobility and disruptive Web 2.0 tools democratize the learning environments, challenging the relations of power between teachers and students (Valentim, 2009). As long as more and more users have the Internet in their pockets, learning becomes more and more mobile and the future is mobile learning 2.0!

In this chapter, we reflect on the emergence of mobile learning 2.0, a new learning environment for the XXI century, made possible by the combination of a new generation of mobile devices with Internet access and the Web 2.0 technologies that allow collaboration, participation, knowledge sharing and construction. As stated by Milrad (2006), "the rapid development of these latest technologies combined with access to content almost from anywhere and every time, allows learners to experience new situations regarding learning in a variety of situations and not only in the school settings" (p. 28).

In this chapter, we will consider new educational scenarios for the development of m-learning 2.0, in particular those made possible with Twitter and m-Flickr.

CONCEPTUALIZING MOBILE LEARNING

The discussion around the establishment of a definition for mobile learning is still emerging and is far from being consensual. In fact, according to Winters (2006), the discussion on the concept of mobile learning can be organized around four broad categories of perspectives:

- the technocentric point of view, dominant in the literature that associates mobile learning to the use of mobile devices;
- the e-learning perspective that considers mobile learning as an extension of e-learning;
- the perspective that regards the place of mobile learning in relation to different forms of formal and informal education;
- the learner-centered approach that emphasizes the linkage between mobile learning and mobile devices to focus on the mobility of the learner.

Table 1. Relationship between the mobility of the learner and the portability of the technology

		Mobility of the learner	
		Conventional classroom	Outside the conventional classroom
Portability of the technology	Fixed technology	No Mobile Learning	Mobile Learning
	Portable technology	Mobile Learning	Mobile Learning

Like Traxler (2007), we believe that m-learning integrates any and all the above features allowing for learning to be delivered "just in time, just enough and just-for-me (...) emerging (m-learning) as an entirely new and distinct concept alongside the mobile workforce and the connected society" (p. 14). New forms of online learning are now possible due to the evolution in the mobile devices that allow companies to sell at affordable prices personal mobile computers that combine in the same device the functions of the phone, camera, media player and multimedia. On the other hand, due to the added facilities in communication (voice, SMS, MMS, video conferencing, Messenger), m-learning includes new forms of collaborative work supported by disruptive technologies that "allow for bridging the on and off campus learning context facilitating real world learning" (Cochrane, 2008, p. 177).

According to Valentim (2009), what defines the originality and gives consistency to mobile learning as an independent area of study, distinct from conventional e-learning, is not the technology but the emergent portability: "The new features are not radio, nor television, nor book, audio, video or digital camera, nor even the computing capacities; what is really new is that, through digital technologies, all those features suddenly emerge integrated and convergent, at the reach in the pocket of any ordinary citizen" (p. 5). Table 1, adapted from Vavoula et al. (2005, p. 22), shows the relationship established between the mobility of the learner and the portability of the technology that defines and individualizes m-learning.

For Sharples et al. (2006), the need to consider the specificity of mobile learning is a natural consequence of the emergence of a new global digital communication society that enables new forms of learning to occur in different contexts, adapted to the requirements both of the individual learner and the community. For the authors, more important than a definition for the concept is the need to define a theory for education in a mobile society, considering both the technologies that support mobility (mobile phones, portable computers and personal audio players) and the mobility of people and knowledge.

According to the authors, such a theory should consider a set of criteria that are important, in our perspective, to understand the essence of the problem. The first requisite is admitting that this form of learning is different from other forms of apprenticeship due to the mobility of the learner; considering these particular features allows for us to view learning from a different point of view (Sharples et al., 2006):

By placing mobility of learning as the object of analysis we may understand better how knowledge and skills can be transferred across contexts such as home and school, how learning can be managed across life transitions, and how new technologies can be designed to support a society in which people on the move increasingly try to cram learning into the interstices of daily life. (p. 2).

The second requisite is the need to consider alternative forms of learning that occur outside the classroom, in sum, all forms of informal and non-formal learning that are organized and personalized according to the learner interests and needs. The notion of Personal Learning Environment (PLE), as conceptualized by Attwell (2007), refers to a kind of e-portfolio extension which shows the abilities and accomplishments of the student and it is from there that the student presents their professional qualifications. It is constructed by an individual and used in everyday life, for learning. It is not an application or a system but a personal assemblage supporting new learning modalities, induced by ubiquitous technologies and social software. According to Henri, Charlier, and Limpens (2008), from the technological point of view, ubiquitous computing allows learning to take place almost everywhere, through wireless and GSM (Global Systems for Mobile Communications) networks and mobile communication devices that are able to access the Internet. Henri, Charlier, and Limpens (2008) consider that new forms of learning can emerge from the use of the PLE:

Because the same technologies are used in the different context of our life, work, home, school, it could be possible to mobilize what has just been learned and apply it in the context it could be used (transfer of knowledge). Additionally, social software, predominant in PLE, represents technological development that allows people to connect and collaborate, and to create and share (p. 3767).

On the other hand, according to Attwell (2007), PLEs have potential for more meaningful learning by facilitating reinvestment of knowledge in different contexts:

There is a major issue in that everyday informal learning is disconnected from the formal learning which takes place in our educational institutions... Personal Learning Environments have the potential to bring together these different worlds and inter-relate learning from life with learning from school and college. (p. 4).

For Valentim (2009), personalization in the access to information is perhaps the most relevant feature of the *homo digitalis*, the one that raises media from the mass condition of information age to the individualized needs of a post information age where *being digital* (Negroponte, 1996, p.173) means access to contents and knowledge at any time and from any place, according to individual needs. “The challenge is not to have access to information but to filter relevant information and technology; more than a diffusion device, is called to play an important role as a provider of efficient criteria of content selection” (Valentim, 2009, p. 6).

The third requisite: A theory for mobile education must consider pre-existing apprenticeship theories that explain successful learning outcomes. Like Milrad (2006), we believe that “in order to design innovative educational practices it is necessary to take an integrative perspective to technology-enhanced learning where pedagogy and learning theory are the driving forces rather than mobile technologies” (p. 30).

According to Valentim (2009), the design of instructional activities with mobile devices must take into account three sets of pedagogical dimensions: individual/constructivist, social/collaborative and contextual/situated. Some authors (Sharples et al., 2006; Cochrane et al., 2009) consider that the Vygostkyan social-constructivist pedagogy approach to learning as well as the Laurillard conversational model (Laurillard, 2002), is the basis to sustain a conceptual framework that guides research and development in the field of ML. Others concentrate the attention on the development of pedagogically designed learning contexts and consider the importance to establish authentic learning environments as a key factor for success on m-learning, according to Herrington & Herrington (2007) and Herrington et al. (2009), the use of authentic contexts (reflecting

Table 2. Learning theories that inspire mobile learning

Learning Theory	Authors	Type of activities
Behaviorism	Skinner, Pavlov	Drill and practice
Contructivism	Piaget, Bruner, Papert	Participative Simulations
Situated learning	Lave, Brown	Case based learning
Collaborative learning	Vygostky	Mobile collaborative learning
Informal learning	Eraut, Engelstrom	Informal learning
Personalized learning	Attwell, van Hammerlen	PLE (Personal Learning Environments)

Table 3. Convergence between constructivism and mobile devices

Constructivist Learning Principles	Mobile devices
Learner centred	Personal
Active	Include data gathering and interactive devices
Constructivist	Wireless communication and easiness of availability to data allow interpretation of results from different sources
Intentional	Ubiquity. Adaptation to different contexts.
Authentic	Exploration of information in real contexts.
Cooperative/collaborative	Connectivity. Support rich multimedia peer to peer and tutor interactions.

the ways we use knowledge in our daily life), authentic activities (complex real problems) and authentic assessment (matched to the ability of the learners). As defended by Valentim (2009), all relevant pedagogical theories under different points of view, inspire the design and development of activities for mobile learning (see Table 2).

From behaviorism, the idea remains that a learning activity must promote changes in learners´ behavior; from constructivism, the learner must have an active role in the learning process; this also means that learning activities must take place in authentic contexts for effective knowledge construction to happen; on the other hand, learning activities should promote collaboration as a social practice and a process where interactions are of fundamental importance; and finally we need to promote activities that support learning to occur outside the traditional learning environments and formal curriculum. On the other hand, an analysis, from the "type of activity" point of view, shows three types of affordances for mobile devices: a) as systems that facilitate dialogue in face-to-face classroom activities (increasing the immediacy of communication and collaboration), b) as instruments that enhance participative simulations, and c) as tools for collaborative data collection outside the classroom, for instance, when students visit a museum or a botanic garden (Valentim, 2009).

According to several authors (Cochrane, 2008; Cochrane et al., 2009; Torrisi-Steele, 2006), the particularity of mobile technologies enhances the development of constructivist learning activities where the learner is in the center of the pedagogical scenario as shown in Table 3 (adapted from Torissi-Steele, 2006).

Finally, the fourth requisite: A theory for mobility in the Web 2.0 era must take into account newer conceptual frameworks in particular those defended by connectivism, a set of ideas that emerged recently coined by George Siemens and

Stephen Downes. According to connectivism, knowledge is distributed across an information network and can be stored in a variety of digital formats (Siemens, 2008a). Since information is constantly changing, its validity and accuracy may change over time, depending on the discovery of new contributions to a specific subject. On the other hand, one's understanding of a subject will also change over time. Connectivism stresses that two important skills that contribute to learning are the ability to seek out current information, and the ability to filter secondary and extraneous information. In other words, "The capacity to know is more critical than what is actually known" (Siemens, 2008b, p. 6). The ability to make quick and wise decisions on the basis of information that has been acquired is considered integral to the learning process and is enormously enhanced if we use mobile devices that allow us to be connected anytime and anywhere. If computers are "tools that help to learn" (Jonassen, 2007), for connectivists, mobile devices can be seen as "tools that help to make significant connections", because to learn is a question of establishing nodes inside the learning ecologies available on the Web (Siemens, 2008b).

As to research, studies on m-learning have increased significantly in recent years both in number and diversity of themes and methodological frameworks. Based on the literature, specially on the works of Trifonova, (2003), Naismith et al. (2004), Kukulska-Hulme & Traxler (2005) and Traxler (2007), six big areas of interest can be identified in the research published on mobile learning: a) the (dominant) technocratic point of view: the use of the innovative technologies in educational settings to demonstrate its potential; b) the e-learning at a small scale: the use of mobile devices in traditional e-learning models; c) learning in a connected classroom: the use of mobile technologies in a traditional face-to-face classroom to explore its potential for collaborative learning activities; d) mobile informal, personalized and situated learning; e) performance enhancement: the use of mobile devices for training; f) mobile learning in rural and remote areas where fixed infrastructures are unavailable. An overall analysis shows that, although research on mobile learning seems promising, it is already in an early stage of development with the dominance of exploratory studies that primary focus on the technological innovations is not anchored in consistent theoretical frameworks that sustain the adoption of mobile technologies in educational contexts.

THE EMERGENCE OF MOBILE WEB 2.0

In Rosen's (2006) opinion, every ten years new technological trends emerge: In the 1970s, mainframe computers appeared; in the 1980s, customer-server technology; in the 1990s, the Internet, and in 2000 onward, Web 2.0 was developed.

For Stephen Downes (2006), Web 2.0 is much more a social than a technological revolution due to the new position and attitude of those accessing and using the network. In fact, the new version of the Web is characterized by: a) focus on the contents; b) independent publication of contents created by the user; c) network effects due to participation-based architecture; d) social ware or collective user intelligence as a result of the contribution and shared experience among users with common interests.

Web 2.0 tools are simple to use resources, do not need installation or constant maintenance, but allow users to publish and share contents with the rest of the academic community and the world. The Web as a platform, with tools that do not require great storage capacity of the hardware, emerges as the great benefit for the development of a new generation in mobile learning that we call mobile-learning 2.0. On the other hand, according to Coutinho and Bottentuit Junior (2009a), Web 2.0 tools,

Encourage new ways of communication, expression and interaction, as well as enrich pedagogical practices, with activities such as: cooperative and collaborative work, writing stimulation, interactive and multidirectional communication, increased ease of use in data storage, creation of online pages, the creation of practice communities. (Coutinho & Bottentuit Junior, 2009, p. 20).

The parallel evolution in mobile devices to the 3G generation phones with technological infrastructures that gather, at the same time, mobility of the learner, ubiquity and connectivity makes it possible for permanent access to the Internet that is carried in the learner's pocket (Moura & Carvalho, 2007; 2008a; 2009) allowing that "learning goes mobile" (Talagi, 2008). In this context, the concept Mobile Web 2.0 emerges to define the interaction between mobile phones and Web 2.0 applications, a powerful alliance that makes sense and supports the enhancement of face-to-face and distance teaching and learning activities by the use of wireless mobile devices (Cochrane, 2008).

For Cohrane (2008), "The smartphone's wireless connectivity and data gathering abilities (e.g. photo blogging, video recording, voice recording, and text input) allow for bridging the on and off campus learning contexts – facilitating real world learning" (p. 177). According to the author, it is the potential for mobile learning to bridge pedagogically designed learning contexts, facilitate learner generated contexts, and content (both personal and collaborative), while providing personalization and ubiquitous social connectedness, which sets it apart from more traditional learning environments. Besides this, the disruptive nature of Web 2.0 technologies and mobile devices (Sharples, 2000, 2002, 2005) enhances the change to learner centered pedagogies, based upon social constructivist pedagogies democratizing the learning context and challenging the relations of power inside the classroom (Moura & Carvalho, 2009).

Valentim (2009, pp. 40-41) systematizes the special attributes of Web 2.0 technologies that made possible the emergence of a new generation in mobile learning:

- Web based tools are suitable for mobile devices that have small memory storage capacities;
- tools like Twitter or Jaipu that operate in convergence with SMS enhance additional conversational facilities and allow the creation of informal virtual communities with similar interests;
- the end on the dependence on new software packages and updates in parallel with the emergence of the philosophy of distribution through widgets (supported by iPhones and Opera Mobile, p.e.), allow easy access to data and applications;
- the possibility to accede to the same contents and functionalities independently from the technological devices (existence of mobile versions of the conventional desktop software, for instance, the Facebook mobile, the Mobile Google Maps, etc.).

PEDAGOGICAL AFFORDANCES OF MOBILE WEB 2.0

However, one must not think that the future of Mobile Web 2.0 is an ocean of opportunities. It is a true challenge for teachers and educators to take advantage of all those potentials for promoting activities that enhance effective learning with mobile devices and Web 2.0 tools. In fact, the establishment of authentic learning environments using mobile Web 2.0 is not easy to carry out as Laurillard (2007) points out,

M-learning technologies offer exciting new opportunities for teachers to place learners in challenging active learning environments, making their own contributions, sharing ideas, exploring, investigating, experimenting, discussing, but they

Table 4. Pedagogical affordances of Mobile Web 2.0

Design of the Activity	Objectives	Mobile devices preferential affordances	Examples of Web 2.0 applications
Assimilation **Dissemination** **Exposition** **Access**	To process, manage and structure information	UMTS, Wi-Fi, TV Streaming, radio	GoogleDocs Zoho Mobile Bookmarking Podcasting Ebooks Blinx Bloove
Adaptation **Reinterpretation**	The environment changes with the learner input Simulations/roleplay	Tactile screen Sensors Electronic bussel	Second Life Mobile Layar Augmented reality Conceptual maps Games
Communication **Discussion** **Reflection** **Sharing** **Collaboration**	Dialogue	Voice and script recognition, video in/out, voice in/out, teleconferencing, push to talk, SMS, microblogging	Eportfolios Blogs/vblogs Wikis OneNote Mobile Winsite Bluepulse Qipir EQQ12 Twitter Woopho
Production **Demonstration** **Elaboration** **Registration**	Learners produce/create something	Photo and audio registry, MMS	Youtube Mobile Flickr Mobile Trackr!
Experience **Discover** **Explore** **Relate**	Interactive activities focalized on problem solving	GPS, RFID, Bluetooth	Layar GMaps Mobile Wherigo mySKY

cannot be left unguided and unsupported. To get the best from the experience the complexity of the learning design must be rich enough to match those rich environments (p. 174).

However, as pointed out by Cochrane (2008), to create successful mobile Web 2.0 learning environments "requires careful planning, appropriate pedagogical design, and plenty of technology support. However the outcomes of student engagement, increased motivation and productivity, and the integration into academic teaching staff's pedagogical toolkits are worth it" (p. 184).

In fact, to design and propose learning environments with mobile Web 2.0 technologies is a true challenge which only depends upon the creativity of the teacher (Cochrane & Bateman, 2010). The pedagogical possibilities are innumerous and, in many cases, mobile devices have preferential affordances to develop learning strategies that can be adapted to a wide array of learning styles and pedagogical contexts, as suggested by Cochrane & Bateman (2010, online) or Valentim (2009, p. 54) based on a review of literature (see Table 4).

For Cochrane, Bateman & Flitta (2009), if "Mobile learning technologies provide the ability to engage in learning conversations between students and lecturers, between student peers, students and subject experts, and students and authentic environments within any context" (p. 143), Web 2.0 tools formatted for use with mobile devices facilitate these activities and open added possibilities for ICT integration into the curriculum.

According to Valentim (2009), mobility is central for the Web 2.0 generation, or even more, it can be seen as part of the emergent Web 3.0 paradigm. In fact, if Web 2.0 was coined by the discovery of the power of collective intelligence, Web 3.0 is the new emerging paradigm where information is contextual and organized through workflow tools as supporting infrastructure. The purpose of this intelligent, new generation Web is to classify information in categories in a standardized way so that it is easy to access required information; discussion is already taking place on search systems that are operated through the human voice or even through the search for similar pictures, by means of digitalization (Teten, 2007).

For Valentim (2009), the semantic Web will become more and more mobile as it increases the number of people who use connected mobile devices. In fact, according to the author, "The ubiquity and omnipresence of mobile devices make them the first truly personal mass media. Web 3.0 recognizes contexts and this is a unique feature essential for education as it allows the development of personal learning environments by the recognition of individual actions and movements in time, made possible by the aggregation of sensors that allow a semantic dialogue optimizing the way we manage time" (Valentim, 2009, p. 42).

EXAMPLES OF MOBILE WEB 2.0 LEARNING SCENARIOS

The number of Web 2.0 tools is nowadays uncountable and every minute a new resource appears requiring increasing attention from teachers and educators in a sense of finding ways of integrating these new resources in their teaching practices (Coutinho, 2009a, 2009b, 2009c; Coutinho & Bottentuit Junior, 2009b). The development of new applications, accessible through mobile devices, motivates the development of innovative educational experiences, which require testing and evaluation, as in both of the cases we come to present involving the use of Twitter and m-flickr.

Twitter's Potentialities

Twitter is a web-based social network in the format of a microblog that allows the user to create a set of short messages (140 characters) which can be updated at anytime, with the use of a computer connected to the Internet or a mobile phone with SMS (short message service). By publishing a message in Twitter, anyone can instantaneously view its content, as well as join the users' webpage, in order to automatically receive all of the updates.

Twitter is not just another socialization website, since it has one differential. This differential resides in the capability to communicate in different forms and manners with the peculiar aspect which few media can provide: real time communication. (Figueiredo & Refkalefsky, 2009, s/p).

This technology appeared in 2006; however, it was only in 2009 that it became popular among Web users. It is mostly used by teenagers, artists, reporters, etc. Its big expansion may be explained by the ease in exchanging messages between its users, since by owning a mobile phone it is possible to send/receive contents directly to/from the Web without the need for Internet access.

According to Martins, Gomes, & Santos (2009), "its use as an educational tool may attract the students' interest in what concerns the interaction and easy content assimilation" (s/p). Among other Twitter educational capabilities we can point out:

- The creation of a discussion forum over a specific subject;
- The ability to send messages with questions to the teacher at any given time;
- The knowledge and digital contents shared with other classroom students;

- The ability to extrapolate the discussion initiated inside the classroom to a wider discussion with all the Web users;
- The possibility to meet people with different points of view, as well as from different geographic places;
- The automatic sending and receiving of messages or notes to all the users associated to a time and space-independent Twitter.

These are only some examples of the use of this tool in educational context. Despite the existence of countless potentialities, there are still few studies implemented in the field. In this sense, we can refer the investigation carried out by Martins, Gomes, & Santos (2009), where they question the potential of this tool as a way to disseminate and exchange contents, by members related to the promotion and advertising discipline, with special focus to teaching in the social communication studies.

Mobile Flickr Potentialities

The Flickr website was developed in 2004 by Ludicorp in Vancouver, Canada, and incorporated by Yahoo in 2005, having as one of its developers Cal Henderson (Cruz, 2008). According to Lisbôa, Bottentuit Junior, & Coutinho (2009), Flickr can be considered as one of the most exemplary tools of Web 2.0, by allowing a high level of interactivity, which grants users the opportunity to comment or cast judgement on the images available there. It also allows categorizing archives through the means of tags, therefore making the search process easier. According to Freitas (2008, online):

Through this website, each user can show his pictures (in private or public), get in touch with different other photographers (amateurs or professionals), share his experiences and post comments over pictures of other members worldwide. The user can rate his photograph as he wishes and can also participate and create groups with specific purposes, joining people who wish to share pictures regarding the same subjects and interests. The groups have the most diverse objectives, which includes both the interest in establishing network contacts and sharing.

What is understood is that the way to save pictures, in digital technologies, is being consolidated in virtual spaces (websites), which combine a storage system with social network publishing and folksonomy, offering a choice of options in order to develop reading and interpretation in different knowledge areas.

Flickr does not have commercial purposes and, through the form it is presented, gathers an infinite number of participants, with the main goal to create a network of relationships between people who have the same interests. As so, any person can participate in this social network and for that it is only required to access the website through the website address (http://www.flickr.com), open an account and then publish the pictures.

More recently, with the development of mobile technologies, Flickr made an adaptation of its website in order to allow access through mobile devices and, in this sense, the mobile Flickr or simply m-flickr, was created. This technology allows the users who own a mobile phone with Internet access to send and access pictures from any place.

According to Moura & Carvalho (2008b,), m-flickr is a resource which "can help to support the development of both individual and collaborative learning activities, sharing images and data collection. The ability to share images with other learners anywhere and at any time is an added value of this service in the field of education" (p. 216).

Considering these principles, the teacher may use m-flickr as a pedagogic resource in the classroom:

- In the History and Geography discipline, the students can describe historical facts through the use of images. The teacher can motivate the students to post comments and considerations, contributing therefore to the development of the knowledge network;

- In the Portuguese language discipline, they can use the images to create textual productions, also promoting a greater diffusion of the students' productions within the school community;
- In the visual and technological education discipline, they can take pictures of a certain space and study the quality aspects of an image;
- In the science discipline, the image may favor the comprehension of concepts, categorizing them according to the theoretical knowledge;
- In mathematics, one can study the geometric forms through pictures taken by the students in their daily routine;
- Visual perception – the students can get to know several places and cultures which are sometimes distant, but which, in other situations, may be part of their own reality;
- To diversify the forms of presentation of the classroom work, with real world images;
- In order to develop abilities and competences in the English language, the teacher may ask the students to comment on the images available in the m-flickr, thus aiming at the practice of writing and interpreting.

This tool's pedagogic potentialities are countless. However, it is up to the teacher, in the scope of the presented reality, to choose the best strategy to use, taking into consideration of the target audience, the goals and the purpose of the curricular subject. The case we present in the next paragraphs is an example of a successful use of the tool in a 10th grade class of Portuguese Literature classes.

The study was developed by Moura & Carvalho (2008b) to evaluate the influence of mobile learning and the engagement of students in the fulfilment of tasks using the mobile phone and the Mobile Flickr application, in Portuguese Literature classes. It regards a fieldwork which had as a main subject the baroque and was carried out in the city of Braga, located in northern Portugal, with a sample of 15 students from a secondary level public school. The task proposed to the students consisted in taking several pictures from the city's different baroque monuments, using their own mobile phones and sending the pictures over to Mobile Flickr through their mobile gadgets. Besides this, the students were asked to prepare information on each of the monuments and discuss the characteristics of the architectural style. The experience revealed the students' involvement and enthusiasm, with all having enjoyed doing the activity, and thus helping the students to become more aware of the mobile phone's potential in the educational process and its implications for teaching and learning. The downsides pointed out by the participants were: a) the expensive nature of the service, even if there was an overall great satisfaction from the students, and b) sending images to be published in the Web, with the use of mobile phones, is still a slow process. These limitations may represent an obstacle to the generalized use of web-based services for mobile phones, but the main benefits of this experience were to create opportunities for the students to develop their artistic capabilities, group work, writing and, most of all, the flexibility in their learning processes. Besides this, it is worth pointing out that the students felt as producers and consumers of their own contents. This feeling of belonging to the information society is, in many cases, difficult to attain in a traditional class.

CONCLUDING REMARKS

The object of the study in this chapter is mobile learning understood as a set of processes to attain knowledge by means of conversation in diversified contexts between persons and personal interactive technologies. The main goal was to discuss and evaluate qualitative changes in the processes of teaching and learning determined by the affordance to mobility of the emergence of the free and easy to use Web 2.0 tools that allow for new forms of communication and content management in educational settings.

We began reviewing mobile learning as an area of study, verifying that it is sustained by the most recognized pedagogical theories, in particular, those that focus on social learning and also the new ideas about the way we learn in a networked world; this is the case for connectivism that values the importance of establishing significant networks for personal learning essential for the success in informal virtual environments.

We discussed the new Internet paradigm called Web 2.0 which offers a new and varied range of online applications that are easy to use and for free. According to this new Web philosophy, users also become producers of the information, distributing and sharing their knowledge and ideas through the Internet, in an easier and faster way. In this context, it is possible to think of new learning scenarios that replace traditional platforms to support teaching and learning (LMS) in more personalized, virtual environments, where, in formal education, students use the same tools that they use every day to communicate in informal environments (Coutinho & Bottentuit Junior, 2009a). We then introduced mobile Web 2.0, a new learning environment for the XXI century, made possible by the combination of a new generation of mobile devices with Internet access and the Web 2.0 technologies that allow collaboration, participation, knowledge sharing and construction. Mobility, portability and ubiquitousness are the three main features of mobile technologies that make them unique for the promotion of effective and quality learning, allowing the creation of virtual environments adapted to the style of each student; it allows the teacher/tutor to have at their disposal a range of free tools for communicating and learning. By building their own personalized space, thanks to ubiquitous technologies and social software, students gain control of their learning. Institutions must then recognize their loss of control over knowledge content, ways of transmission, learning process and validation. They have to accept the fact that ownership of learning is moving towards the students. Educational systems should not ignore this phenomenon but rather, try to find ways to appreciate the learning that takes place outside the institution and recognize its contribution to personal and professional development (Coutinho & Bottentuit Junior, 2009a).

In conclusion, it is in the hands of teachers and educators the responsibility to explore alternative classroom methodologies as the ones we presented using Twitter and m-Flickr; those initiatives come to shorten the gap between the classroom and the real world while improving the students' motivation and the learning outcomes. Schools need to open doors to new scenarios, in either formal or non-formal education, provided by the mobility offered by a new generation of mobile devices that break time and space boundaries and allow learning to occur in accordance with each individual's cognitive level, as well as from any geographical place.

REFERENCES

Attwell, G. (2007). Personal learning environments-the future of e-learning? *eLearning Papers, 2*(1). Retrieved April 24, 2008, from www.elearningpapers.eu

Bottentuit, J. B., Jr., & Coutinho, C. P. (2007). Virtual laboratories and m-learning: Learning with mobile devices. In *Proceedings of International Multi-Conference on Society, Cybernetics and Informatics (WMSCI)* (pp. 275-278). Orlando, FL, EUA.

Bottentuit, J. B., Jr., & Coutinho, C. P. (2008). The use of mobile technologies in higher education in Portugal: An exploratory survey. In C. Bonk, M. M. Lee & T. Reynolds (Eds.), *Proceedings of E-Learn 2008 Conference on E-Learning in Corporate, Government, Healthcare & Higher Education* (pp. 2102-2107). Las Vegas, NV.

Castells, M. A. (2004). *Galáxia Internet. Reflexões sobre Internet, negócios e sociedade*. Lisboa, Portugal: Fundação Calouste Gulbenkian.

Cochrane, T. (2008). Mobile Web 2.0: The new frontier. In *Hello! Where are you in the landscape of educational technology? Proceedings ascilite Melbourne 2008* (pp. 177-186). Retrieved April 24, 2009, from http://www.ascilite.org.au/conferences/melbourne08/procs/cochrane.pdf

Cochrane, T., & Bateman, R. (2010). Smartphones give you wings: Pedagogical affordances of mobile Web 2.0. *Australasian Journal of Educational Technology 26*(1), 1-14. Retrieved June 24, 2010, from http://www.ascilite.org.au/ajet/ajet26/cochrane.html

Cochrane, T., Bateman, R., & Flitta, I. (2009). Integrating mobile Web 2.0 within tertiary education. In *Proceedings of the V-MICTE 2009 - V International Conference on Multimedia and ICT in Education* (pp. 1348-1353).

Coutinho, C. (2009b). Using blogs, podcasts and Google sites as educational tools in a teacher education program. In G. Richards (Ed.), *Proceedings of World Conference on E-Learning in Corporate, Government, Healthcare, and Higher Education 2009* (pp. 2476-2484). Chesapeake, VA: AACE. Retrieved July 24, 2009, from http://hdl.handle.net/1822/9984

Coutinho, C. P. (2009a). Challenges for teacher education in the learning society: Case studies of promising practice. In H. H. Yang & S. H. Yuen (Eds.), *Handbook of research on practices and outcomes in e-learning: Issues and trends* (pp. 385-401). Hershey, PA: Information Science Reference/ IGI Global. Retrieved April 24, 2009, from http://hdl.handle.net/1822/9981

Coutinho, C. P. (2009c). E-learning 2.0: Challenges for lifelong learning. In I. Gibson (Ed.). *Proceedings of the 20th International Conference of the Society for Information Technology and Teacher Education, SITE 2009,* (pp. 2768-2773).

Coutinho, C. P., & Bottentuit, J. B. Jr. (2009a). From Web to Web 2.0 and e-learning 2.0 . In Yang, H. H., & Yuen, S. H. (Eds.), *Handbook of research on practices and outcomes in e-learning: Issues and trends* (pp. 19–37). Hershey, PA: Information Science Reference - IGI Global.

Coutinho, C. P., & Bottentuit, J. B., Jr. (2009b). Literacy 2.0: Preparing digitally wise teachers. In A. Klucznick-Toro, et al. (Eds.), *Higher Education, Partnership and Innovation (IHEPI 2009)* (pp. 253-261). Budapest, Hungary: PublikonPublishers/IDResearch, Lda.

Cruz, S. (2009). Blogue, YouTube, Flickr e Delicious: Software social. In A. A. Carvalho (org.). *Manual de ferramentas da Web 2.0 para professores*. Lisboa, Portugal: Direcção-Geral de Inovação e de Desenvolvimento Curricular do Ministério da Educação.

Downes, S. (2006). E-learning 2.0 at the e-learning forum. In *E-Learning Forum.* Canadá: Institute for Information Teachnology. Retrieved April 24, 2008, from http://www.teacher.be/files/page0_blog_entry38_1.pdf

Figueiredo, P., & Refkalefsky, E. (2009). Twitter: Uma nova forma de se comunicar? In *XXXII Congresso Brasileiro de Ciências da Comunicação. Sociedade Brasileira de Estudos Interdisciplinares da Comunicação*: Curitiba. Retrieved April 13, 2010, from http://www.intercom.org.br/papers/nacionais/2009/resumos/R4-1316-1.pdf

Freitas, G. P. (2008). Configurações da fotografia contemporânea em comunidades virtuais. In *XXXI Congresso Brasileiro de Ciências da Comunicação*. Natal. Retrieved December, 8 2009 from http://www.intercom.org.br/papers/nacionais/2008/resumos/R3-1125-1.pdf

Henri, F., Charlier, B., & Limpens, F. (2008). Understanding PLE as an essential component of the learning process. In J. Luca & E. R. Weippl (Eds.), *Proceedings of the 20th World Conference on Educational Multimédia Hypermedia & Telecommunications, EDMEDIA 2008* (pp. 3766-3770). Vienna, Austria: University of Vienna.

Herrington, A., & Herrington, J. (2007). *Authentic mobile learning in higher education*. Paper presented at the AARE 2007 International Educational Research Conference, Fremantle, Australia.

Herrington, J., Herrington, A., Mantei, J., Olney, I., & Ferry, B. (Eds.). (2009). *New technologies, new pedagogies: Mobile learning in higher education.* Wollongong, Australia: Faculty of Education, University of Wollongong. Retrieved May 13, 2010, http://ro.uow.edu.au/newtech

Jonassen, D. H. (2007). *Computadores, ferramentas cognitivas - desenvolver o pensamento crítico nas escolas*. Porto, Portugal: Porto Editora.

Kukulska-Hulme, A., & Traxler, J. (2005). *Mobile learning: A handbook for educators and trainers*. Londres, UK: Routledge.

Laurillard, D. (2002). *Rethinking university teaching: A conversational framework for the effective use of learning technologies* (2nd ed.). Londres, UK & Nova Iorque, NY: Routledge Falmer. Retrieved October 25, 2009, from http://www.questiaschool.com/read/103888453

Laurillard, D. (2007). Pedagogical forms of mobile learning: Framing research questions . In Pachler, N. (Ed.), *Mobile learning: Towards a research agenda* (*Vol. 1*, pp. 153–175). London, UK: WLE Centre, Institute of Education.

Lisbôa, E. S., Bottentuit Junior, J. B., & Coutinho, C. P. (2009). Uso do Flickr em contexto educativo. In: VI Congresso Brasileiro de Ensino Superior a Distância (ESuD), 2009, São Luís. *Anais do VI Congresso Brasileiro de ensino Superior a Distância (ESuD).* São Luís - Maranhão: Universidade Estadual do Maranhão.

Martins, E., Gomes, I., & Santos, L. (2009). O Twitter como ferramenta no ensino e atuação de profissionais de publicidade e propaganda. In *XXXII Congresso Brasileiro de Ciências da Comunicação.* Sociedade Brasileira de Estudos Interdisciplinares da Comunicação: Curitiba. Retrieved October 13, 2009, from http://www.intercom.org.br/papers/nacionais/2009/resumos/R4-3861-1.pdf

Milrad, M. (2006). How should learning activities using mobile technologies be designed to support innovative educational practices? In M. Sharples (Ed.), *Big issues in mobile learning* (pp. 28- 30). Report of a workshop by the Kaleidoscope Network of Excellence Mobile Learning Initiative. University of Nottingham.

Moura, A., & Carvalho, A. (2007). Das tecnologias com fios ao wireless: Implicações no trabalho escolar individual e colaborativo em pares . In Dias, P., Freitas, C. V., Silva, B., Osósio, A., & Ramos, A. (Eds.), *Actas da V Conferência Internacional de Tecnologias de Informação e Comunicação na Educação: Challenges 2007*. Braga: Universidade do Minho.

Moura, A., & Carvalho, A. (2008a). *Mobile learning: Teaching and learning with mobile phone and podcasts*. In 8th IEEE International Conference on Advanced Learning Technologies, 2008 (ICALT 2008) (pp. 631-633). Santander, Spain.

Moura, A., & Carvalho, A. (2008b). Mobile learning with cell phones and mobile Flickr: One experience in a secondary school. In I. Arnedillo-Sánchez, & P. Isaías (Eds.), *Proceedings IADIS Conference Mobile Learning 2008* (pp. 216-220). Algarve, Portugal.

Moura, A., & Carvalho, A. (2009). Peddy-paper literário mediado por telemóvel. *Educação, Formação & Tecnologias, 2*(2), 22-40. Retrieved May 29, 2010, from http://eft.educom.pt

Naismith, L., Lonsdale, P., Vavoula, G., & Sharples, M. (2004). *Literature review in mobile technologies and learning*. Report 11: FUTURELAB SERIES.

Negroponte, N. (1996). *Ser Digital*. Lisboa: Editorial Caminho.

Rosen, A. (2006). Technology trends: e-Learning 2.0. *The e-Learning Guild's Learning Solutions E-Magazine*. Retrieved May 24, 2008, from http://www.readygo.com/e-learning-2.0.pdf

Sharples, M. (2000). The design of personal mobile technologies for lifelong learning. *Computers & Education, 34,* 177-193. Retrieved April 20, 2010, from http://www.eee.bham.ac.uk/sharplem/Papers/handler%20comped.pdf

Sharples, M. (2002). Disruptive devices: Mobile technology for conversational learning. *International Journal of Continuing Engineering Education and Lifelong Learning, 12*(5/6), 504–520. doi:10.1504/IJCEELL.2002.002148

Sharples, M. (2005). Learning as conversation: Transforming education in the mobile age. In *Proceedings of Conference on Seeing, Understanding, Learning in the Mobile Age* (pp. 147-152). Budapest, Hungary.

Sharples, M., Taylor, J., & Vavoula, G. (2006). *A theory of learning for the mobile age*. Retrieved October 13, 2009, from http://kn.open.ac.uk/public/document.cfm?docid=8558

Siemens, G. (2008a). *Connectivism: A learning theory for today's learner*. Retrieved March 11, 2009, from http://www.connectivism.ca/about.html

Siemens, G. (2008b). *Learning and knowing in networks: Changing roles for educators and designers*. Paper presented at the IT Forum. Retrieved October 10, 2009, from http://it.coe.uga.edu/itforum/Paper105/Siemens.pdf

Talagi, S. (2008). Learning goes mobile. In *Western Leader* (p. 4).

Teten, D. (2007). *Web 3.0: Where are we headed?* Retrieved January 13, 2010, from www.teten.com/assets/docs/Teten-Web-3.0.pdf

Torrisi-Steele, G. (2006). *The making of m-learning spaces.* Paper presented at the mLearn 2005, Cape Town, South Africa. Retrieved May 13, 2009, from https://olt.qut.edu.au/udf/OLT2006/gen/static/papers/Torrisi-Steele_OLT2006_paper.pdf

Traxler, J. (2007). Defining, discussing and evaluating mobile learning: The moving finger writes and having writ. *The International Review of Research in Open and Distance Learning, 8*(2). Retrieved January 23, 2009, from http://www.irrodl.org/index.php/irrodl/article/view/346

Trifonova, A. (2003). *Mobile learning - review of the literature*. (Technical Report DIT-03-009, Informatica e Telecomunicazioni, University of Trento). Retrieved March 9, 2010, from http://eprints.biblio.unitn.it/archive/00000359

Valentim, H. (2009). *Para uma compreensão do mobile learning: Reflexão sobre a utilidade das tecnologias móveis na aprendizagem informal e para a construção de ambientes pessoais de aprendizagem*. Unpublished Master Dissertation, Universidade Nova de Lisboa, Portugal.

Vavoula, G., Sharples, M., Scanlon, E., Lonsdale, P., Jones, A., & Sharples, M. (2005). Report on literature on mobile learning, science and collaborative activity. *Kaleidoscope*. Retrieved October 25, 2009, from http://hal.archives-ouvertes.fr/hal-00190175/en

Winters, N. (2006). What is mobile-learning? In M. Sharples (Ed.), *Big issues in mobile learning* (pp. 5-9). Report of a workshop by the Kaleidoscope Network of Excellence Mobile Learning Initiative. University of Nottingham.

KEY TERMS AND DEFINITIONS

Flickr: Flickr can be considered as one of the most interesting Web 2.0 tools that allows a high level of interactivity, which grants users the opportunity to categorize archives through the means of tags or to comment or cast judgement on the images posted on the website.

Mobile Flickr: With the development of mobile technologies, Flickr made an adaptation of its website in order to allow access through mobile devices and, in this sense, mobile Flickr or simply m-flickr, was created.

Mobile Web 2.0: The concept of Mobile Web 2.0 emerges to define the interaction between mobile phones and Web 2.0 applications, a powerful alliance that makes sense and supports the enhancement of face-to-face and distance teaching and learning activities by the use of wireless mobile devices.

Mobile-Learning: A set of processes to attain knowledge by means of conversation in diversified contexts between persons and personal interactive technologies.

Twitter: It is a web-based social network in the format of a microblog that allows the user to create a set of short messages (140 characters) which can be updated at anytime, with the use of a computer connected to the Internet or a mobile phone with SMS (Short Message Service).

Web 2.0: The term Web 2.0 was first coined by Tim O'Reilly in 2004 to refer to a second generation in the history of Web-based communities of users and a range of special services such as social networks, blogs, wikis, podcast that encourage collaboration and exchange of information between users.

Web 3.0: Web 3.0 is used to describe the evolution of the use and interaction in the network through different paths. This includes the transformation of the network in a database, a move towards making content accessible by multiple non-browser applications, the thrust of artificial intelligence technologies, the semantic web, the Geospatial Web, or Web 3D.

Chapter 14
Exploring the Pedagogical Affordances of Mobile Web 2.0

Thomas Cochrane
Te Puna Ako (Centre for Teaching and Learning Innovation), Unitec, New Zealand

ABSTRACT

This chapter explores the potential of Mobile Web 2.0 to enhance tertiary education today, outlining both research-informed principles, as well as providing case study examples of Mobile Web 2.0 participants' experiences of how the use of Mobile Web 2.0 within a pedagogical framework has transformed both students' and lecturers' conceptions of teaching and learning.

INTRODUCTION

This chapter is based upon the author's research that covers three years of action research mLearning (mobile learning) projects encompassing five different courses, forming five case studies spanning from one to three years of implementation and refinement, and involved thirteen mLearning projects undertaken between 2007 and 2009 with a total of 280 participants (Cochrane & Bateman, 2010a). The learning contexts included: Bachelor of Product Design (2006 using Palm Lifedrive, 2008 using Nokia N80, N95, 2009 using Nokia XM5800, N95, N97), Diploma of Landscape Design (2006 using Palm TX, 2007 using Nokia N80, 2008 using Sonyericsson P1i, 2009 using Dell mini9 netbook), Diploma of Contemporary Music (2008, 2009 using iPod Touch, iPhone 3G), Bachelor of Architecture (2009 using Nokia XM5800 and Dell Mini9 netbook), and the Bachelor of Performing and Screen Arts (2009 using Dell Mini9 netbook and Nokia XM5800). The aim of the research was to investigate the potential of Mobile Web 2.0 tools (with a focus upon smartphones coupled with mobile formatted Web 2.0 social software) to facilitate social constructivist learning environments across multiple learning contexts (both formal and informal).

The research used a participatory action research methodology (Swantz, 2008) and based its pedagogical decisions upon the foundation

DOI: 10.4018/978-1-60960-613-8.ch014

of social constructivist learning theories (Piaget, 1973; Vygotsky, 1978).

The research captured the learning journeys of the researcher and participants as they moved from initial skepticism to personal appropriation of the new technologies, to the ontological shifts required for integrating the unique affordances of these Mobile Web 2.0 technologies into their pedagogical practice and courses, enabling collaborative learning environments that bridge multiple contexts.

The research led to the development of an intentional community of practice model (Langelier, 2005; Wenger, White, Smith, & Rowe, 2005) for lecturer professional development and scaffolding student learning, established a pedagogical design framework, identified critical success factors, and developed an implementation strategy for the integration of mLearning within tertiary education.

The research adds the insights of a longitudinal study to the relatively new body of knowledge around mLearning. What began as an investigation of the affordances of Web 2.0 in 2007 developed into a wide exploration of Mobile Web 2.0 within a variety of learning contexts. The success of these projects led to the implementation of integrating Mobile Web 2.0 technologies (based on an explicit social constructivist pedagogy) across the institution.

This chapter summarises the potential of Mobile Web 2.0 in tertiary education from the researcher's perspective and experiences, overviewing a range of current freely available Web 2.0 services. Two case studies are used to illustrate the integration of Mobile Web 2.0.

BACKGROUND

This section briefly outlines some of the core concepts and the author's position on mLearning and Mobile Web 2.0.

Web 2.0

While definitions of Web 2.0 are difficult to pin down, it is their similar characteristics that link these diverse web services. "Ultimately, the label "Web 2.0" is far less important than the concepts, projects, and practices included in its scope" (Alexander, 2006, p. 33). McLoughlin defines Web 2.0 as "a second generation, or more personalised, communicative form of the World Wide Web that emphasises active participation, connectivity, collaboration and sharing of knowledge and ideas among users" (McLoughlin & Lee, 2007, p. 665).

Pedagogy 2.0

Recent years have seen many attempts to reconceptualise pedagogical approaches within tertiary education (JISC, 2009b; Laurillard, 2001). These have been driven by the emergence of new learning theories based broadly upon constructivist and social constructivist foundations, and the development of new learner-centred technologies that facilitate these newer pedagogies (Alexander, et al., 2006; JISC, 2009a). For example, the appropriation of Web 2.0 tools within a social constructivist pedagogy facilitates what has been termed "Pedagogy 2.0" (McLoughlin & Lee, 2008a). McLoughlin advocates the exploration of the potential of the alignment of Web 2.0 tools and emerging learning paradigms based loosely upon social constructivism such as "navigationism", and "connectivism".

The affordances of these technologies, coupled with a paradigm of learning focused on knowledge creation and networking, offer the potential for transformational shifts in teaching and learning practices, whereby learners can access peers, experts, the wider community and digital media in ways that enable reflective, self-directed learning (McLoughlin & Lee, 2008b, p. 649).

Similarly, Herrington has proposed that mobile technologies can facilitate "Authentic Learning" (J. Herrington, Mantei, Herrington, Olney, & Ferry, 2008).

Focusing even more explicitly on empowering independent learners, Luckin et al (2008) propose the concept of Learner Generated Contexts (LGC) as a potential framework for technology based learning based on the Vygotskian concept of 'Obuchenie'. Though not explicitly limited to mobile learning, the concept focuses upon learning within learners own environments that new technologies facilitate. 'Obuchenie' blurs the distinction between teaching and learning, creating a two-way dyadic interaction within the Zone of Proximal Development. Luckin et al see a reconceptualisation of the level of influence the teacher plays in these contexts, and attempt to breakdown the classical PAH continuum (Pedagogy – Andragogy – Heutagogy) (see Table 1).

While the researcher is not advocating a radical reconceptualising of educational pedagogy on the scale that is proposed by Luckin et al, the researcher sees similarities and useful alignment of our pedagogical approaches with 'Pedagogy 2.0', 'Authentic Learning' and some of the PAH continuum principles. The key point of difference is in the role that the researcher assigns to the lecturer within the formal and informal learning environments. The researcher views the input and facilitation of the lecturer as a critical success factor in implementing Mobile Web 2.0 technologies, and would agree with Laurillard's position that states "mLearning, being the digital support of adaptive, investigative, communicative, collaborative, and productive learning activities in remote locations, proposes a wide variety of environments in which the teacher can operate" (Laurillard, 2007, p. 172). However, the role of the lecturer is significantly changed. The focus moves from teacher-directed to student-centred, where students create accounts on free Web 2.0 sites and then invite their lecturer and peers to collaborate within these environments, turning the control of the learning environment beyond the domain of the teacher-directed learning management system (LMS) to focus upon student-generated content and student-generated contexts.

mLearning

mLearning technologies provide the ability to engage in learning conversations between students and lecturers, between student peers, students and subject experts, and students and authentic environments within any context. It is the potential for mobile learning to bridge pedagogically designed learning contexts, facilitate learner generated contexts, and content (both personal and collaborative), while providing personalisation and ubiquitous social connectedness, that sets it apart from more traditional learning environments. Mobile learning, as defined in this research, involves the use of wireless enabled mobile digital devices (Wireless Mobile Devices or WMDs) within and between pedagogically designed learning environments or contexts. From an Activity Theory perspective, WMDs are the tools that mediate a wide range of learning activities and facilitate collaborative learning environments (Uden, 2007).

Table 1. The PAH continuum, from Luckin et al (Luckin, et al., 2008, p. 10)

	Pedagogy	Andragogy	Heutagogy
Locus of Control	Teacher	Learner	Learner
Educational Sector	Schools	Adult education	Doctoral research
Cognition Level	Cognitive	Metacognitive	Epistemic
Knowledge Production Context	Subject understanding	Process negotiation	Context shaping

Mobile Web 2.0

The WMD's wireless connectivity and data gathering abilities (for example: photoblogging, video recording, voice recording, and text input) allow for bridging the on and off campus learning contexts – facilitating "real world learning", disrupting traditional instructivist teaching models and facilitating a move along the PAH continuum to social constructivist learning paradigms.

PEDAGOGICAL AFFORDANCES OF MOBILE WEB 2.0

Adoption and integration of technology into educational environments need to be based upon a pedagogical design framework that aligns the affordances of the chosen technologies with the chosen pedagogical framework (Cochrane & Bateman, 2010b).

Successful technology integration is a sociological issue, intimately connected to institutional cultures and practices, to social groups (formal and informal), and to individual intention, agency and interest. Most importantly, appropriate use of technology in teaching requires the thoughtful integration of content, pedagogy, and technology. (Mishra, Koehler, & Zhao, 2007, p. 2)

The addition of a new technology reconstructs the dynamic equilibrium between all three elements forcing instructors to develop new representations of content and new pedagogical strategies that exploit the affordances (and overcome the constraints) of this new medium. Similarly, changing pedagogical strategies (say moving from a lecture to a discussion format) necessarily requires rethinking the manner in which content is represented, as well as the technologies used to support it. (Mishra, et al., 2007, p. 8)

Experience and feedback from the research participants (2007 to 2009) has shown that the focus should be on the affordances of Wireless Mobile Devices (WMDs) that are most suitable for the small screens and slower text entry, as well as those affordances that are unique to WMDs (For example: the built-in geotagging, media recording capabilities, and communication tools). In particular, it is the WMD's potential to bridge multiple learning contexts that facilitate rich interactions between formal and informal social constructivist learning environments. As Laurillard notes, "The intrinsic nature of mobile technologies is to offer digitally-facilitated site-specific learning, which is motivating because of the degree of ownership and control." (Laurillard, 2007, p. 157).

Mobile Web 2.0 Design Framework

The design framework for each of the projects is shown in Table 2. This framework was developed iteratively over the life of the research, which began in 2006 with two test projects that informed the practical implementation of the subsequent projects in 2007 to 2009. The framework table format is based loosely on that suggested by Sharples et al (2009), emphasizing that the starting point of the design process is the learning practice and chosen pedagogical framework, which then informs the appropriate choice of mediating technologies.

Evaluation of Example Mobile Web 2.0 Services

This section overviews some of the currently available Mobile Web 2.0 services that can be utilised within a social constructivist pedagogy.

The researcher has focused upon utilising freely available Web 2.0 services that are easily accessible via smartphones. The smartphone's constant connectivity and built-in media capturing affordances allow students to capture, share, and critique ideas and continue learning conversations within virtually any context. The following concept map (Figure 1) illustrates this process with some of the 'core' Web 2.0 tools used in the research projects, which are expanded upon in the following section.

Google Mobile

Google provides a gateway into the Google Mobile services (http://mobile.google.com) via a phone's web browser. iGoogle (http://www.google.com/ig/i) is a customisable mobile Google homepage.

Pedagogy: Links to mobile formatted software tools that support social constructivist pedagogies.

I. Maps

Google Maps (http://maps.google.com), a free world-wide mapping service is optimised for use on mobile devices. Additionally, most smartphones now include an integrated GPS for geotagging of photos and videos and geolocation via mapping services such as Google Maps. Geolocation adds an extra layer of information to mobile captured content and, along with a built-in compass, provides the foundation for Augmented Reality application interaction.

Pedagogy: Context awareness and sharing of geolocation data.

II. Calendar

Google Calendars (http://calendar.google.com) can be shared between groups of people via invitation. Google Calendars use an open format that provides interoperability between many calendar systems – for example iCal on Mac OSX.

Pedagogy: Time scheduling and collaboration of group activities.

III. Reader

RSS enables subscribing, tracking and sharing of online activities. RSS provides a link between all Web 2.0 media sites. Google Reader (http://reader.

Table 2. MLearning project design framework

Learning Practice	Mediating Circumstances		
Social Constructivism	**Context**	**Technology**	**Agent**
Lecturer Community of Practice	Lecturer professional development, pedagogical brainstorming	Face to face Scaffolded using LMS Smartphone Web 2.0 services	Lecturers as peers, with researcher as technology steward
Student and lecturer Community of Practice	Pedagogical integration and technical support	Face to face Scaffolded using LMS Smartphone Web 2.0 services	Students as peers, Lecturer as guide and pedagogical modeler, with the researcher as technology steward
Collaboration	Group projects	Social networking, Collaborative documents	Google Docs, student peers
Sharing	Peer commenting and critique	Web 2.0 media sites, eportfolio creation	RSS, student peers, lecturer
Student content creation	Student individual and group projects	Smartphone with camera and microphone, content uploaded to Web 2.0 sites	Student and peers
Reflective	Journal of learning and processes, recording critical incidents	Web 2.0 hosted Blog	Personal appropriation, formative feedback from lecturer
Learning Context Bridging	Linking formal and informal learning	Smartphone used as communication tool and content capturing	Student interacting with context, peers, and lecturers

Table 3. Affordances of smartphones mapped to social constructivist activities

Activity	Overview	Examples	Pedagogy
Video Streaming	Record and share live events	Flixwagon, Qik http://www.qik.com Knocking, Livestream	Real-time Event, data and resource capturing and collaboration.
Geotagging	Geotag original photos, geolocate events on Google Maps	Flickr, Twitter, Google Maps http://tinyurl.com/5a85yh	Enable rich data sharing.
Micro-blogging	Post short updates and collaborate using micro-blogging services	Twitter http://tinyurl.com/2j5sz3	Asynchronous communication, collaboration and support.
Txt notifications	Course notices and support	Txttools plugin for Moodle and Blackboard txt and twitter polls: http://www.pol-leverywhere.com/ http://twitter.polldaddy.com http://twtpoll.com/	Scaffolding, learning and administrative support
Direct image and video blogging	Capture and upload images and video of ideas and events	Flickr, YouTube, Vox	Student journals, eportfolios, presentations, peer and lecturer critique.
Mobile Codes	2D Codes scanned by cameraphone to reveal URL, text, etc.	QR Codes, Datamatrix 2D Codes http://tinyurl.com/af2u6d	Situated Learning – providing context linking
Enhanced Student Podcasts	Remote recording of audio, tagged with GPS and images, etc.	AudioBoo	Situated and Collaborative Learning – providing context linking
Augmented Reality	Overlaying the real world with digital information	Wikitude Layar	Situated Learning and Metacognition
Social Networking	Collaborate in groups using social networking tools	Vox groups, Ning, peer and lecturer comments on Blog and media posts http://tinyurl.com/4uz6rj	Formative peer and lecturer feedback.

Figure 1. Mobile Web 2.0 concept map

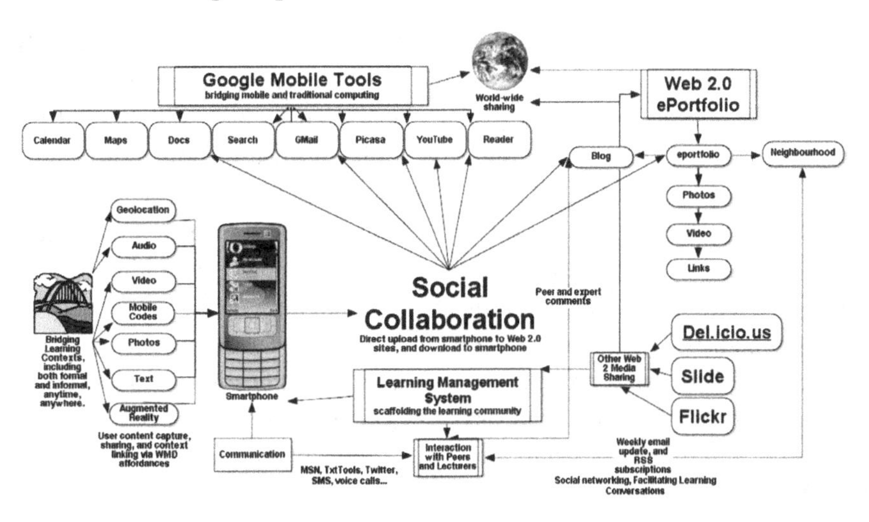

google.com) is a mobile formatted web based RSS reader, and there are also RSS client applications for synchronizing Google Reader subscriptions via PC, Mac or mobile.

Pedagogy: Collaboration, collation, categorising and sharing of multiple sources of information.

IV. Picasa

Dedicated image sharing repositories such as Flickr (http://www.flickr.com) and Picasaweb (http://picasaweb.google.com) offer interactive features beyond image repositories – including interactive slideshows and the ability to annotate and tag student captured images. These are linkable and embeddable in most Blogging systems. Direct mobile upload to online image sharing sites can be achieved via either mobile application clients, or email. Picasaweb mobile is supported via mobile media sharing systems such as Shozu or Pixelpipe.

Pedagogy: Event, data and resource capturing and collaboration. Creativity.

V. GMail

GMail (http://gmail.com) provides a free email account that can be used on almost any Internet capable device. A GMail account also opens free access to all other Google web services. The Google Java application optimises GMail for a wide range of cellphones.

Pedagogy: Communication and collaboration.

VI. Docs

Google Docs (http://docs.google.com) is Microsoft Word, Excel and PowerPoint compatible. Documents can be uploaded, shared and edited by a group. They are viewable online in a web browser without MS Office. Docs can be created on mobile devices by emailing the document to a private Google Docs address. To edit uploaded documents, you need a full PC web browser or a full version of mobile applications such as 'QuickOffice' on your smartphone - a mobile version of MS Office.

Pedagogy: Documentation, reflection, critique, description, and collaborative document publishing.

VII. YouTube

YouTube (http://www.youtube.com) is currently the most popular video sharing site. The mobile version (http://m.youtube.com) supports viewing of videos online in the mobiles web browser, or via a downloadable Java client for specific phones. Uploading mobile videos to YouTube is achieved via email attachments or as a Shozu destination.

Pedagogy: Event, data and resource capturing and collaboration. Creativity.

VIII. Search

Google's mobile search feature enables voice activated web searches for quick access to information on the go.

Pedagogy: Information literacy, flexible information access and evaluation.

IX. World-wide Sharing

Blog posts and online media can be shared with a world-wide audience via RSS feeds or URLs from either VOX or the Google tools. This enables the creation of world-wide virtual learning communities.

Pedagogy: Collaboration, peer support and critique.

X. Interaction with Peers and Tutors

The core support element of the mLearning projects is a weekly "community of practice" investigating the use and integration of the smartphones and Web 2.0 tools involving: the technology steward, the course tutors, and the students. Each trial "learning community" is also supported by the "neighbourhood" social networking feature of Vox, and the use of instant messaging for facilitating communication and a sense of social presence.

Pedagogy: These tools facilitate context independent learning conversations.

Learning Management System

Moodle is an example of a mobile friendly Learning Management System, hosted on a production level Unitec server. Course notes, discussion forums, and various activities can be hosted on Moodle. Learning management systems are controlled by the institution and courses are administered by the course lecturers. In the researcher's projects, students' content is hosted outside of Moodle on Web 2.0 site accounts, while Moodle is used as a tutorial space for scaffolding the technology support for the projects.

Pedagogy: Scaffolding and support.

Blogging

A blog post (including media) can be uploaded directly to most blog hosts using mobile blogging apps on smartphones, or media forwarding services such as Shozu (http://www.shozu.com) or Pixelpipe, or emailed to a user's blog email upload address.

Pedagogy: Developing critical and reflective thinking, journaling.

Smartphone

Students and teaching staff are provided with a 3G smartphone paying for their own 3G data and voice call usage. Internet connectivity is also available via the Unitec WiFi network while on campus. This provides faster, free web access while on campus. The smartphone's wireless connectivity and data gathering abilities (For example: photoblogging, video recording, voice recording, and text input) allow for bridging the on and off campus learning contexts - facilitating "real world learning".

The core activity of each project is the creation and maintenance of a reflective Blog as part of a course group project. Additionally, a variety of mobile friendly Web 2.0 tools are used in conjunction with the smartphone. The choice of mobile device for each project is based on the best fit of features with the key requirements of each course. Previous projects identified the importance of a ubiquitous connection to the Internet for student productivity across multiple contexts, and the preference of students and tutors to carry a single device (i.e. a cellphone); hence preference was given to smartphones over WiFi capable PDAs. Common specifications required include: WiFi capability for free web access while on campus, 3G for fast web access off campus, a built-in camera, media playback, alternative text entry capability, support for key Web 2.0 applications. Windows Mobile devices were not considered based on their small marketshare and inherent "uncoolness" for students. Palm smartphones had been trialed initially in 2007 but had been rejected by students because of the poor quality of the built-in camera, 'clunky' form-factor, and aging OS. Budget was another factor, limiting the cost of the device to $700NZ each. To keep the cost of the devices down, the smartphones were purchased 'unlocked' through parallel importers.

Communication

Instant Messaging (IM) is a synchronous communications technology, with the most popular IM service being MSN. There are many mobile IM clients available. Fring (http://www.fring.com) is a free Instant Messaging and Skype client for most mobile phones. It allows messaging between the most popular IM systems. It works best over a WiFi connection, or good 3G connection.

Microblogging is a cross between SMS texting, blogging, and instant messaging. Microblogging is an asynchronous, collaborative communication technology, suited to use on mobile devices. The most popular microblogging service is currently Twitter. Twitter usage had exponential growth during 2008, with an increase of 752% to over 3 million users world-wide.

Pedagogy: Communication and collaboration.

Social Collaboration

Mobile Media Sharing

Mobile media sharing services provide web based portals for bridging Web 2.0 media services. Two examples are Pixelpipe and Shozu (http://www.shozu.com). These services link all your online mobile Blog and media sites together via either (for example) the Shozu client application, or an email sent to go@m.shozu.com

Pedagogy: Collaboration, sharing and collation of student generated content.

ePortfolio

An example of a mobile friendly ePortfolio was VOX (http://www.vox.com). Vox included media sharing (video, audio, documents, images, links) and linking (YouTube, Flickr) as well as social networking. The untimely shut-down of Vox in mid 2010 demonstrates the volatile nature of freely hosted Web 2.0 services. Similar eportfolios can be constructed from a collage or mashup of several Web 2.0 services.

Pedagogy: Collaborative sharing of media and peer critique, also form the basis for a career portfolio.

Social Networking

Many Web 2.0 services allow users to define user groups and give secure access to content. Collaboration can be enhanced by automatic email updates or auto announcing new content via services such as Twitter, facilitating a community environment.

Almost all smartphones now include a built-in camera that is capable of capturing still images and video. Most smartphones also include a built-in GPS (Global Positioning Service) that works via satellites to provide longitude and latitude information for geotagging and geolocation. This facilitates geotagging original photos, and the ability to geolocate events on Google Maps, adding a location dimension to captured images and video. Web 2.0 services that support geotagged photos include Flickr and Picasaweb.

The built-in camera on smartphones can record video and audio at up to almost DVD quality. This facilitates students recording events, interviews, and reflections with a visual dimension, and sharing these online via a variety of mobile friendly video sites such as YouTube. Video streaming applications such as Qik and Flixwagon allow real-time sharing of video directly from smartphones to these web-based services. Qik and Flixwagon then archive the video stream for later viewing, sharing and commenting. Additionally, video streaming sites integrate with other mobile Web 2.0 technologies such as Twitter - creating an automatic announcement on Twitter regarding a live video stream that a student's Twitter followers could then watch in almost real-time. Qik and Flixwagon also feature the ability to forward video streams to a user's YouTube account for sharing on that service as well. Qik supports the association of geolocation data with video streams, providing a Google Maps link to the actual location of the recorded event.

The built-in microphone of smartphones can be used to record audio and then upload that audio file to an online Blog or other Web 2.0 site that supports audio. Most Web 2.0 sites support the uploading on audio files in either mp3 or .wav formats. Some support a popular mobile audio format, .amr. Podcasting is a popular form of audio recording that has an associated RSS feed for subscribing to new audio recordings. For example, students could record themselves reflecting or reporting on their progress in an assignment or project, or they could record an interview with an expert in the field.

Bridging Learning Contexts

It is the potential for mobile learning to bridge pedagogically designed learning contexts, facilitate learner generated contexts and content (both personal and collaborative), while providing personalisation and ubiquitous social connectedness, that sets it apart from more traditional learning environments. Mobile learning, as defined in this project, involves the use of wireless enabled mobile digital devices (Wireless Mobile Devices or WMDs) within and between pedagogically designed learning environments or contexts. From an activity theory perspective, WMDs are the tools that mediate a wide range of learning activities and facilitate collaborative learning environments (Uden, 2007).

Pedagogy: Collaborative sharing of media and peer critique, learner generated contexts.

DESIGNING AND IMPLEMENTING MOBILE WEB 2.0 IN TERTIARY EDUCATION

Key issues (that have been identified across the researcher's thirteen mobile web 2.0 projects 2007-2009) for integrating mobile web 2.0 within an education course include:

- The integration of the mobile web 2.0 tools into the course assessment criteria
- The authentic use of the mobile web 2.0 tools – that is, they are not just added as a 'gimmick'.
- Lecturer modeling of the pedagogical use of the mobile web 2.0 tools
- Creating the sense of a learning community around the integration of the mobile web 2.0 tools
- Providing adequate pedagogical and technological support for the lecturers and students

These are illustrated and explored in the following two case study examples from the researcher's mobile web 2.0 projects. Explorations of several of the researcher's other mLearning case studies can be found in various journal papers (Cochrane, 2010; Cochrane & Bateman, 2009; Cochrane, Flitta, & Bateman, 2009).

Case Study 1: Diploma of Contemporary Music

The 2009 mLearning project within the Diploma of Contemporary Music was informed by the lessons learnt from the 2008 trial. A compilation of 2008 student reflections as Vodcasts (Online video recordings) of the 2009 mLearning project is available on YouTube: http://nz.youtube.com/watch?v=0It5XUfvOj.

During 2008, no assessment tasks were directly related to the use of the iPhones or iPod Touch's, and this resulted in varying commitment to the project by the students. While all iPhone recipients regularly used the device, there was limited use for directly course-related activities. This suggested that, while the students appropriated the use of the tools into their personal and informal learning, they had not been convinced (neither modeled by the lecturers) of the potential for the iPhones and associated activities to be useful in their formal learning environment. It also suggested that students are more likely to respond to tasks for which they receive credit. It became clear that the iPhone project needed to be embedded in a course, with clearly related assessment tasks, for the students to participate more fully in it. In particular, 2009 projects were designed to investigate the use of MySpace, student created podcasts, and microblogging as authentic mobile learning environments within the context of music delivery, promotion and critique.

The 2009 project (See Table 4) was explicitly linked to two courses, one within the second year of the Diploma of Contemporary Music, the other within the first year of the course with second year students as peer mentors. Thus the integration of mLearning was staged across the two years of the course, and the use of Mobile Web 2.0 tools were integrated into the course assessment. MLearning was explicitly integrated into the Web Technologies paper (PASA5011) during semester one of the second year of the Diploma of Contemporary Music course. All students were issued with

Table 4. Outline of 2009 Diploma of Contemporary Music mLearning project

Course: Diploma of Contemporary Music 2009	
Participants	• 24 students • 2 Course Lecturers • Technology Steward (Thom Cochrane - CTLI)
Mobile Technology	12 students using iPhone, participants responsible for 3G data, voice & txt costs. 12 students using iPod Touch – during Semester2.
Pedagogical Model	From Pedagogy to Andragogy
Pedagogical Focus	1. (5011) An investigation of the current and future uses of web 2.0 technologies in music production and distribution. Students research and report on various technologies using a weekly podcast/vodcast that is peer critiqued by the other students in the course. 2. (4006) Recording and peer critique of student performances.
Community of Practice	Weekly throughout the entire course
Support LMS	Blackboard plus an institutionally hosted Wiki
Deliverables	An assessed online Blog/eportfolio documenting and showcasing students' design processes and forming the basis of a collaborative hub with worldwide peers and potential employers/clients. And the weekly use of instant messaging, microblogging, and VODCasts.
YouTube Links	• Project Summary http://www.youtube.com/watch?v=hLNNTK1_wGQ • Lecturer2 Reflections http://www.youtube.com/watch?v=o9p4i23CsPE • Student Reflections http://www.youtube.com/watch?v=5wbryYTmW88
Blog Links	• Course Tutorial Wiki http://ctliwiki.unitec.ac.nz/index.php/IphoneTutorials • Example student Blog http://rima803.vox.com/ • Example student AudioBoo http://audioboo.fm/profile/ting019 • Example student Group Blog http://groupb.groups.vox.com/
Course Project Outlines	1. Environmental Recording Assignment http://docs.google.com/Doc?docid=0Adkx7n-UKqvBZGNocjRyZ2dfNDNkenRwbTdqOQ&hl=en_GB 2. MySpace Assignment http://docs.google.com/Doc?docid=0Adkx7n-UKqvBZGNocjRyZ2dfNDJkZ2s5N2ZjbQ&hl=en_GB 3. 4006 Performance Groups http://docs.google.com/Doc?docid=0Adkx7n-UKqvBZGNocjRyZ2dfNDFmOXczanhjaw&hl=en_GB
Timeframe	March 2009 through to July 2009 for PASA5011. July to November 2009 for PASA4006.

iPhones for use within the course throughout 2009, and were also encouraged to personalise the use of the iPhone into their daily routines. Internet access was available for free via the campus WiFi network, but students and staff were responsible for any voice and 3G data costs accrued. The focus of the semester one project was on the Contemporary Music students using iPhones as tools to record and share environmental sounds from a variety of off-campus contexts, as well as creating online profiles on Vox (http://www.vox.com) and MySpace (http://www.myspace.com), evaluating the use of new technologies for music generation, sharing, marketing, and distribution. Thus the iPhones facilitated both learner-generated content (Bruns, 2007) and learner-generated contexts (Cook, Bradley, Lance, Smith, & Haynes, 2007; Luckin, et al., 2008). Several assessed projects within the course involved the direct use of the iPhone and web 2.0 tools, as described in the summarised course outline below.

Implications of Case Study1: Diploma of Contemporary Music 2008 to 2009

The Diploma of Contemporary Music mLearning project developed from an initial exploration of the potential of mLearning to engage students and enhance the course to an example of successful course integration and student adoption and appropriation of mLearning. During the first iteration of the mLearning project, students and lecturers were enthusiastic and engaged by the tools, but skeptical as to the potential impact on the course and learning outcomes. The second iteration of the mLearning project integrated the mLearning tools into the course assessment leading to adoption and appropriation by the students beyond personal and social use, leveraging the learning context bridging (Vavoula, 2007) affordances of mobile web 2.0 for facilitating authentic (A. Herrington & Herrington, 2007) course-related learning environments beyond the classroom. This case study also demonstrates the need for significant time for lecturer pedagogical reflection for the necessary ontological shifts (Chi & Hausmann, 2003; Hameed & Shah, 2009) in their pedagogical conceptions to be able to integrate mLearning authentically.

Case Study 2: Bachelor of Performing and Screen Arts

This project focused upon an investigation of the potential of mobile web 2.0 technologies within the field of Film and Television within the Bachelor of Performing and Screen Arts (PASA). The PASA mLearning integration was focused on the context of the mLearning tools themselves as key new technologies that are becoming important in reinventing and democratizing the recording and distribution of film that will have significant impact on the industry. The tools themselves were thus the focus of learning as well as used to record students' learning journeys, thus acting as mediators (Uden, 2007) and bridges of external learning contexts (Vavoula, 2007). Topics covered by the mLearning project included: mobile video streaming and sharing, collation and broadcasting mobile video using Livestream or UStream, creating an online identity, and associated business practices. The course lecturer created a Vox group, and all resources for the project were shared with the class via this group page (http://unutechsy309.groups.Vox.com/), including links to several Google Docs. Table 5 summarises the 2009 PASA mLearning project.

Implications of Case Study2: Bachelor of Performing and Screen Arts 2009

The Performing and Screen Arts mLearning project was one of the most ambitious of the mLearning projects with regards to the use and exploration of the mobile technologies. However, its implementation suffered from the relatively short time the lecturers had for personally appropriating the mLearning tools themselves, and timetabling limitations led to a significant change in the community of practice support model. While not personally modeling (A. Herrington, Herrington, & Mantei, 2009; J. Herrington & Oliver, 2000) the use of the mobile web 2.0 tools to a high level, the course lecturers nevertheless created an atmosphere of high expectations of the students that created an energetic 'buzz' among them, facilitating experimentation and collaboration around the use of the tools. While there was a lack of course-focused community facilitated by the WMD implementation, there was a very high level of personal appropriation of the WMDs by the participating students. The students found the portability and ubiquitous connectivity of the smartphones empowering for both accessing course content and their social networks. This case study therefore highlights the importance of the development of a regular supportive learning community, and the positive impact of high expectations from the lecturers on the participating students.

Student Feedback

The PASA students demonstrated a high level of personal, social and emotional attachment to the smartphones, exhibiting a reluctance to return the smartphones at the end of the project. The connectivity of both the netbooks and smartphones

Table 5. Outline of 2009 Performing and Screen Arts mLearning project

Course: Bachelor of Performing and Screen Arts, third year Film and TV class 2009	
Participants	• 25 students • 4 Course Lecturers • Technology Steward (Thom Cochrane - CTLI)
Mobile Technology	Dell Mini9 3G netbook, plus Nokia XpressMusic 5800 WiFi smartphone (or similar), participants responsible for 3G data, voice and txt costs.
Pedagogical Model	From Pedagogy to Andragogy
Pedagogical Focus	Film and TV major students investigate the current and future uses of web 2.0 technologies in performing arts film production and distribution. Students research and report on various technologies using a weekly podcast/vodcast that is peer critiqued by students in the course. Students experiment with live video streaming and collation of video using Livestream.com. The focus is upon students developing an understanding of the importance of a quality online profile and presence in the emerging crowd-source web 2.0 environment.
Community of Practice	Six introductory COP sessions at the start of the course
Support LMS	Moodle
Deliverables	An assessed online Blog/eportfolio documenting and showcasing students' design processes and forming the basis of a collaborative hub with worldwide peers and potential employers/clients. Scripting, shooting, editing and presentation of a mobile video short film.
YouTube Links	• Introduction to the assessed student project http://www.youtube.com/watch?v=00d-t0F9AzY • Student reflections on the use of the WMDs http://www.youtube.com/watch?v=jEA7EEcAQCA
Blog Links	• http://unutechsy309.groups.vox.com/ • http://karenperedo.vox.com/ • http://helloagnes.vox.com/
Course Project Outlines	1. Assessment Outline http://docs.google.com/Doc?docid=0ATo8wcQiO76XZDI3Z2QzZl8yNGdmNjdxY2Ru&hl=en 2. Project Workshops Outline http://docs.google.com/Doc?docid=0ATo8wcQiO76XZDI3Z2QzZl8yOWNucDk5NWM1&hl=en_GB
Timeframe	March 2009 through to July 2009 with Lecturers. Student projects begin Semester 2, 2009.

was highly appreciated by the students, who had very limited connectivity options within their course previous to the mLearning project. Some examples of PASA student feedback on the 2009 mLearning project are given below.

I find with the WMD you have a phone, a camera, a notebook, a flash drive, a pen to write with, music and entertainment (PASA student 2009 survey feedback).

It worked really well, but you get really used to the system and it's going to be sad to give the devices back (PASA student 2009 survey feedback).

I was always online at Uni, work and at home. I could check my mail while lounging on the couch or lying in bed. As a camera student, we don't have PCs at school, so I used it a lot (PASA student 2009 survey feedback).

We are actually sharing tips and teaching each other how to use these new technologies...:) it has kind of united us even more... all for the same cause (PASA student blog post July 2009).

I'm currently sitting in bed, watching Entourage on my PC on the desk and typing this with my new netbook, while idly playing Zelda on my Nintendo DS. I could also be writing this post on my phone and then doing a mobile upload, but it's easier to type on one of these. So I've been thinking, is this too much technology? Do I really need all of these things? Now that I'm twittered and blogged and Facebooked and Flickred and Qiked and everything, I feel a little overwhelmed... So do I really need a smartphone, a netbook, a PC, and everything else that goes with it? I certainly like it... So now I can take photos on my DSLR, upload them to my netbook, put my sim card in that and upload photos to the web with 3G, add them to my blog and then send the link around with twitter. CRAZINESS! (PASA student blog post July 2009).

The contemporary music students' appropriated a wide range of the iPhone's affordances both into their daily lives and into their course workflow. The portability, connectivity, and wealth of music-related applications for the iPhone were all highly rated by students. A selection of representative 2009 student feedback on the mLearning project integration is included in the following paragraphs.

As I went about my daily life I took my iPhone everywhere with me so that when I heard something that I thought I could use I could just record using Cycorder, which was really useful. This was very interesting because I was more aware of the sounds around me, other times I usually block them out, distracted by my own thoughts. ... I found musical elements in these sounds, through dynamics of engines to the rhythm of enthusiasm towards voting and politics, and the beeping of the horn of course (2009 Student1 blog post).

I've found the iPhone really useful. I use it a lot for surfing the net, and music related applications, and games, checking my emails, and just the ease of using something so small to do so much: Txting, making calls, all on one device, videoing, and using the recording apps, such as iTalk. I used to do a lot of MSN, but I use Twitter more now. I don't use Blackboard on it much, as it's not really iPhone friendly, and you have to do a lot of zooming to navigate around Blackboard. ... I've used it to record environmental sounds, record video, use Qik, geotagging, check my emails for the course, NetNewsWire for all the RSS feeds of the rest of the classes online sites, and Twitter has been great for keeping in contact with people, and Fring – especially for direct contact with Thom whenever I needed help, and accessing the web wherever there is WiFi access (2009 Student2 reflection).

There are huge benefits for any student using the iPhone – WiFi access, all the different applications and games you can get on the iPhone... These devices keep us up to date with each other. I use the iPhone for blogging, recording lectures and rehearsals, so I can take these home and review them and find ways to better our sound. I know how to use the Internet now much more than just sending emails. People have been amazed at what I can do with the iPhone – photos, videos, internet access, music applications, communicating, and even using it as a musical instrument – I have used the iPhone as a flute using the Ocarina application during live performances on stage (2009 Student3 reflection).

Lecturer Feedback

Lecturers were asked to record Vodcast reflections on the impact of the mLearning project within the course for 2009. Examples are transcribed in the following two paragraphs. Overall the 2009 lecturer feedback was very positive and evidenced a progression in their understanding of the pedagogical potential of mLearning within their courses.

I've found the iPhone really useful using Cycorder for recording student performances and then upload through YouTube onto their Vox blog so they can then actually review their performances and see what they've done (Part-time Lecturer, 2009).

I'm something of a skeptic when it comes to using the iPhones to assist with course work. So the goal for me was to integrate them into the coursework and not use them for the sake of it, but actually try and use them in a way where it would be the best way to do things. And I think we've seen to some extent, and particularly with the recording project that the students have been working on recently, that the technology of the iPhone can be very useful for their learning, and I think we're also starting to see the students working in very different ways than what we've seen before (Contemporary Music Lecturer, 2009).

I can't say enough about your contribution to our Year 3 New Technologies mobile learning project this year. You facilitated it seamlessly, laying the initial groundwork by up-skilling the staff – all the while imbuing your training with the social-constructivist applications of the gear. This provided an initial context for these new communication tools, with which the Screen Arts staff involved shall always associate and use them (PASA lecturer, 2009).

Pedagogical Strategies

Curriculum Integration of Mobile Web 2.0

A key strategy to facilitate a move along the PAH continuum is curriculum integration of mobile web 2.0. The case studies illustrate that curriculum integration must focus on the unique affordances of mobile web 2.0 in order to create authentic learning environments. To achieve this, curriculum integration must start with the learning practice that is to be achieved (As illustrated in Table 1), aligning and choosing appropriate mobile web 2.0 affordances with this goal. Following such a design framework will ensure that the technology is not the primary focus, or that good pedagogy is retrofitted to technology.

Modeling the Pedagogical Use of the WMDs and Social Software

Lecturers must model the use of the mobile web 2.0 tools within their own daily workflows and within authentic course-related contexts. The various mLearning projects undertaken have illustrated that pedagogical integration of mLearning into a course or curriculum requires a paradigm shift on behalf of the lecturers involved, and that this takes significant time. Hameed & Shah (2009) describe this process as a "cultural re-alignment". An intentional Community Of Practice model (Langelier, 2005) has been found to be effective for guiding and supporting the mobile web 2.0 projects. This comprises weekly "technology sessions" (Community of Practice) with small groups of lecturers facilitated by an appropriate 'technology steward' (Wenger, White, & Smith, 2009; Wenger, et al., 2005).

Stage and Scaffold the Learning Load

Mobile web 2.0 integration into a course produces and requires significant rethinking of lecturer pedagogies and assessment procedures. To minimise the level of technological load and scaffolding required by the students (and lecturers), the implementation of mobile web 2.0 should be staged and scaffolded using a select range of activities over significant timeframes. Thus beginning the introduction of web 2.0 integration into the first year of a course (in multi-year courses) will prepare students for higher-level context bridging in subsequent years of their course.

Table 6. Example mLearning roll-out timeframe

Deliverable	Timeframe	Outcome
Establish weekly COP with lecturers and technology steward. Establish support requirements (with IT Services and Telco).	Semester 1	Staff develop competency with mLearning. Staff develop pedagogical mLearning activities based on social constructivist pedagogies.
mLearning projects with staff and students. Implementation of the mLearning activities within each course and assessment.	Semester 2	Increased student engagement. Flexible delivery. Facilitating social constructivist pedagogies and bridging learning contexts.
Lecturers publish and present case studies based on project implementation.	End of Semester 2 and beginning of Semester 3	Conference, Journal publications and symposia presentations.

Implementation Strategies

Implementation Model

Based upon the researcher's experiences, in order to achieve an explicit move to a social constructivist learning environment using mobile web 2.0 tools, a staged, and scaffolded approach has been adopted (illustrated below in Table 6). This staged approach allows the bridging of the PAH (Pedagogy, Andragogy, Heutagogy) continuum (Luckin, et al., 2008), and the embedding of mobile web 2.0 affordances that support each stage. Additionally, as the life-span of mobile computing is generally shorter than that of desktop computing, a staged roll-out of WMD computing for students involved in three year long courses can be achieved to minimise the redundancy of the student-owned WMDs. Lecturer professional development and technological support have been found to be critical in facilitating the pedagogical focus of this roll-out.

A staged integration of mLearning (mobile web 2.0) across the three years of a programme can be structured as follows in Table 7 below.

Table 7. Scaffolding the roll-out of mobile web 2.0 throughout various course levels

Stage	Web 2.0 Tools	MLearning Tools	Indicative Student Course Related Costs (NZ)	Course Timeframe	PAH Alignment
Level 1	Social Collaboration with peers and lecturer. Student generated content.	Use of student-owned netbook or mid-range smartphone, LMS and basic web2.0 sites	Netbook $700 Internet paid access $250	1 year Certificate programmes, or first year of longer programmes	Pedagogy (Lecturer directed)
Level 2	Social collaboration with peers and 'authentic environments'. Context Aware	Student-owned laptop and/or mid-range smartphone	Laptop cost $750 ($1500 spread over 2 years) And/or smartphone $750 Internet paid access $250	Second year of two year or longer programmes	From Pedagogy to Andragogy (Students become the content creators)
Level 3	Context independent. Student generated contexts.	Student-owned laptop and/or high-end smartphone	Laptop cost $750 ($1500 spread over 2 years) And/or smartphone $750 Internet paid access $250	Third year of programme	From Andragogy to Heutagogy (Students become independent learners)

FUTURE RESEARCH DIRECTIONS

While the research has sought to produce transferable principles and strategies to enhance tertiary education using mobile web 2.0, it is ultimately bound by the limits of the contexts of the learning communities that it is embedded in (the five case studies are based in the 'creative arts and industries' fields), and the current affordances of the available mobile web 2.0 technologies. The mobile web 2.0 projects have so far used a model of providing a common smartphone for the students and lecturers within a course. The students and lecturers involved have been encouraged to use the smartphones as if they owned them for the period of the projects. This approach was used to seed the concept and provide proof of concept results. However, to create a sustainable approach, the goal going forward is to move to a student-owned model, where students purchase a smartphone that meets specifications outlined by the course requirements – much as many institutions currently require students to purchase a specified laptop computer to ease support requirements. As the cost of appropriate smartphones and 3G data costs drop, the purchase cost may be sustainably subsidized by institutions in lieu of other course related costs that the mobile web 2.0 paradigm replaces. However, it is yet to be seen whether there can be transferability of the research outcomes based upon an institution supplied or specified WMD and mLearning projects based upon student chosen and owned WMDs.

The technological goal-posts of mobile web 2.0 are rapidly changing, and new integrated smartphone affordances continue to provide new ways of communicating, collaborating and enhancing learning. An example for future research is the rise of augmented reality applications for smartphones and integration with web-based services. The challenge is to implement these new technologies from a sound pedagogical basis.

A limitation of the participatory action research methodology of the research is the significance of the input of the researcher as the technology steward for the projects. The partnerships developed between the researcher and the participants (particularly the lecturers) have been critical in supporting and providing direction for the projects. It is yet to be seen whether the approach can be transferred to other mLearning contexts involving a different technology steward.

CONCLUSION

The combination of the user content creation and sharing capabilities of web 2.0 with the ubiquitous connectivity and unique context aware affordances of smartphones provides a rich tool for facilitating social constructivist learning environments. This chapter has outlined a research-informed approach to implementing mobile web 2.0 in tertiary education. While a lot can be achieved with a standard cameraphone combined with web 2.0, the cost of smartphones with built-in GPS, compass, and quality camera continues to reduce, bringing these tools into the pockets of more students. Harnessing the power of these tools within education can transform teaching and learning in unique ways. Achieving this involves a design framework, lecturer development and support, and student scaffolding and technology support.

REFERENCES

Alexander, B. (2006, March/April). Web 2.0: A new wave of innovation for teaching and learning? *EDUCAUSE Review*, *41*, 33–44.

Alexander, B., Brown, M., Doogan, V., Johnson, L., Levine, A., & Sabean, R. (2006). *The Horizon report 2006 edition*. Austin, TX: The New Media Consortium.

Bruns, A. (2007, March 21-23). *Beyond difference: Reconfiguring education for the user-led age.* Paper presented at the ICE3: Ideas in cyberspace education: digital difference, Ross Priory, Loch Lomond.

Chi, M., & Hausmann, R. (2003). Do radical discoveries require ontological shifts? In Shavinina, L., & Sternberg, R. (Eds.), *International handbook on innovation* (*Vol. 3*, pp. 430–444). New York, NY: Elsevier Science Ltd. doi:10.1016/B978-008044198-6/50030-9

Cochrane, T. (2010). Mobilizing learning: Intentional disruption. Harnessing the potential of social software tools in higher education using wireless mobile devices. *International Journal of Mobile Learning and Organisation*, *3*(4), 399–419. doi:10.1504/IJMLO.2009.027456

Cochrane, T., & Bateman, R. (2009). Transforming pedagogy using Mobile Web 2.0. *International Journal of Mobile and Blended Learning*, *1*(4), 56–83. doi:10.4018/jmbl.2009090804

Cochrane, T., & Bateman, R. (2010a). *Reflections on 3 years of mlearning implementation (2007-2009)*. Paper presented at the IADIS International Conference Mobile Learning 2010.

Cochrane, T., & Bateman, R. (2010b). Smartphones give you wings: Pedagogical affordances of mobile Web 2.0. *Australasian Journal of Educational Technology*, *26*(1), 1–14.

Cochrane, T., Flitta, I., & Bateman, R. (2009). Facilitating social constructivist learning environments for product design students using social software (Web2) and wireless mobile devices. *DESIGN Principles and Practices: An International Journal*, *3*(1), 15.

Cook, J., Bradley, C., Lance, J., Smith, C., & Haynes, R. (2007). Generating learner contexts with mobile devices. In Pachler, N. (Ed.), *Mobile learning: Towards a research agenda* (pp. 55–73). London, UK: WLE Centre, Institute of Education.

Hameed, K., & Shah, H. (2009). *Mobile learning in higher education: Adoption and siscussion criteria.* Paper presented at the IADIS International Conference on Mobile Learning 2009, Barcelona, Spain.

Herrington, A., & Herrington, J. (2007). *Authentic mobile learning in higher education.* Paper presented at the AARE 2007 International Educational Research Conference, Fremantle, Australia.

Herrington, A., Herrington, J., & Mantei, J. (2009). Design principles for mobile learning. In Herrington, J., Herrington, A., Mantei, J., Olney, I., & Ferry, B. (Eds.), *New technologies, new pedagogies: Mobile learning in higher education* (pp. 129–138). Wollongong, Australia: Faculty of Education, University of Wollongong.

Herrington, J., Mantei, J., Herrington, A., Olney, I., & Ferry, B. (2008). *New technologies, new pedagogies: Mobile technologies and new ways of teaching and learning.* Paper presented at the ASCILITE 2008, Deakin University, Melbourne, Australia.

Herrington, J., & Oliver, R. (2000). An instructional design framework for authentic learning environments. *Educational Technology Research and Development*, *48*(3), 23–48. doi:10.1007/BF02319856

JISC. (2009a). Effective practice in a digital age. Retrieved from http://www.jisc.ac.uk/publications/documents/ effectivepracticedigitalage.aspx

JISC. (2009b). *Higher education in a Web 2.0 world.* Retrieved from http://www.jisc.ac.uk/publications/documents/heweb2.aspx

Langelier, L. (2005). *Working, learning and collaborating in a network: Guide to the implementation and leadership of intentional communities of practice*. Quebec City, Canada: CEFIRO (Recherche et Études de cas collection).

Laurillard, D. (2001). *Rethinking university teaching: A framework for the effective use of educational technology* (2nd ed.). London, UK: Routledge.

Laurillard, D. (2007). Pedagogical forms of mobile learning: Framing research questions . In Pachler, N. (Ed.), *Mobile learning: Towards a research agenda* (pp. 33–54). London, UK: WLE Centre, Institute of Education.

Luckin, R., Clark, W., Garnett, F., Whitworth, A., Akass, J., Cook, J., et al. (2008). *Learner generated contexts: A framework to support the effective use of technology to support learning.* Retrieved 5 November, 2008, from http://api.ning.com/files/Ij6j7ucsB9vgb11pKPHU6LK-MGQQkR- YDVnxruI9tBGf1Q-eSYUDv- Mi-l6uWqX4F1jYA1PUkZRXvbxhnxuHusyL1l-RXVrBKnO/ LGCOpenContextModelning.doc

McLoughlin, C., & Lee, M. (2007). *Social software and participatory learning: Pedagogical choices with technology affordances in the Web 2.0 era.* Paper presented at the Ascilite 2007, ICT: Providing Choices for Learners and Learning, Centre for Educational Development, Nanyang Technological University, Singapore.

McLoughlin, C., & Lee, M. (2008a). Future learning landscapes: Transforming pedagogy through social software. *Innovate: Journal of Online Education, 4*(5), 7.

McLoughlin, C., & Lee, M. (2008b). *Mapping the digital terrain: New media and social software as catalysts for pedagogical change.* Paper presented at the ASCILITE Melbourne 2008, Deakin University, Melbourne.

Mishra, P., Koehler, M. J., & Zhao, Y. (Eds.). (2007). *Faculty development by design: Integrating technology in higher education.* Charlotte, NC: Information Age Publishing.

Piaget, J. (1973). *To understand is to invent.* New York, NY: Viking Press.

Sharples, M., Crook, C., Jones, I., Kay, D., Chowcat, I., & Balmer, K. (2009). *CAPITAL year one final report (Draft).* University of Nottingham.

Swantz, M. L. (2008). Participatory action research as practice . In Reason, P., & Bradbury, H. (Eds.), *The SAGE handbook of action research: Participative inquiry and practice* (2nd ed., pp. 31–48). London, UK: SAGE Publications.

Uden, L. (2007). Activity theory for designing mobile learning. *International Journal of Mobile Learning and Organisation, 1*(1), 81–102. doi:10.1504/IJMLO.2007.011190

Vavoula, G. (2007). *Learning bridges: A role for mobile technologies in education.* Paper presented at the M-Learning Symposium. Retrieved from http://www.wlecentre.ac.uk/cms/index.php?option=com_content&task=view&id=105&Itemid=39

Vygotsky, L. (1978). *Mind in society.* Cambridge, MA: Harvard University Press.

Wenger, E., White, N., & Smith, J. (2009). *Digital habitats: Stewarding technology for communities.* Portland, OR: CPsquare.

Wenger, E., White, N., Smith, J., & Rowe, K. (2005). Technology for communities . In Langelier, L. (Ed.), *Working, learning and collaborating in a network: Guide to the implementation and leadership of intentional communities of practice* (pp. 71–94). Quebec City, Canada: CEFIRO.

KEY TERMS AND DEFINITIONS

3G: Third generation mobile 'broadband'.

Andragogy: Student-centred (Adult) learning.

Blog: Weblog: Online journal.

Community Of Practice (COP): A peer support group with a common interest and practice.

Eportfolio: Collection of online media.

Heutagogy: Self-directed learning.

LMS: Learning Management System: For example Blackboard or Moodle.

Mlearning: Mobile learning.

Mobile Web 2.0: Web 2.0 sites optimised for the affordances of mobile devices.

Pedagogy: Teacher-directed learning.

RSS: Rich Site Summary: For subscribing to update information to web 2.0 sites.

Smartphones: Mobile phones with an extensible operating system.

Social Constructivism: A social theory of learning that postulates students learn via social interaction, and personal experimentation and investigation.

Web 2.0: Interactive, customisable web services, facilitating user-generated content.

WiFi: Wireless Ethernet connectivity.

Wiki: Editable collaborative web page.

WMD: Wireless Mobile Device.

Zone of Proximal Development (ZPD): The difference between what a learner can learn on their own and what they can learn with the guidance of an expert.

Chapter 15
Open Source Implementation of Mobile Pair Programming for Java Programming Class

Lee Chao
University of Houston-Victoria, USA

ABSTRACT

Pair programming has been used to improve the learning of programming by many computer science educators. Implementing pair programming on the mobile learning platform can significantly improve the availability and collaboration between pair programming partners. This chapter explores a set of open source mobile technologies that can be used to implement mobile pair programming. It shows that mobile pair programming can be implemented entirely with open source or free software.

INTRODUCTION

As mobile technology advances, mobile learning has been adopted by more and more educational institutions. Instructors and students of universities and schools have developed various effective solutions to implement mobile learning in many academic fields. For non-technical academic fields such as English, history, communication, and so on, the implementation of mobile learning requires less effort. On the other hand, in lab intensive academic fields such as computer science, biology, and so on, the implementation of mobile learning has to consider how to implement lab activities through a mobile network. The mobile learning designer needs to consider the usability, compatibility, and affordability of a mobile network infrastructure when implementing the lab activities required by the curriculum. The designer is required to thoroughly understand the lab activities' requirements and the technologies used to carry out the mobile learning. To illustrate

DOI: 10.4018/978-1-60960-613-8.ch015

some strategies and methods for implementing lab activities over a mobile network, this chapter considers the implementation of mobile pair programming with open source and free products.

The discussion starts with the investigation of the behaviors of novices in learning programming. This chapter briefly reviews the literature of research in teaching beginning programming classes. One of the methods suggested by these researchers to improve learning programming is pair programming. Several articles discussed the use of pair programming and gave the evaluation results. Most of the pair programming studies have been done in the lab environment. A pair of students works together in front of one computer. In a mobile learning environment, it is challenging for two students to do the same. Currently, it is hard to find research on implementing pair programming in a mobile learning environment, especially with open source products. This situation leads to the discussion in this chapter.

This chapter gives an overview of mobile pair programming. Originally, pair programming is a method used by professional programmers to efficiently develop computer software. This chapter briefly discusses how pair programming can also improve programming skills. It explains the methods of pair programming and how pair programming is used in a programming class. This chapter also examines the difficulties in implementing mobile pair programming.

This chapter discusses how mobile pair programming is implemented with open source or free software. It introduces a list of open source and free software that can possibly be used in a mobile pair programming project. The list includes the software on the server side, the software on the client side, and the software for collaboration. The server side software includes the server operating system, integrated development environment (IDE), software development kit (SDK) for mobile devices, and virtualization software. The client software includes the mobile device emulator, Internet voice service, Web browser, messaging service client and social service client software. For the implementation of pair programming, this chapter also introduces several collaboration software and screen sharing software. It briefly describes the features and functionalities of the open source and free software.

This chapter illustrates the implementation of pair programming through a case study project. The software used in this project is either open source or free software and can be freely downloaded from the Web. The project includes the implementation of the server operating system Ubuntu, the virtual machine VMware, Eclipse IDE, Android Application SDK, VNC screen sharing, the integrated voice service Nimbuzz, the social service software Skype, and the Android mobile device emulator. The project demonstrates software selection strategies, installation procedures, and the configuration of software to meet the pair programming requirements.

BACKGROUND

Due to its ease of use, versatility, security, cross-platform capabilities, and multimedia features, Java as a programming language has been widely used in computing related curriculums (Klawonn, 2008). Java is often the choice for the courses such as introduction to programming, multimedia applications, Web development, operating systems, human computer interfaces, and so on.

On the other hand, in computing and technology education, learning programming posts some challenges to students who are new to programming. Programming has been identified as one of the great challenges in computing education (McGettrick et al., 2005). The challenge of programming has contributed to the relatively low enrollment and retention rate in the computing related programs at higher education institutions (Denning, 2004). Robins et al. (2003) described the characteristics of the novice programmers as below.

- Without thinking structurally, novices work on programming tasks through a line-by-line approach.
- When trying to understand the code in a computer program, novices are poor at tracking/tracing the code. They have a poor grasp of the order of code execution.
- Novices lack strategies to map a problem to a computer program. They have difficulty applying relevant knowledge.
- For novices, it takes a longer learning curve to get to know about object-oriented programs. It is difficult for novices to mentally represent the control structures and functions in an object-oriented program.
- Novices tend to focus on specific language features, but fewer program planning/designing.

Each programming language has its own functionalities and features. During a lecture, students learn about these functionalities and features of a programming language. However, it is difficult for novices to apply these functionalities and features in problem solving. They often do not have strategies on how to apply what they have learned.

By understanding the behavior of novices, researchers have developed different teaching strategies to improve the learning of programming skills. For example, Spohrer and Soloway (1989) suggested that we improve teaching by illustrating control flows, exhibiting design principles, and displaying program states. Gárcia-Mateos and Fernández-Alemán (2009) recommended using a web-based automatic evaluation system to continuously evaluate students' activities, instead of the final exam. Mennila (2007) suggested a similar evaluation method that evaluates students' understanding of the program code as a whole, evaluates their understanding of individual constructs, and finds out how they consider the difficulty levels of different programming topics. Fernández-Alemán (2009) recommended sequential access models to allow students to deduce algorithmic schemes for implementing iterative algorithms. In the research conducted by Nagappan et al. (2003), pair programming is used to improve the learning of programming. Through pair programming, students can understand the code better and can achieve better retention than learning programming solo.

MOBILE PAIR PROGRAMMING

Before being applied to the learning of programming, pair programming was one of the approaches used in the development of software. During a software developing process, two programmers are involved in the project. One programmer creates the computer code and the other thinks structurally about the implementation of the future software while reviewing the code. The programmer working with the code is called the driver. The other one who is reviewing the code is called the navigator. The two programmers regularly switch their roles during the software development process. Software developers involved in pair programming can share their knowledge and strategies on mapping a problem to a computer program. By teaming up as a group, the software developers make fewer mistakes and have much better chance to complete the software development assignment before the deadline.

Recently, pair programming has been extended to the Web based environment. Two programmers in different locations work together through a shared desktop or a collaborative enabled programming editor (Kappe, 2007). F Steimann, J Gößner, U Thaden (2009) proposed an experimental design strategy on the evaluation of mobile based pair programming.

Pair programming has some limitations, especially in the mobile learning environment. The students who participate in pair programming need to work together for an extended period of time. This requires the batteries of mobile devices to last long enough to support the entire learning process. Unlike professional programmers, students are less disciplined. Some students may not able to keep up with pair programming schedule. If students from different continents carry out pair programming, they may face even more challenges. The time zones may be many hours apart, the performance of the mobile networks on both sides may be uneven, there may be a cultural difference between the two learners; and the cost for long distance dialing may add another factor to the difficulty.

MOBILE PAIR PROGRAMMING TECHNOLOGIES

To implement the distance pairing, two students should be able to access a remote desktop or server computer with their mobile devices. They can do so through a Wi-Fi network or a 3G mobile network managed by a wireless phone company. Mobile devices that are capable for pair programming are Smartphones, PDAs, MIDs, netbook computers, and notebook computers. Since not all the students in a pair programming project own a Smartphone, they can use a notebook or netbook computer to remotely access a virtual machine which is used to host the Integrated Development Environment (IDE) for writing, editing, and testing the programming code. The virtual machines may be hosted by a server and run in a computer lab on campus and managed by IT service teams. Once a virtual machine is created with open source or free virtual servers, the open source operating system Ubuntu Linux can be installed as the guest operating system of the virtual machine. Ubuntu is known for its sophisticated GUI interface and its support for wireless and mobile networks. Other open source Linux operating systems can also get the job done.

Various open source or free software is available for supporting pair programming in teaching programming languages. For a software development project with one of many programming languages, Eclipse is the one of the well-known IDE packages.

To be able to create applications that are able to run on an Android Smartphone, one needs to install the Android Software Development Kit (SDK) plug-in on Eclipse. To implement Web based pair programming, open source software such as Sangam or Saros can be used to deliver the driver's activities to the computer on the navigator's end.

For real time communication during a pair programming process, one needs Internet voice service software which allows the two students to communicate with each other. The Internet voice service technology such as Skype Lite allows the partners in a pair programming process to talk to each other. By using the Internet voice service, the partners in a pair programming group can make affordable local and international calls from their mobile devices.

For sharing the driver's screen, one can consider using the Virtual Network Computing (VNC) technology. The VNC technology allows the student who acts as the driver to remotely edit the code and the student who acts as the navigator to share the same desktop image. Some of the VNC software is specifically designed to run on mobile devices.

For Web project management and debugging, one can find software such as Redmine that may help. Some of the Web project management jobs can also be done with a Web browser such as Firefox.

In the following, a brief overview of these technologies is given. The overview describes some open source and free software that can possibly be used to implement mobile pair programming. The overview starts with the server side technologies.

Ubuntu GNU/Linux: Ubuntu (Ubuntu, 2009) is the open source GNU/Linux operating system software. It is sponsored by Canonical Ltd., a private company from South Africa. Ubuntu provides three editions, server, desktop, and mobile. It is designed to handle a wide range of computing tasks, from supercomputers to hand-held mobile devices. It is designed to fit in various computer technologies. Therefore, it is widely supported by computer manufacturers. On the server side, it is supported by IBM, Sun, Dell, and so on. For desktop and notebook computers, it is supported by companies such as Dell and Lenovo. It is also supported by a wide range of Intel based mobile devices. In recent years, Ubuntu Linux has been gaining attraction among users.

The strong support for the GUI based desktop is one of the attractive features provided by Ubuntu. The GUI support is important for the beginning programming class since most of our students are new to the Linux operating system. They are not familiar with the Linux commands. When compared with other Linux distributions, Ubuntu is relatively easy to use.

For mobile learning, Ubuntu is an ideal operating system since it supports 3G mobile networks. It is also one of the best Linux operating systems in supporting Wi-Fi networks. With these kinds of support, students can smoothly move from a Wi-Fi network to a 3G mobile network and vice versa. This feature is ideal for those students who are on the move. When a Wi-Fi network is available, students should use the Wi-Fi network for communication so that they can save the valuable 3G mobile network minutes. When a Wi-Fi network is not available, they need to switch to a 3G mobile network for communication. Ubuntu can automatically detect and manage the 3G connection. Ubuntu also works well with the 3G modem, Smartphone, Bluetooth, and other mobile technologies.

Ubuntu is regularly updated. Every 6 months, it is upgraded so that it can keep up with the fast developing PC industry. Since mobile devices are changing rapidly and there are so many different types of them, it is important to regularly upgrade the operating system with new drivers and applications.

Ubuntu provides strong support for portable devices such as CDs, DVDs, and USB flash drives. Ubuntu allows users to use a USB flash drive for operating system installation. It allows users to back up the current operating system to a USB flash derive and restore the operating system from the USB flash drive. Ubuntu can also write the Linux operating system on a live CD and use the CD for live hard disk installation without restarting the computer. Ubuntu Mobile and Embedded Edition is developed on the Low-Power Intel Architecture (LPIA). To support low-power mobile devices, Ubuntu provides the Personal Package Archive (PPA) service which is used to build and publish packages.

Ubuntu works well with multimedia content. It handles high quality video clips, radio broadcasts, and VoIP phones. Even for mobile devices such as Mobile Internet Devices (MIDs), Ubuntu Mobile and Embedded Edition can replay video and sound with small sized memory and disks. It also delivers fast animations and clear graphics. In addition, Ubuntu includes many utilities for multimedia content editing such as photo editing and media authoring.

Ubuntu has a large user base. According to a desktop Linux survey (Vaughan-Nichols, 2007), the Ubuntu family was number one in usage among the Linux desktop operating systems during the year 2007. It took 30 percent of the Linux desktop operating system market share according to the survey. With the large user base, it is relatively easy to get solutions for various technical problems.

When used as a server operating system, Ubuntu includes the LAMP (Linux, Apache, MySQL and PHP) package. It can automatically install and configure the LAMP package and make it ready to use. This feature makes the server installation quick and easy. It avoids the risk of misconfiguration. For operating system management, it provides management tools that can be used for searching, calendaring, Web form spell checking, phishing detection, network monitoring, and system security. In addition, Ubuntu includes a variety of application software such as the latest e-mail server and Web browser, the office suite,

OpenOffice.org, the instant messenger Pidgin, and the graphic editor GIMP.

Virtual Machine: Using software to implement a device such as a computer or a hard drive is called virtualization (Smith and Nair, 2005). The software that is used to implement a computer is called a virtual machine (VM), which includes the virtual hard drive, virtual memory, virtual network interface card, and so on. Virtual machines need to be hosted by a server computer which allows multiple users to log on simultaneously. While running on the server computer, the virtual machine uses the physical memory and hard drive. A server computer should have enough RAM and hard drive space to host multiple virtual machines. The reasons to use virtual machines instead of physical computers are:

- Virtual machines can run independently and pose no security risk to the host computer. Since students need to remotely access computers through the Internet, it is risky to expose the physical computers to the public. With virtual machines, students can even get the administrative permission without a security risk so that they are able to configure connections to mobile devices.
- Using virtual machines can significantly reduce cost. There is no need to purchase new computer systems.
- The management of virtual machines is much easier than that of physical machines. For each new semester, each student can be assigned a copy of a virtual machine. By the end of the semester, the administrator can simply delete the copy. In case of crash, a new copy of the virtual machine can be used to replace the crashed virtual machine; there is no need to reinstall the operating system and applications.

In general, virtual machines run more slowly than physical machines. This disadvantage may not be a real problem since most students' projects do not require much computing power.

Several open source or free virtual machine products are available. The free version of VMware (2009) includes VMware Sever used to create virtual machines and VMware Player used to run virtual machines. These two free packages are adequate for the needs of pair programming.

Xen is open source virtualization software that is designed to run on a host computer powered by a UNIX like operating system (Citrix Systems, 2009). Xen virtual machines are able to run as fast as the host computer. It can also adjust its usage of resources to adapt to the resources available on the host computer. When the host computer is busy, Xen can reduce its usage of physical resources.

Eclipse: Eclipse is an Integrated Development Environment (IDE) package (The Eclipse Foundation, 2009). Multiple programming languages such as Java, C, C++, COBOL, Python, Perl, PHP, and others are supported by Eclipse. To develop software with a specific programming language, the developer needs to install a plug-in specially designed for that programming language. Eclipse provides a graphical user interface (GUI) environment called Eclipse Workbench for viewing, editing, executing, and debugging the programming code. The Eclipse Workbench consists of an editor menu bar, perspective, tool bar, and view. These Workbench components have the following functionalities:

- *Editor*: An editor can be used to display, insert, delete, and modify the programming source code. To reduce syntax errors, keywords and commands are displayed in various colors.
- *Menu Bar*: A menu bar is used for the management of files, editing, projects, windows, help documents, and property configuration.
- *Perspective*: A perspective is a container hosting the editor, resource view, debugging view, and message console. By pro-

viding the necessary information, the perspective offers an environment for the programmer to work on his/her project.

- *Tool Bar*: A tool bar contains the commonly used commands to perform tasks such as searching, printing, and saving.
- *View*: A view is an area where a specific piece of information can be displayed, such as code, error message, and resource. Through the view, the programmer can interact with the code or configure the properties of a project.

In addition to the Workbench, Eclipse also provides built-in classes to help programmers to create GUI and to run plug-ins.

Sangam: As an Eclipse plug-in, Sangam allows two programmers to share the same Eclipse Workbench (Ho, Raha, Gehringer, & Williams, 2009). Sangam provides a server called Syncshare Server which is built to run within the Eclipse development environment. Syncshare allows multiple users to create their own accounts on the server. Two students who participate in pair programming need to connect to the same server. After both students in pair programming are connected to the server, the student who is acting as the driver begins to work on the code. Sangam collects the events generated by the activities done by the driver and runs the same events on the navigator's computer. In such a way, the drive is actually working on the navigator's computer as well.

Saros: Saros is another Eclipse plug-in designed for distributed collaborative editing and pair programming (Freie Universität Berlin, 2010). Saros distributes the changes made by the driver to all other participants in group programming. Saros uses the XMPP/Jabber server to initiate a session. The Extensible Messaging and Presence Protocol (XMPP) is a protocol designed for instant messaging and online presence detection. Jabber is another name for XMPP. Openfire is one of the open source XMPP servers. Once a driver creates a session, the driver invites the other partner to join the session. Unlike Sangam, Saros supports a multi-driver platform and supports high quality audio conferencing for collaboration.

Virtual Network Computing (VNC): One of the technologies used in remote pair programming is VNC which an open source desktop sharing technology developed by the Olivetti Research Laboratory. VNC allows two or more programmers to share the same desktop screen. After the programmers at different locations log on to a VNC server, each of their computer screens will display the same desktop. VNC transmits the keyboard and mouse movements on one programmer's desktop to other programmers' desktops.

The VNC technology can now be used in the mobile computing environment (Rennison & Andrews, 2008). The mobile edition of VNC developed by RealVNC allows users to access the desktop on a PC from mobile devices and allows PC users to remotely manage mobile devices. To run RealVNC, the RealVNC server should be installed on the desktop or notebook computer where Eclipse is installed. On an Android Smartphone, install and start the VNC client called Remote VNC Pro. Once the connection is established to the RealVNC server, both students in pair programming can share the Eclipse Workbench.

Many other VNC solutions are also available for mobile devices. In fact, the Android emulator supports a version of the VNC server. Open source software such as android-vnc-viewer (Ohloh, 2010) can be used to connect an Android Smartphone to the VNC server.

Internet Voice Service: There are dozens of companies that provide the VoIP and P2P technology based Internet voice service. The Internet voice service allows a PC user to make free calls to another PC user or make low cost calls to other landline or mobile phone users. Companies such as Skype, Truphone, Sipdroid, Nimbuzz, and fring provide the free Internet voice service software specifically designed for Android mobile devices.

Skype Lite is free software for voice communication through the Internet (Skype Lite, 2009). It can be installed on Android Smartphones and other Android mobile devices. Through Skype, a student in pair programming can check if his/her partner is online. If the partner is online and ready to chat, the student can call his/her partner or contact the partner through instant messaging. If an Android mobile device has the mobile calling plan and data plan, Skype Lite can use the local phone system or mobile network for communication. Unlike PC users, the call from one mobile device to another mobile device is not free. It requires users to use their valuable Skype credits.

Truphone allows Android based mobile devices to communicate through VoIP (Truphone, 2010). It is free for calls from one Truphone user to another Truphone user. Once Truphone is installed on a mobile device, it uses Wi-Fi connectivity to route calls thought the Internet. The partners in a pair programming group do not have to use their valuable mobile phone minutes to discuss programming code. Truphone also supports native language multi-communication. Truphone has a specific version that is designed for Android mobile devices. It can be downloaded from Android Market.

Sipdroid (Sipdroid, 2009) is VoIP client software for Android based mobile devices. To make a VoIP call, Sipdroid needs to use the Session Initiation Protocol (SIP) service offered by companies such as PBXes (PBXes, 2009). The protocol SIP is used to establish a session between callers in an IP network. By using Sipdroid, students in a pair programming project can make phone calls and do text messaging through a 3G or wireless network. Sipdroid integrates the mobile phone service and VoIP service so that users do not need to manage two separate accounts. Students can download Sipdroid from Android Market.

Nimbuzz (Nimbuzz, 2010) is free Internet service software that provides VoIP, instant messaging, and presence services. Through Nimbuzz, students can integrate the mobile service and popular social networks such as Facebook and MySpace, as well as messaging services such as Skype, Windows Live Messenger, Yahoo! Messenger, and Google Talk into a single account. Since Nimbuzz integrates popular messaging and social service networks, Nimbuzz users can communicate with all their contacts who may have registered in different services. The Nimbuzz mobile service allows students to make free calls to each other on the Internet. Students can also use Nimbuzz to make affordable calls through a landline or mobile network. The phone backup service called Phonebook allows students to back up their phone books on their Smartphones. In case a student loses his/her phone, the student can get the contact information from the Phonebook service. Nimbuzz can be freely downloaded from Android Market.

Like Nimbuzz, fring (fring, 2009) is another free service integration software package which allows students to interact with their friends who are registered in different social networks and messaging services. The users of fring form a mobile Internet community. Like Sipdroid mentioned earlier, fring allows students to use a third party SIP service such as Skypeout/Skypein to call other landline or mobile phones. To avoid the reconfiguration of the Wi-Fi connection when a mobile user enters a different Wi-Fi network zone, fring automatically connects a Smartphone to an available Wi-Fi network. Once a mobile device is connected to a Wi-Fi network, fring automatically switches the 3G connection to the Wi-Fi connection to save some valuable 3G network minutes. From Android Market, students can also download fring to their Android enabled mobile devices.

Redmine: Redmine is open source software for Web application management (Redmine, 2009). By using Redmine, programmers can assemble multiple pieces of code to form a large project. It can also be used to keep the schedule and daily progress for pair programming.

MOBILE PAIR PROGRAMMING PROJECT

To be able to pair two students so that they can work on a programming assignment together, pair programming should be created. Based on the requirements and availability, different sets of mobile pair programming technologies can be used. For the project presented in this section, one may need to install and configure open source or free software such as the Android software development kit (SDK), Eclipse, Eclipse Android Plug-in, and RealVNC server and client. The following discusses the strategy and design for implementing pair programming.

Project Description: In this beginning-level online Java programming class, more than 60 percent of the students are currently working part time or full time. Some of them may need to travel. They may not have formal training on Java programming. Since it is an online class, the students will complete all their programming assignments online. To improve the teaching and learning of Java programming in such a diversified environment, the computer information system department wants to try the pair programming approach.

At the beginning of the class, the students are divided into groups based on their geographic areas (where they live) and their GPAs, two students in each group. If it is possible, the instructor may try to group two students so that their GPA values are about same as those of other groups. Students within the same group may not come from the same geographic area. Some group members may travel around during the semester. Each group is assigned about ten programming assignments per semester. The students are asked to schedule some time blocks so that both students can participate in the pair programming. The students in each group are asked to alter their pair programming roles for different assignments. For the students to be able work as distant pairs, the IT service department must create a pair programming environment so that the students in the same group can work on the programming together.

The infrastructure for the pair programming is created on the client-server architecture. The client side includes mobile devices or computers used by the students. The server side includes virtual machines which are used to host an integrated development environment (IDE) for writing, editing, and debugging Java code. To be more efficient, the students are required to design computer programs for their assignments and create the first draft of Java code on paper before participating in pair programming.

Also, to motivate the students and to make the learning more fun, the students in the class will be asked to create some simple applications for their mobile phones. Due to the first-rate usability, flexibility, and availability of the Android operating system, the Android based Smartphone is selected as the mobile device to be used in the class. For students who do not own an Android Smartphone, they can use the Android Smartphone emulator to complete their assignments.

There are many ways to implement the pair programming project. The following is an example that illustrates the implementation for the university environment. This example shows an implementation scheme that helps faculty members to implement pair programming for their own classes on virtual machines that are used to host Eclipse. Ubuntu Linux is the operating system used by the virtual machines and their host computers. On the student side, Smartphones or other Wi-Fi enabled mobile devices are used to remotely access the virtual machines. Students who do not own a Wi-Fi enabled mobile device can use the Android emulator to complete their assignments.

Server Implementation: On the server side, assume that the computer used as a server is physically linked to the Internet. The operating system of the server machine can be Ubuntu Server Edition or any other server operating system depending on the university's preference. The size of memory and the hard disk should be large enough so that it can host at least dozens of pair programming groups. The server machine must be a member of the university's network domain so that students' logons can be authenticated by the domain controller on the university's network. After the server machine is installed and configured, the server administrator creates one virtual machine for each pair programming group. On each virtual machine, the guest operating system can be an open source Linux based operating system such as Ubuntu Desktop Edition or another type of operating system. After the desktop operating system is installed, it is time to install and test pair programming related software such as Eclipse, the Android SDK, and the RealVNC server. The following describes how the pair programming software is used in this project and gives strategies for the installation and configuration. Our first step is to install the Android SDK.

Java Installation: Since Android uses Java as the programming language for application development, Java should also be installed on the virtual machines. To do so, one needs to download Ubuntu's apt-get tool and install the packages, sun-java6-bin, sun-java6-jre, and sun-java6-jdk. After these packages are installed, one needs to configure the environmental variables. Ubuntu may also have a built-in version of GNU Java which has poor performance. One needs to edit the JVM configuration file to make sure that Sun's Java is used as default.

Android SDK Installation: In the programming class, the students are asked to develop applications for Android Smartphones with the Java programming language. To make sure that the programs written by the students work properly, it requires the students to verify the code on the Android Smartphones. Also, although the Android Smartphones are the mobile devices to implement mobile pair programming, not all the students own an Android Smartphone. In such a case, on their own desktop, laptop, or netbook computers, the students can install the Android SDK which includes the Android emulator. They can use the Android emulator to access the server for pair programming or to verify their code. Assume that Ubuntu Linux is used as the operating system of the virtual machines. The Android SDK can be downloaded from the Web site: http://linux.softpedia.com/get/Programming/Interpreters/Android-32340.shtml

The downloaded file is a compressed file in the tgz format. After the file is downloaded, extract the tgz file into the home directory.

Android SDK Configuration: The configuration of the Android SDK follows the procedure below:

- Start the Android SDK in the extract folder and run all the available updates.
- Then, create an Android virtual machine and add some hardware such as a GPS or camera to the Android virtual machine.
- After the virtual machine is created, start the Android phone emulator shown in Figure 1. The emulator can be used to test the applications created for the Android Smartphone.

Through the Android Smartphone emulator, the students are able to access the Internet (Figure 2). The ability to access the Internet is important for the remote pair programming project.

Eclipse Installation and Configuration: To create applications for an Android Smartphone, Eclipse is the application development environment for writing, editing, and testing the programming code. If the operating system is Ubuntu Linux, the procedure of installation can be briefly described as the following.

- First, download the file eclipse-java-galileo-linux-gtk.tar.gz. After the file is downloaded, unzip the file.
- Set the path and permission to make Eclipse executable in the home directory.
- For the user's convenience, add the Eclipse icon to the panel on the Ubuntu desktop.
- Launch Eclipse as shown in Figure 3.

Figure 1. Android phone emulator

Figure 2. Accessing the Internet from Android phone

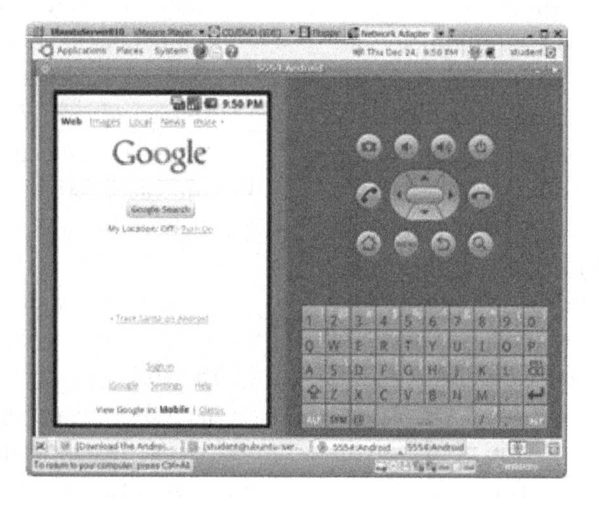

Figure 3. Launching Eclipse from Panel

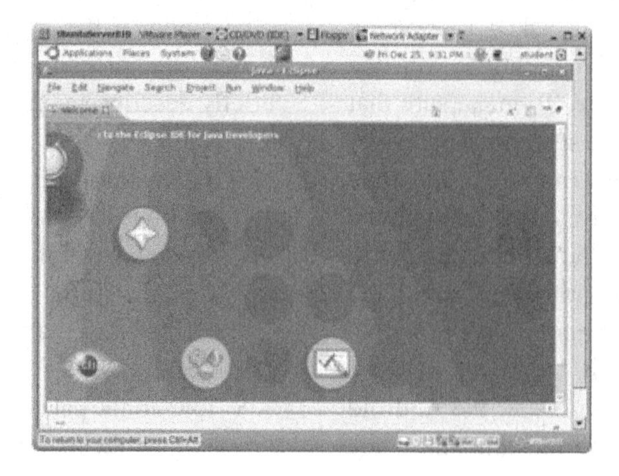

Figure 4. Linking ADT plug-in to Android SDK

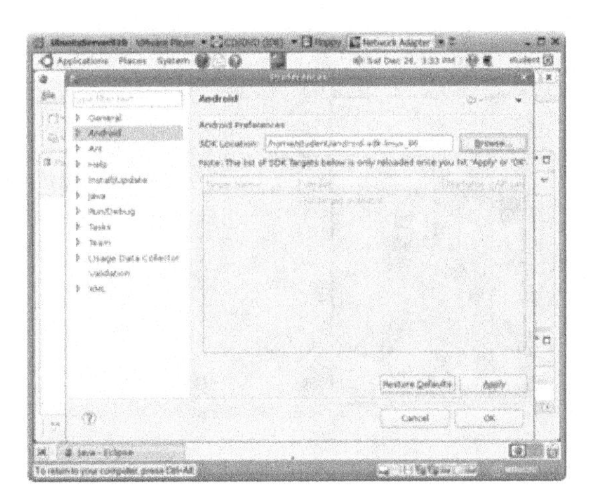

Eclipse is now ready to be used for developing Java applications.

Android Development Tools (ADT) Plug-in Installation: After Eclipse is installed, the next task is to add some Android development tools to Eclipse. These tools will help students to create Java code for Android based mobile devices. The installation procedure is illustrated below:

- After Eclipse is started, update Eclipse with the files from the Web site: http://dl-ssl.google.com/android/eclipse. Then, restart Eclipse.
- The next step is to link the ADT plug-in to the Android SDK folder as shown in Figure 4.

Now, Eclipse is ready. The students can use it create their Java projects for the Android Smartphones.

Creation of Java Project for Android Smartphone: To test if Eclipse is working properly, one can create a simple project called Hello Android. Enter some Java code as shown in Figure 5 and run the project.

Figure 5. Java project

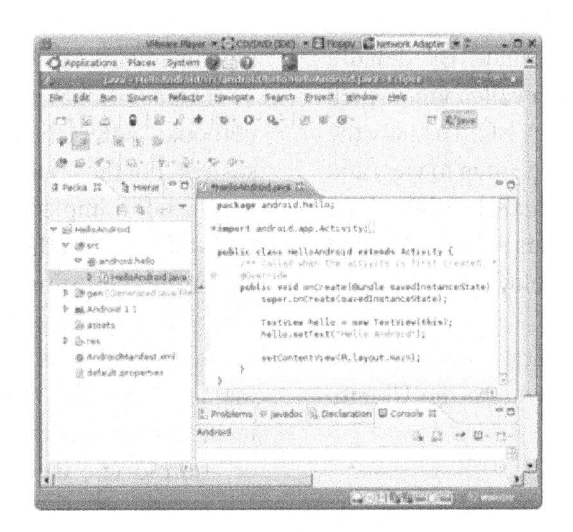

Figure 6. Android application Hello Android

After the Java code is successfully executed, Eclipse can save the application project of Hello Android to the Android Smartphone or Android emulator. In Figure 6, the application Hello Android is displayed on the Android emulator.

The above procedure shows that Eclipse has been successfully installed on the Ubuntu virtual machine and it is ready to be used to create Android applications. The next task is to make sure that the students are able to share the virtual machine's desktop during pair programming.

VNC Installation and Configuration: VNC is the technology that allows the students to share the screen of Eclipse. If Ubuntu Linux operating system is used to power the virtual machine for pair programming, the installation and configuration of VNC are fairly easy since several VNC packages are included in Ubuntu. By default, Ubuntu includes a VNC server called vino which is the VNC server for Gnome VNC. To allow the students in a pair programming group to view and control the desktop of the virtual machine where Eclipse is installed, VNC needs to be configured in the procedure illustrated below:

- The VNC server configuration can be done from the Preference menu on the Ubuntu desktop.
- When a Ubuntu Linux powered notebook computer is used as the VNC client, if needed, use the apt-get command to install the following packages:

xvnc4viewer
xtightvncviewer
tightvnc-java
vnc-java

- After these packages are installed, run the command:

vncviewer -fullscreen xxx.xxx.xxx.xxx:0

where xxx.xxx.xxx.xxx is the virtual machine's IP address. Figure 7 shows the desktop of the virtual machine which is displayed on an Ubuntu client machine named as UbuntuCleint810.

This solution is good for the students who have netbooks with Ubuntu Linux as the operating system. For those students who use Smartphones, they need to implement the VNC client on their Smartphones or other mobile devices.

Implementation of VNC on Smartphone: The remote pair programming can also be implemented by using Smartphones or PDAs. To share the desktop of the virtual machine, one needs to install the VNC client software, such as RealVNC or android-vnc-viewer, on the mobile devices.

Figure 7. Desktop sharing

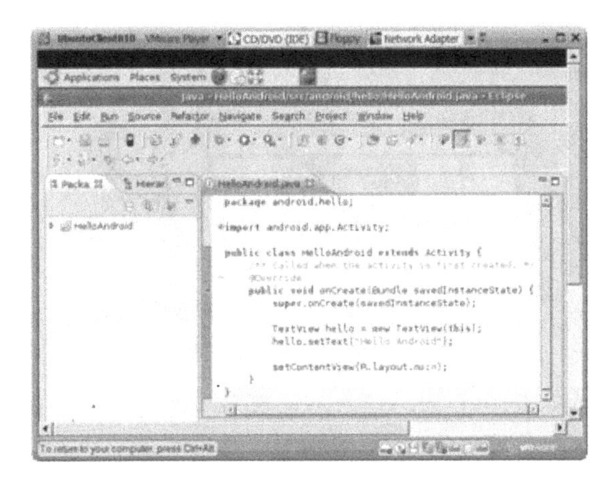

After the VNC client software is installed, it can be used to connect to the VNC server created on the virtual machine in the previous step. Now, two students can share the Eclipse screen on mobile devices as shown in Figure 8.

Implementation of Internet Voice Service: To discuss problems in Java code, the students can communicate with their partner in the same group with their cell phones, or use short messaging. The discussion of Java code can sometimes take a long time. Not all the students have an unlimited talk or messaging plan for their cell phones. In such a case, using the Internet voice service such as Skype or Truphone is an affordable alternative. As an example, the messaging integration software Nimbuzz is used to illustrate the implementation. The procedure of implementing the Internet voice service is as below:

- From Android Market, download and install Nimbuzz.
- After Nimbuzz is installed on an Android based mobile device, Nimbuzz will prompt the user to select the supported messaging service or social network service. For example, the students can choose Skype, or many other similar services.
- After a student logs on to the Skype account, the student's contact information will be displayed on the Smartphone's screen as shown in Figure 9.

Figure 8. Screen sharing on mobile devices

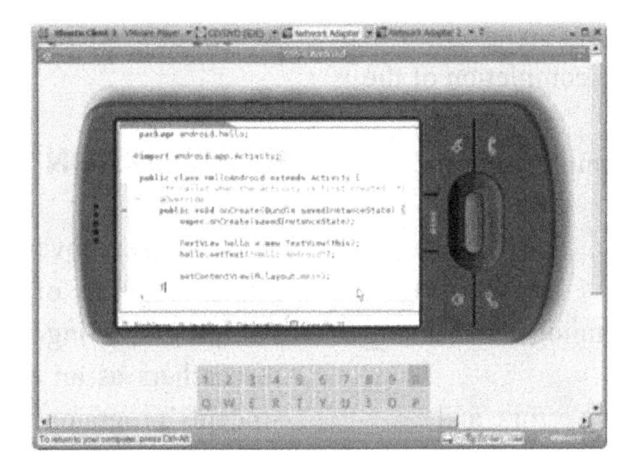

Figure 9. Skype contacts

- After the connection of the selected contact is established, the user will be prompted to choose an activity such as making a phone call or sending text messages.

In such a way, the students in a pair programming group can begin their discussion about the Java code.

Implementation of Class Evaluation: The class evaluation may include the evaluation of the instructor, the students' achievement, course materials, and the usage of pair programming. For the evaluation of pair programming, it may include the students' test scores and the survey. The survey may include questions related to the following aspects:

- The time spent on jointly designing computer programs,
- The time used to jointly code and test the computer programs,
- The average time for the completion of the assignments,
- The size of the programming code and the difficulty level,
- Scheduling problems and technical problems,
- The number of programming errors being detected and corrected,
- The type of programming errors, and
- Proposed improvement.

The students are required to keep a log about the time spent and errors occurred during the pair programming process so that they will be able to answer these questions at the completion of each assignment.

To evaluate the students' understanding of the assignments, multiple choice mid-term and final exams are given to assess the effect of learning. The students are required to take the exams individually to prevent the situation where one student does all the programming while the other student remains passive.

The above is an example for demonstrating the implementation of remote pair programming. It only represents one possible solution. There are many other ways to implement mobile pair programming. The implementation of pair programming should be based on the learning requirements and the available technologies. For a more systematical approach, readers may refer to the book by Cockburn and Williams (2001).

CONCLUSION

This chapter covers the topics related to the implementation of mobile pair programming. Pair programming has been considered by many researchers as an effective method to improve students' programming skills. The implementation of pair programming in a mobile learning environment posts challenges. Unlike traditional pair programming, students are not able to meet face to face. This chapter explores a set of open source mobile technologies that can be used to implement mobile pair programming. It explains how mobile pair programming can be implemented entirely with open source or free software. From the case study project, one can see that the implementation of pair programming can be done by using free downloaded software from the Internet. By using open source and free software, the educational institutions can significantly reduce the cost in developing mobile learning. In addition to low cost, open source and free software is rich in features and functionalities, so it can play a major role in implementing the mobile learning infrastructure.

REFERENCES

Citrix Systems. (2009). *Xen*. Retrieved August 15, 2009, from http://www.xen.org

Cockburn, A., & Williams, L. (2001). *The costs and benefits of pair programming*. Boston, MA: Addison-Wesley Longman Publishing.

Denning, P. J. (2004). The field of programmers myth. *Communications of the ACM, 47*(7), 15–20. doi:10.1145/1005817.1005836

Fernández-Alemán, J. L. (2009). Deducing loop patterns in CS1: A comparative study. In *Proceedings of the 2009 Ninth IEEE International Conference on Advanced Learning Technologies* (pp. 247-248).

Freie Universität Berlin. (2010). *Saros - distributed collaborative editing and pair programming*. Retrieved January 17, 2010, from https://www.inf.fu-berlin.de/w/SE/DPP

fring. (2009). *fring*. Retrieved December 6, 2009, from http://www.fring.com

Gárcia-Mateos, G., & Fernández-Alemán, J. L. (2009). A course on algorithms and data structures using online judging. *ACM SIGCSE Bulletin, 41*(3), 45–49. doi:10.1145/1595496.1562897

Ho, C., Raha, S., Gehringer, E., & Williams, L. (2009). *Sangam - a distributed pair programming plug-in for Eclipse*. Retrieved January 15, 2010, from http://collaboration.csc.ncsu.edu/laurie/Papers/Sangam.pdf

Kappe, D. (2007). *Pair programming with VNC*. Retrieved August 6, 2009, from http://www.pathf.com/blogs/2007/09/pair-programmin

Klawonn, F. (2008). *Introduction to computer graphics*. Godalming, UK: Springer-Verlag London Limited.

McGettrick, A., Boyle, R., Ibbett, R., Lloyd, J., Lovegrove, G., & Mander, K. (2005). Grand challenges in computing: Education - A summary. *The Computer Journal, 48*(1), 42–48. doi:10.1093/comjnl/bxh064

Mennila, L. (2007). Novices' progress in introductory programming courses. *Informatics in Education, 6*(1), 139–152.

Nagappan, N., Williams, L., Ferzli, M., Wiebe, E., Yang, K., Miller, C., & Balik, S. (2003). Improving the CS1 experience with pair programming. *ACM SIGCSE Bulletin, 35*(1), 359–362. doi:10.1145/792548.612006

Nimbuzz. (2010). *About Nimbuzz Mobile*. Retrieved January 18, 2010, from http://nimbuzz.com/en/mobile

Ohloh. (2010). *Android-vnc-viewer*. Retrieved January 17, 2010, from http://www.ohloh.net/p/android-vnc-viewer

PBXes. (2009). *PBXes*. Retrieved December 6, 2009, from https://www1.pbxes.com/index_e.php

Redmine. (2009). *Redmine*. Retrieved December 6, 2009, from http://www.redmine.org

Rennison, P., & Andrews, A. (2008). *RealVNC goes mobile*. Retrieved January 17, 2010, from http://www.realvnc.com/company/news/mobile.html

Robins, A., Rountree, J., & Rountree, N. (2003). Learning and teaching programming: A review and discussion. *Computer Science Education*, *13*(2), 137–172. doi:10.1076/csed.13.2.137.14200

Sipdroid. (2009). *Sipdroid*. Retrieved January 18, 2010, from http://sipdroid.org

Skype lite. (2009). *Skype launches on Android platform and more than 100 Java-enabled mobile phones*. Retrieved January 18, 2010, from http://about.skype.com/2009/01/ skype_launches_on_android_plat.html

Smith, D. E., & Nair, R. (2005). The architecture of virtual machines. *Computer*, *38*(5), 32–38. doi:10.1109/MC.2005.173

Spohrer, J. C., & Soloway, E. (1989). Novice mistakes: Are the folk wisdoms correct? In Soloway, E., & Spohrer, J. C. (Eds.), *Studying the novice programmer* (pp. 401–416). Hillsdale, NJ: Lawrence Erlbaum.

Steimann, F., Gößner, J., & Thaden, U. (2009). *Proposing mobile pair programming*. Retrieved December 6, 2009, from http://www.projectory.org/publications/MobilePairProgramming.pdf

The Eclipse Foundation. (2009). *Home*. Retrieved December 6, 2009, from http://www.eclipse.org

Truphone. (2010). *Truphone*. Retrieved January 18, 2010, from http://www.truphone.com

Ubuntu. (2009). *Ubuntu*. Retrieved December 6, 2009, from http://www.ubuntu.com

Vaughan-Nichols, S. J. (2007). *2007 desktop Linux survey results revealed*. Retrieved December 1, 2007, from http://www.desktoplinux.com/news/NS8454912761.html

VMware. (2009). *VMWare*. Retrieved August 6, 2009, from http://www.vmware.com

KEY TERMS AND DEFINITIONS

Eclipse: It is a comprehensive multi-language software development environment. It consists of an IDE and an extensive plug-in system.

Integrated Development Environment (IDE): It is a software development environment where the software developer can write, edit, and test the programming code.

Internet Voice Service: It is a service that allows a PC user to make free calls to another PC user or make low cost calls to landline or mobile phone users.

Software Development Kit (SDK): It is a set of tools used by programmers to develop application software.

Virtual Machine: It is the software that is used to implement a computer including the virtual hard drive, virtual memory, virtual network interface card, and so on.

Virtual Network Computing (VNC): It is a GUI based remote desktop accessing software.

Voice over Internet Protocol (VoIP): It is a transmission technology that delivers voice communications over IP networks.

Chapter 16
Accessing Remote Laboratories from Mobile Devices

Pablo Orduña
DeustoTech – Tecnológico Fundación Deusto, Spain

Javier García-Zubia
University of Deusto, Spain

Diego López-de-Ipiña
University of Deusto, Spain

Jaime Irurzun
DeustoTech – Tecnológico Fundación Deusto, Spain

ABSTRACT

Remote Laboratories constitute a first order didactic resource in engineering faculties. Its use in mobile devices to increase the availability of the system is a challenge highly coupled to the requirements established by each experiment. This work presents the main strategies for adapting a Remote Laboratory to mobile devices, as well as the experience of a real Remote Laboratory, WebLab-Deusto, in this adaption. These strategies are analyzed and compared in order to detail what strategy is more suitable under certain situations.

INTRODUCTION

A Remote Laboratory is a software and hardware system that enables students to use real experiments physically located in a university. This way, students can access real experiments 24 hours a day, 7 days a week, even including holidays, from anywhere with access to the Internet. Given that the experiments are real, the Remote Laboratories have the opportunity to probe this fact to the students (i.e., with a webcam), so the students don't lose the feeling that they are doing exactly what they would do in a hands-on-lab session.

As an example, WebLab-Deusto, a Remote Laboratory developed in the University of Deusto, counts with different experiments. In the FPGA experiment, the student works with the Xilinx IDE at home, develops a program for a FPGA device. Once the student has finished programming it, he can connect to the Remote Laboratory, send the program, which is programmed in a real FPGA,

DOI: 10.4018/978-1-60960-613-8.ch016

Figure 1. Sample screen of WebLab-Deusto

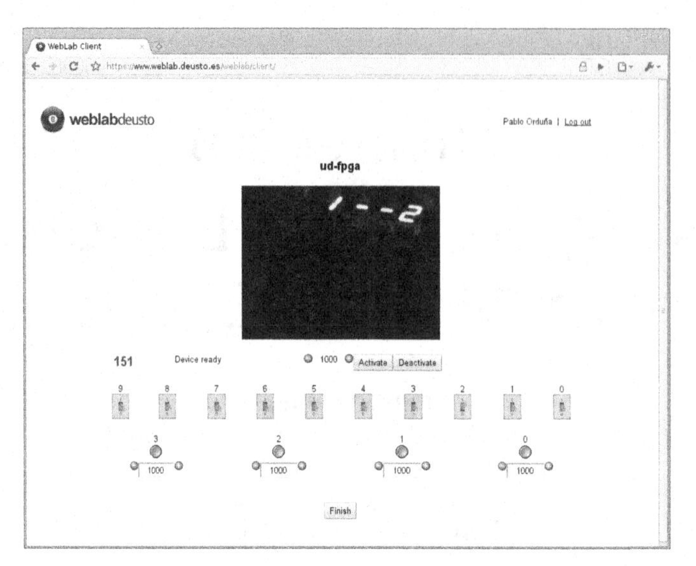

and finally interact with it with the panel of Figure 1. During a limited amount of time, the user will be able to modify some inputs of the device and see through a webcam the produced outputs. The students are not limited in how many times they can use it, but the amount of time for each use is limited so the device can be reused by other students.

Remote Laboratories also provide an efficient performance of the devices for the university. Many experiments do not require a continuous connection to the device. For instance, in the FPGA experiment described above, the device is only used by the student during the small amount of time in which the device is programmed and tested. This amount of time is enough for testing if the code was correct or what failed. After this time, the user can go back to the IDE and try to correct the problems, and test again connecting to the experiment. If somebody else is using the device, the latter user will enter in a queue of users and will stay there until all the users that entered before him have finished. If there are many users, the university can put several copies of the same experiment, so the students will automatically use different experiments and the mean time in queue will decrease. Due to this transparent and efficient sharing among different remote students, the Remote Laboratory might achieve higher levels of throughput per device than traditional laboratories.

As an example of a real scenario, during the course 2008-2009 in the University of Deusto the CPLD experiment had 74 users, and the FPGA experiment had 23 users. Figure 2 represents the occurrences of each position in a queue for the

Figure 2. Relation between positions in a queue and how many occurrences of those positions in CPLD and FPGA experiments of WebLab-Deusto during the course 2008-2009

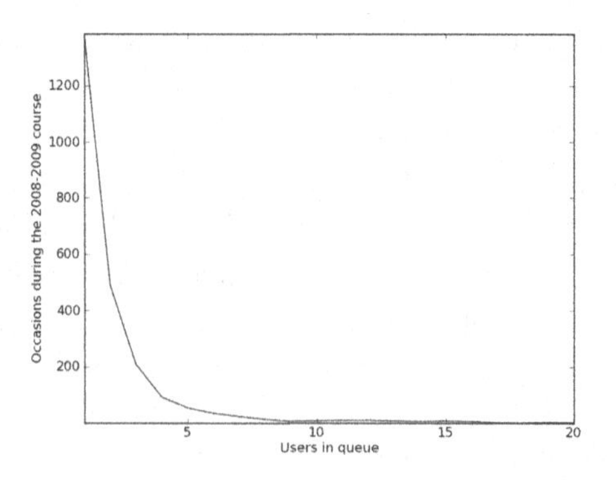

two experiments during that course. This is, if a student entered in a queue in position 5, Figure 2 represents one occurrence for position 5, and no occurrence for lower positions, even if he stays in the queue and achieves those lower positions. It also represents only the positions in the queue, from 0, which means that the student will use an experiment as soon as a device becomes free, to 19, which was the maximum achieved position. Therefore, the figure does not represent the number of students that directly used the experiment without being in queue, which was the majority. Furthermore, during this course there was a single FPGA device and a single CPLD device, so since WebLab-Deusto balances the load of users among the available devices, it is predictable that the distribution would have varied considerably if more devices had been placed. However, it is still remarkable that given that the maximum session time was 200 seconds, a single device enabled students to wait less than 10 minutes 87% of all cases where the user entered in a queue, getting a higher percent if we counted the users that didn't even enter in the queue.

Section 2 details the characteristics of Remote Laboratories, presenting the challenges on the adaption of Remote Laboratories to mobile devices and showing the WebLab-Deusto architecture as an example of Remote Laboratory. Section 3 describes the available strategies for adapting a Remote Laboratory to a mobile device, compares them and shows the strategy used in WebLab-Deusto. Finally, Section 4 details the future research directions and Section 5 presents the conclusions.

CHARACTERISTICS OF REMOTE LABORATORIES

Figure 3 shows how a simple Remote Laboratory is composed by three main layers:

- Client software placed in the student's computer
- Server software placed in the university, connected to the experiment
- Physical experiment placed in the university

Each layer can later be divided into complex components. For instance, the second layer can be composed of several servers deployed in a university for security reasons so as to distribute the management of experiments by the experts on the experiments. The third layer can also provide high complexity, designing experiments that can handle several users concurrently at the hardware level, experiments that self test that everything works even under adverse situations, such as a

Figure 3. Initial view of a Remote Laboratory

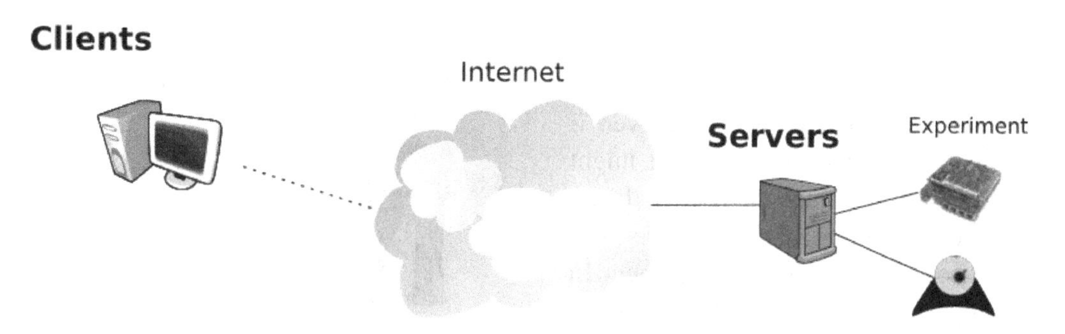

robot handling its own battery, or experiments checking that the inputs sent by the user will not damage the devices.

However, while these layers are briefly addressed in other sections, the focus of this chapter is to detail how to access these experiments from mobile devices, and thus on the first layer.

What Experiments May Be Used from Mobile Devices?

There is a wide range of mobile devices, and the experiments will only work on the subsets of them. The more powerful the mobile device is, the more suitable it usually becomes for supporting an experiment.

However, there are experiments that simply can not work on a mobile device. For instance, the FPGA experiment described in the first section requires the user to use an IDE which is not available in a mobile device. It is difficult that this experiment can be used with a mobile device. There are two approaches to consume this type of experiment:

1. Providing the IDE software in a remote machine. The Carinthian University of Applied Sciences counts with an experiment that uses a CPLD. In order to provide the required software for the student, the student accesses directly to the remote machine with the software installed. This way, the student can efficiently use the IDE software without having to install it, from anywhere on the Internet. This approach could be applied to the scenarios related to mobile devices, where the student might be able use a mobile device to interact through a remote desktop with the required IDE. However, even if this approach could be applied, it might be difficult to use since the IDE GUIs are designed for desktop machines.
2. Being a secondary user of the experiment. In the sample use described above, the student works on his own, without collaborating with other users. However, this is not always the case. In SecondLab, a Second Life based client for an experiment in WebLab-Deusto, a microrobot is remotely programmed by a student from Second Life (Figure 4). Other students can be in the same room in Second Life watching the same results without interacting with the experiment. This master/slave usage of the experiment could be adapted to mobile devices: a student could have developed the program sent to the FPGA and then other students could see what it does from a mobile device, or they could even interact with the experiment during the session managed by the master student.

Sample Remote Laboratory: WebLab-Deusto

WebLab-Deusto is an Open Source (GNU GPL v2 at the moment of this writing; BSD in the next release) Remote Laboratory, developed by the University of Deusto. WebLab-Deusto is not an experiment itself, but a framework on top of

Figure 4. WebLab-Deusto microrobot experiment used from Second Life with SecondLab

Figure 5. WebLab-Deusto architecture

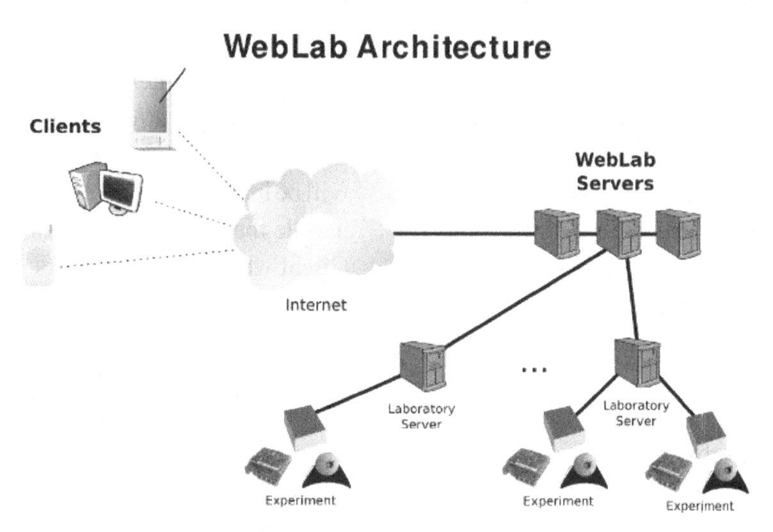

which experiments can be developed (see Figure 5). Therefore, there are two separate development roles:

1. WebLab-Deusto Developer
2. Experiment Developer

The first type of developer must be an IT expert that will deal with low level issues, such as scalability, communications, authentication, authorization, session management, scheduling, user tracking, user management, logging, or handle complex deployments.

However, the second type of developer should not deal with these issues and should only focus on building the experiment itself. In order to do so, the Experiment Developer will build two components:

1. **Server side component:** it will validate the user input and translate it to the devices.
2. **Client side component:** it will show the user interface of the experiment. The user will interact with this user interface, which will send commands to the server side component and will show the responses.

In order to make the development of these components easy, WebLab-Deusto provides an API for developing clients and another API for developing servers. These APIs are based on commands: the client will call a method for sending a command (which will be a string such as "turn switch 1 on"), and the framework will internally send that the command to the server, receive a response and provide it to the client. The Experiment Developer will handle the serialization and deserialization of the commands.

Since the background of the different Experiment Developers is diverse, as well as the tools they have experience with, WebLab-Deusto supports a range of development frameworks for both the client and server development. In the latest release (3.9.0), WebLab-Deusto provides libraries for developing the server in:

- C
- C++
- Java
- .NET
- LabVIEW
- Python

These libraries have an interface that the developer must implement, with four simple methods:

- **Initialize:** a new user is going to use the experiment
- **Send command:** the current user sends a string, and a response is expected
- **Send file:** the current user sends a file, and a response is expected
- **Dispose:** the current user has finished using the experiment

WebLab-Deusto provides libraries for developing the client in:

- Java Applets
- JavaScript
- Adobe Flash

These libraries have methods for retrieving current configuration, as well as for sending commands and gathering responses from the experiment server.

Once both sides have been developed and deployed, the user will load the WebLab-Deusto web page. The user will log in, and the WebLab-Deusto server will handle the authentication and authorization. The list of experiments available for this user will be shown, and the user will select an experiment from this list. The user will perform a reservation and the client developed by the Experiment Developer (Experiment Client) will be initialized. The WebLab-Deusto client will handle the queue and will notify the Experiment Client when the usage starts. From that point, the user will interact with the user interface developed in the Experiment Client. The provided API will handle sessions, and the WebLab-Deusto client will manage that the session doesn't expire while the web page is loaded. The WebLab-Deusto client will also notify the Experiment Client that the time has finished and that all the client resources, such as timers, should be disposed, and finally the WebLab-Deusto client will unload the Experiment Client.

Therefore, as detailed in Figure 6, the framework already handles the common requirements of Remote Laboratories, providing the Experiment Developer with a flexible system on top of which develops the experiments using simple APIs. These common requirements are (Orduña, et al., 2009):

Figure 6. WebLab-Deusto stack

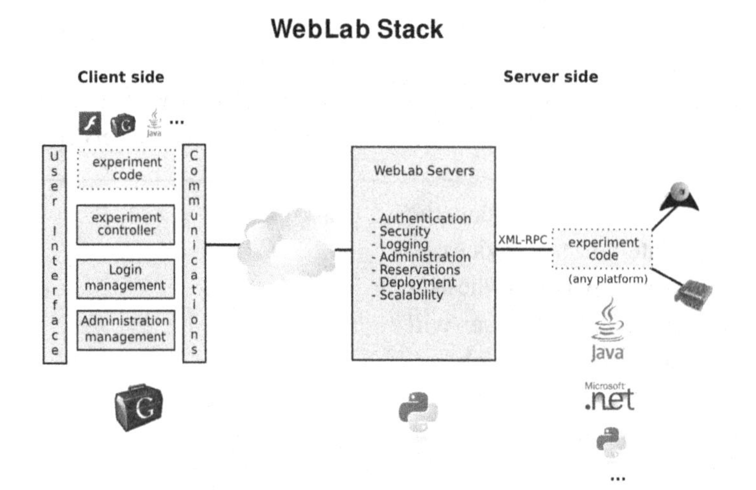

- **Authentication and authorization:** the framework handles authentication and authorization.
- **Security:** the framework wraps the database management and the communications (i.e., the framework can be configured to use SSL without changing the client).
- Logging and user tracking: the framework stores who used what experiments, when, and what was done during that session. If a user sends a malicious file, the administrators can later find it since it has been stored in a different node.
- **Reservations:** scheduling models are provided by the system, so without changing the client the administrator can grant access to the experiments to different users with different priorities and assigned time.
- **Deployment:** the experiments can be placed in different rooms in the university while the core servers are placed in a different place (such as the IT services). The experiments don't need to have a public IP address or a port open to the Internet, since the framework proxies the requests.
- **Scalability:** the core servers can be spread among different machines to balance the load of users.
- **Administration:** administration tools are provided to manage the rest of the requirements.

Since WebLab-Deusto 4.0M1, the system has provided two versions of the client side: the desktop version and the mobile version (see Figure 7). While they reuse most of the code (controller, communications, serialization, configuration management), the mobile version reimplements the user interface, adapting it to smaller screens. Both versions are developed in pure JavaScript, not relying on any plug-in. The client also gives the chance to the Experiment Developer to provide two different Experiment Clients for the same experiment: the regular one and the one adapted for mobile devices. It is up to the Experiment Developer to implement only one of them or both of them, and to deal with the problems that can arise in mobile development, described in detail in the following section.

Figure 7. On the left side, the desktop version of the WebLab-Deusto client. On the right side, the mobile version running on an Android 1.6 simulator

BUILDING MOBILE-ENABLED REMOTE LABORATORIES

There are two main approaches for porting an application to mobile devices: adapting a web version to the constraints of mobile devices and building a native implementation in each supported mobile phone. This section details both approaches and compares them.

Building Native Remote Laboratories in a Mobile Device

Nowadays, most mobile devices provide development frameworks on top of which third-party developers can build applications. The added value of this is clear: the functionality of the mobile device becomes flexible, since new applications can be built using the capabilities of the mobile device.

However, the range of development frameworks has become wide: Symbian, Android, iPhone, Windows Mobile, LiMo, BlackBerry or Maemo (now merged with Moglin in MeeGo) are examples of mobile operating systems. Applications available for these operating systems are usually native applications developed in their own SDK, which is only supported by each operating system. For instance, a native application for Symbian, developed using Symbian C++, will only run on Symbian.

An approach to avoid this is to use Java ME (formerly known as J2ME). Java ME was built under the Java slogan "Write once, run anywhere": different mobile devices come with a Java Virtual Machine (JVM), which will run applications compiled for Java ME. The base system does not count with modules for graphics or access to hardware, but different JSRs are specified and different mobile device providers will implement these specifications. This way, a new feature (i.e., access to accelerometers of the device) that was not defined when the base system was defined, can later be defined and the mobile vendors will implement it in those devices that have accelerometers. Then, application developers can specify that their applications rely on a set of specifications, so the applications will only run on those devices that implement them. This system is extensible, so the application can run as long as the requirements are a subset of the features provided by the mobile device.

The availability of Java ME in mobile devices is not so universal since it is not available in some of the new mobile device operating systems. From the list of operating systems above, iPhone and Android do not support Java ME. There are efforts to implement a JVM in Android or to be able to compile a Midlet to an Android application, but it is not supported and it is not transparent for the user. A JVM can be installed in Maemo, however it does not come with it, and it is not available in the default application repositories. Windows Mobile might come with a JVM, but it is not supported by Microsoft itself and it is not considered as reliable as Compact Framework by the industry: as an example, Google Mail provides a Java ME application that specifically states that does not support Windows Mobile, and it provides a web version for this platform. Finally, Symbian, BlackBerry, and LiMo phones do support Java ME.

Another approach to reuse code among two platforms is to reuse device independent logic code, and it will work as long as both platforms can use the programming languages. This is, logic code that does not rely on graphics, sound, location, calendar, network, and even threads, can be reused between two development frameworks. For instance, ZXing (a Barcode Scanner) is developed in Java. The core of the project, which manages to decode a barcode inside an "image" and generates some results, relies on a set of interfaces that will later be implemented by other parts of the project, such as the Java ME implementation or the Android implementation of these interfaces. With this isolation of the logic code, most of the code is reused and only the user interface requires to be reimplemented.

Nevertheless, the restriction of using the same programming language is still a constraint for this approach: Android and Java ME use Java as the programming language or C/C++, while Symbian native applications use C/C++ or Python, iPhone applications use Objective C or C/C++, Windows Mobile applications use .NET or C/C++, LiMo applications use C/C++, Blackberry applications use Java, and Maemo applications use C/C++ or Python, so ZXing cannot reuse code directly with other development platforms. If the code was developed in C/C++, then it could be reused among more platforms, but there are certain security constraints in some platforms that might invalidate this approach. Android provides a security manager that tells the user what are the permissions for a given application. This security manager can work as long as it works with managed code, and C/C++ code is unmanaged code. Therefore, the user could use an application that reuses code in C++ as long as he accepts that the application will run with higher privileges than regular applications.

The advantage of using a native technology is that it can use all resources that the mobile device provides through the used SDK. If the mobile device supports it, the application may use 3D graphics, retrieve the user's position, access the accelerometers, the camera, use Bluetooth, interact with files and handle disk storage, access the mobile calendar or contacts, or even play music and videos, while mobile web browsers usually do not provide these features to web applications.

The main question, however, is how useful these features are in Remote Laboratories:

- 3D graphics might be used to provide an immersive scenario that could improve the interaction with the experiment, as is already done with WonderLand (Scheucher, et al., 2009a) or Second Life (Garcia-Zubia, et al., 2010).
- Access to accelerometers can improve the usability of the system, especially if the experiment requires some movement.
- Handling disk storage is useful for retrieving results of an experiment and for providing scripts that the remote experiment might require. This would not be supported in iPhone or Android, since the user cannot directly handle files.
- Accessing user's calendar could assist the user by showing the user's events in the same screen of the booking system.
- Notifying the user that the experiment enqueued is now ready to be used could be useful, so the user does not need to check if it is his turn every few seconds.

However, other features such as retrieving the user's position, Bluetooth or camera are not features subject to be exploited by a Remote Laboratory.

Building Web Remote Laboratories in a Mobile Device

Support for web applications in mobile devices has increased during the last few years. With the arrival of Web 2.0 and Cloud computing, it became necessary to support complex web applications in mobile devices. Also, with the progress on mobile device development, new mobile device operating systems are direct ports of desktop or server operating systems: the iPhone OS is derived from Mac OS X, while LiMo, Maemo and Android are derived from different flavours of Linux. Nowadays, mobile devices count with different modern web browsers. Maemo comes with MicroB, based on gecko (the engine used by Mozilla Firefox), while iPhone, Android, Symbian and LiMo come with browsers based on WebKit (the engine used by Safari and Google Chrome), and Windows Mobile is supported by Opera Mobile. It is actually claimed that around 40% of iPhone users browse the web more often from their iPhone than their own desktop (Sauer, 2009), which means that mobile devices have become a proper way to support the web.

However, web applications usually need to be adapted for mobile devices. This adaption requires three changes:

1. Provide a proper layout. Developers should think what is actually going to be used from a mobile device, and how the user may see it on a small screen. For instance, newspapers tend to provide a vertical panel where each news item is represented in a row with a single sentence, so the user can quickly see what news item is more interesting and click on it. Each row acts as a button, so it becomes easy to click it with a touch screen.
2. Provide the required contents. Developers should think what contents are going to be migrated to the mobile version. Users might look at the mobile version as a complement to the desktop version, so it becomes normal that some features are not present.
3. Avoid plug-ins. Many web applications provide features that are based on plug-ins such as Java applets, Adobe Flash or Microsoft Silverlight. These plug-ins are not available in most devices, and it is difficult that they become available there, due to the resources required for the plug-in developer to port the plug-in to the wide range of mobile platforms.

Remote Laboratories, however, tend to rely on plug-ins, such as Java Applets or Adobe Flash, as detailed by Gravier, et al. (2008). To establish two examples, in MIT iLabs (iLabs, 2010), Java has been the preferred client development environment (Harward, et al., 2008), and the BTH VISIR (VISIR, 2010) project use Adobe Flash (Gustavsson, et al., 2007). While some devices support Adobe Flash (such as those based on Maemo), relying on a plug-in will exhaustively decrease the number of supported users.

The capabilities of a web page are also increasing nowadays. Most of the webs we know are based on the current web standards and plug-ins. However, the next release of HTML, HTML5, introduces important new functionalities to the web. While these functionalities are not yet stable, some of them are already implemented by major web browsers. These features include:

- Audio and video
- Canvas
- Geolocation
- Storage and databases
- New forms

Some widespread web applications, such as YouTube or Google Maps, already provide contents using HTML5. YouTube supports video through HTML5 instead of depending on Flash if the user explicitly agrees. Google Maps use HTML5 geolocation capabilities to show the user where he is by pressing a small button. This button will only appear if the web browser supports geolocation.

The most interesting point is that, since mobile web browsers are based on modern web browsers, some mobile web browsers already provide these functionalities. For instance, both iPhones and Android devices can already handle geolocation, as well as Windows Mobile by installing a plug-in.

WebLab-Deusto: A Web Remote Laboratory Working in a Mobile Device

In December 2006, the University of Deusto presented how WebLab-Deusto 2.0 could be accessed from mobile phones such as Symbian devices with the Opera Mobile web browser (López-de-Ipiña, et al., 2006).

During the year after, the Remote Laboratory was rewritten, and in September 2007 this new version (WebLab-Deusto 3.0) started to be used with students. The client of WebLab-Deusto 2.0

Figure 8. WebLab-Deusto 2.0 working under a Nokia 6630 device with Opera Mobile

had been written in plain JavaScript. However, WebLab-Deusto 3.0 (WebLab-Deusto, 2010) is completely written in Google Web Toolkit (as well as in the current -3.9.0- and the next version -4.0-).

Google Web Toolkit -GWT- is an open source technology, developed by Google, that provides an API that can be used from the Java programming language, and the toolkit will translate it to JavaScript code. This way, a developer can use a Java IDE such as Eclipse; (it actually includes a plug-in for Eclipse), and write and test code in Java, and finally compile it to JavaScript. This Java code can only use libraries provided by the GWT API; (for instance, the classes for managing files are not available since the result is compiled to JavaScript, which does not support files). The following are remarkable features that GWT provides:

- **Fast development cycle:** in the development mode, any change to the code will be updated as soon as the web browser refreshes the web page. With the Eclipse plug-in, the developer can debug the code as a regular Java application and can use JUnit to create unit tests.
- **Deferred binding:** GWT does not generate a single JavaScript file with all the resulting code, but a set of JavaScript codes. For instance, it compiles a version for each major web browser, optimizing all the low level calls for each web browser. It also compiles a version for each supported language. Therefore, if it supports 6 web browsers and 4 languages, it will generate 24 different versions of the same web, transparently for the developer, and each optimized for the user.
- **Size optimizations:** GWT can compile in the "obfuscated mode" if requested, which reduces considerably the size of the generated scripts (given that they do not conserve the original Java names). It also supports a feature called "code splitting", which is that the developer can select parts of the code and explicitly mark them so they will be separated as different JavaScript files and will only be downloaded if a certain condition matches. For instance, each experiment of WebLab-Deusto is compiled in a different file, so the code and resources of an experiment are not downloaded if the user does not use that experiment. The same applies to the different themes available (such as the "mobile" theme and the "desktop" theme).
- **Cache management:** GWT can include resources (such as images or CSS files) in the JavaScript files. With this system, as soon as the current fragment of code is downloaded, it is ready to be rendered, without requiring some more time for downloading the images. Additionally, since this JavaScript file will be cached by the web browser, next time the browser will only perform a request to the server to check if

Figure 9. At the left, ud-logic running on a desktop. At the right, the same experiment adapted to a mobile screen

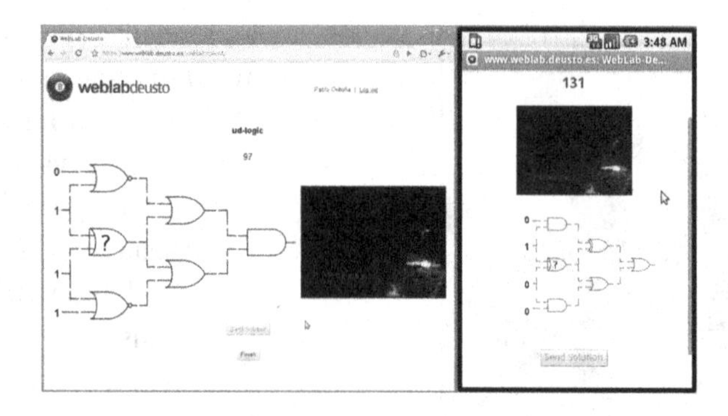

the JavaScript file was changed, without having to check if every single image was changed too.

GWT also enables the developer to easily reuse code among different layouts, which makes it very suitable for implementing a mobile version. GWT has been used for building the Google Wave client, and Lars Rasmussen, Head of Engineering in Google Australia and one of the originators of Google Wave, claimed that "[using GWT] we estimate that the extra requirement for serving this [Google Wave] both into desktop browsers and mobile browsers only adds about 5% extra engineering effort" during his presentation in the Google I/O 2009 conference (Rasmussen, 2009).

As detailed in the previous section, WebLab-Deusto currently provides two versions (actually two themes, which a third-party could reimplement to adapt WebLab-Deusto to its institution) of the common, experiment independent user interface: a desktop version and a mobile version. The framework enables the Experiment Developer to implement two different user interfaces of its experiment: one for desktop themes and another for mobile themes. Therefore, it is up to the Experiment Developer to write the client of the experiment in the suitable technology. If the Experiment Developer chooses to use Adobe Flash or Java Applets, and does not provide another implementation for mobile versions, then that experiment will hardly be supported in mobile devices. However, if the Experiment Developer only implements one version in JavaScript or GWT, then it will probably run on mobile devices, although it might not have been adapted to smaller screens.

As an example, Figure 9 shows a single experiment (ud-logic), which has been adapted to the mobile version, by using smaller images and reorganizing the layout of the experiment.

Comparing Both Approaches

The decision of implementing a native version or not depends mostly on if it is affordable. In general, the web approach requires adapting it once to mobile screens and then it will work on most modern mobile devices. Depending on the technologies and on the design used for developing the Remote Laboratory client, providing this web version does not require much effort. However, if it was built on top of a plug-in such as Java Applets, Adobe Flash or LabVIEW, it might require to be completely rewritten.

On the other side, providing a version of the application for each mobile platform is expensive to build and expensive to maintain. Every time there is a new feature, all the versions should be updated. Google Mail, for instance, provides a mobile web interface, and Google provides an enhanced version that is integrated in the operating system so as to perform notifications or to add the "send through Google Mail" option in the gallery. However, this enhanced version is only for Android, Blackberry and Symbian, while it relies on the web version for iPhone, Windows Mobile and other mobile devices. Facebook also provides a mobile web version supported by most mobile phones, but at the same time it targets a wide range of mobile platforms with native versions, including Symbian, Android, Windows Mobile, iPhone, Palm and Blackberry.

The main difference between both projects is that Facebook requires more integration with the mobile device: it is usually used for sharing pictures that the user may take from the mobile phone. The geolocation is also important in order to tag these pictures. Notifications are also crucial in Facebook and may not be mapped to other protocols, while users of not supported platforms of Google Mail can use a regular mail client to get notifications.

Therefore, the decision of implementing the native version or the web version will rely on if the requirements of the experiment are so important that they justify the amount of work required for implementing it for the most common mobile platforms that are used by the students. This decision must be made for each experiment.

FUTURE RESEARCH DIRECTIONS

As previously stated, the main areas where the use of Remote Laboratories from mobile devices could become especially interesting are two: 3D Virtual Environments and Collaborative Environments.

The interest in 3D Virtual Environments is increasing, expecting that 80% of active Internet users will have a second life in a virtual world by the end of 2011 (Gartner, 2009). Applied to Remote Laboratories, Collaborative Virtual Learning Environments are starting to be present (Scheucher, et al., 2009b; García-Zubia, et al., 2010). Since modern mobile devices add hardware support for 3D, mobile devices become a suitable tool to spread 3D Virtual Environments on top of which deploy experiments that the user can use intuitively.

Regarding Collaborative Environments, different users may interact with the same experiment in a collaborative way, in a master/slave way or in a fully collaborative way. This could increase the potential of the Remote Laboratories' use in mobile environments: if a teammate is present in a laptop on the Internet, the technical drawbacks of mobile devices might be blurred.

While both areas provide interesting results in Remote Laboratories, they have not yet been exploited in mobile devices.

CONCLUSION

Using the experience in the software development of Remote Laboratories and the adoption of Remote Laboratories in mobile devices since 2006, the paper has presented and analyzed strategies to implement a mobile enabled Remote Laboratory.

The first strategy is using web technologies, so they may reuse existing code available in the desktop version of the Remote Laboratory, although some adaption is still required. Using this strategy, the number of target mobile devices is maximized.

The second strategy is using native mobile technologies. While the control on the device is higher, being able to access the camera or drawing 3D graphics, the application must be rewritten and maintained in each target mobile platform.

The adoption of the first strategy in WebLab-Deusto is detailed, and finally both strategies are compared. The use of one or the other depends on the affordable cost of development and on the requirements of the experiment. This situation is similar to the decision of using lightweight technologies such as AJAX or using more powerful but plug-in requiring technologies such as Adobe Flash or Java Applets in regular Remote Laboratories, already addressed in the literature (García-Zubia, et al., 2009).

REFERENCES

García-Zubia, J., Irurzun, J., Angulo, I., Orduña, P., Ruiz-de-Garibay, J., & Hernández, U. ... Sancristobal, E. (2010). SecondLab: A remote laboratory under Second Life. In *Proceedings of the IEEE Engineering Education Conference 2010* (pp. 351-356).

García-Zubia, J., Orduña, P., López de Ipiña, D., & Alves, G. (2009). Addressing software impact in the design of remote labs. *IEEE Transactions on Industrial Electronics, 56*(12), 4757–4767. doi:10.1109/TIE.2009.2026368

Gartner Research Inc. (2009). *Gartner says 80 & of active Internet users will have a "second life" in the virtual world by the end of 2011*. Retrieved on June 12, 2010, from http://www.gartner.com/it/page.jsp?id=503861

Gravier, C., Fayolle, J., Bayard, B., Ates, M., & Lardon, J. (2008). State of the art about remote laboratories paradigms - foundations of ongoing mutations. *International Journal of Online Engineering, 4*(1). Retrieved from http://online-journals.org/index.php/i-joe/article/view/480.

Gustavsson, I., Zackrisson, J., Håkansson, L., Claesson, I., & Lagö, T. (2007). The VISIR project - an open source software initiative for distributed online laboratories. In *Proceedings of the Remote Eng. And Virtual Instrumentation Conference 2007*.

Harward, V. J., del Alamo, J. A., Lerman, S. R., Bailey, P. H., Carpenter, J., & DeLong, K. ... Zych, D. (2008). The iLab shared architecture: A Web services infrastructure to build communities of Internet accessible laboratories. *Proceedings of the IEEE, 96*(6), 931-950.

iLabs. (2010). *Computer software*. Available from http://ilab.mit.edu/wiki

López-de-Ipiña, D., García-Zubia, J., & Orduña, P. (2006) Remote control of Web 2.0-enabled laboratories from mobile devices. In *Proceedings of the 2nd IEEE International Conference on e-Science and Grid Computing 2006* (p. 123).

Orduña, P., García-Zubia, J., Irurzun, J., Sancristobal, E., Martín, S., & Castro, M. ... González. J. M. (2009). Designing experiment agnostic remote laboratories. In *Proceedings of the Remote Engineering and Virtual Instrumentation Conference 2009*.

Rasmussen, L. (2009, May 28). *Google Wave developer preview at Google I/O 2009* [Video file]. Retrieved from http://www.youtube.com/watch?v=v_UyVmITiYQ

Sauer, F. (2009, October 19). *The enterprise apps in your pocket*. [Web log post]. Retrieved from http://googlewebtoolkit.blogspot.com/2009/10/enterprise-apps-in-your-pocket.html

Scheucher, B., Belcher, J., Bailey, P. H., dos Santos, P., & Gütl, C. (2009b). Evaluation results of a 3D virtual environment for Internet-accessible physics experiments. In *Proceedings of the International Conference of Interactive Computer Aided Learning 2009*.

Scheucher, T., Bailey, P. H., Gütl, C., & Harward, V. J. (2009a). Collaborative virtual 3D environment for Internet-accessible physics experiments. In *Proceedings of the International Conference of Remote Engineering and Virtual Instrumentation 2009*.

VISIR. (2010). *Computer software*. Retrieved from http://openlabs.bth.se

WebLab-Deusto. (2010). *Computer software*. Retrieved from http://www.weblab.deusto.es

Section 3
Open Culture and Ubiquitous Education

Chapter 17
Supporting Distance Users of Mobile Learning Technology

Yong Liu
Åbo Akademi University, Finland

Hongxiu Li
Turku University, Finland

ABSTRACT

With the rapid deployment of mobile devices, mobile learning emerges as a promising approach giving rise to a wide spectrum of new education possibilities. It serves as an effective conduit to deliver education to civilians of all social-economic levels, in particular the learners previously unreachable from traditional education systems, such as problem teenagers, social employees, and ageing people. Hence, unlike traditional education approaches, it is considered to be a good alternative to deal with the challenges posed by demographic shifts and social transformation. The purpose of this chapter is to: (i) identify the theoretical and technological underpinnings for delivering mobile learning to the distance learner, and (ii) discuss the possible learner communities that can benefit from mobile learning technology, with regard to their unique learning requirements and features.

INTRODUCTION

Increasingly, information and communication technologies, or ICTs, have started to permeate nearly every aspect of our lives. It not only dramatically alters the way we communicate, work and run businesses, but also gradually changes the way people access training and education. In particular, advance in broadband wireless network technology today enables mobile devices to transmit text, voice, video and animated images independent of time and location. This establishes a concrete technical basis for translating mobile learning from theory into actual practice.

The potential and impact of mobile learning are further enhanced in consideration of a worldwide proliferation of the mobile phone. A report from Portio Research (2007) predicts that the global

DOI: 10.4018/978-1-60960-613-8.ch017

mobile penetration rate will surpass 50 percent in 2008, and further 1.5 billion new mobile phone users are expected to bring the overall penetration rate to 75 percent by 2011, in which 65 percent of new consumers will come from the Asia Pacific Region. The statistic is further confirmed by a recent report released by Euromonitor (Banjanovic, 2010), which indicates 4.0 billion mobile phone subscriptions in the world in 2008. In some parts of the world, such as Western Europe, the figure has already hit 100% since 2007. The worldwide penetration of mobile devices indicates that the number of potential users of mobile learning services has far exceeded the amount of students within the current education systems.

As mobile devices are becoming more and more sophisticated and affordable, they are increasingly equipped by ordinary consumers. As a result, it comes as no surprise that sooner or later people would begin to look for new ways to activate learners, in particular those with academic ambitions but reluctant to or can't enroll in the formal education systems. A Europe-wide mobile learning project—m-learning, for instance, has been launched for the purpose of educationally disadvantaged young adults, such as teen dropouts and unemployed. In addition to common students, it is clear that a number of new learner communities could benefit and be involved, and become an indispensable part of the future mobile learning landscape. For audience, this chapter seeks to draw a brief picture of mobile learning in terms of its theoretical and technological underpinnings, and identify its potentials regarding a diversity of users.

ENABLING MOBILE LEARNING IN SOCIAL CONTEXTS

Knowledge has an inherent nature to mobilize in concert with people's increasingly mobile lifestyle. Research indicates that learning activities happen frequently in daily lives. It can take place as long as people hope to start and adapt their activities to enable educational behavior and outcomes. Vavoula (2005) conducted a study on everyday adult learning episodes in which 161 learning episodes were reported from 15 participants in a research period of two weeks. Of the total 161 learning episodes, 51% of them took place at learners' home or workplace, while 21%, 6%, 5% and 2% of episodes happened respectively in a workplace outside the office, at places of leisure, outdoors and in a friend's house (Vavoula, 2005). Other locations took 14%, including places of worship, the doctor's surgery rooms, cafes, hobby stores, and in cars. In addition to this, 48% of mobile episodes were found to be associated with work. Note that only 1% of the self-reported episodes occurred on public transport, indicating that there may be a chance to explore learning opportunities for people to utilize unproductive travelling time. These findings indicated that there are many learning episodes in daily lives where mobile learning can probably be involved and lend a helping hand.

Further, among all the learning episodes, mobile learning will be favored if a learner is situated in the 'right' scenario. Mobile learning can be advantageous, particularly when a learner is on the move or at a 'non-place'. The term 'non-place' refers to places such as airport terminals, waiting halls and hotels (Kynäslahti & Seppälä, 2003), where people are physically immobile but mobile in logic. Also, mobile learning facilitates learning activities where a learner is in a stable scenario, such as learning in class, or in a situation where a learner wants to avoid moving, e.g., a patient following a daily prescription and diagnosis at home when the doctor is working in the hospital. At home, a bed or a sofa is the most often mentioned place by mobile device owners (Hujala, Kynäslahti & Seppälä, 2003), which shows a potentially ideal location for mobile learning. What is more, mobile learning is effective for just-in-time learning or learning in urgent situations, such as first aid (Kynäslahti, 2003).

In addition, a number of studies reveal that mobile technologies have many unique advantages to support teaching and learning activities. Savill-Smith & Kent (2003), during a review of the published literature on the use of palmtop computers for learning, stated that palmtop computers can "assist students' motivation, help organizational skills, encourage a sense of responsibility, help both independent and collaborative learning, act as reference tools, and can be used to help track students' progress and for assessment" (p. 4). Similarly, Corbeil & Valdes-Corbeil (2007, pp. 54) summarized the benefits of using mobile learning as follows:

- Great for people on the go.
- Anytime, anywhere access to content.
- Can enhance interaction between and among students and instructors.
- Great for just-in-time training or review of content.
- Can enhance student-centered learning.
- Can appeal to tech-savvy students because of the media-rich environment.
- Support differentiation of student learning needs and personalized learning.
- Reduce cultural and communication barriers between faculty and students by using communication channels that students like.
- Facilitate collaboration through synchronous and asynchronous communication.

Based on an analysis of 12 international case studies, Kukulska-Hulme & Traxler (2005) summarized the reasons to use mobile learning in teaching and learning activities, including:

Access

- Improving access to assessment, learning materials and learning resources
- Increasing flexibility of learning for students
- Compliance with special educational needs and disability legislation

Changes in teaching and learning:

- Exploring the potential for collaborative learning, for increasing students' appreciation of their own learning process, and for consolidation of learning
- Guiding students to see a subject differently than they would have done without the use of mobile devices
- Identifying learners' needs for just-in-time knowledge
- Exploring whether the time and task management facilities of mobile devices can help students to manage their studies
- Reducing cultural and communication barriers between staff and students by using channels that students like
- Wanting to know how wireless/mobile technology alters attitudes, patterns of study, and communication activities among students

Alignment with institutional or business aims:

- Making wireless, mobile, interactive learning available to all students without incurring the expense of costly hardware
- Delivering communications, information and training to large numbers of people regardless of their location
- Blending mobile technologies into e-learning infrastructures to improve interactivity and connectivity for the learner
- Harnessing the existing proliferation of mobile phone services and their many users. (Traxler & Kukulska-Hulme, 2005, pp. 3-4)

The benefits of mobile learning abound. Generally, it not only can engage learners from different backgrounds, enable more effective learning activities, but also support a shift of current education system and teaching style for a better performance. Note that, as mobile learning is still in its initial stage, its benefits have not yet been fully addressed.

Technological Underpinnings for Realizing Mobile Learning

Since learning practices are mobile in terms of location and time, technologies that support learning should be mobile as well (O'Malley et al., 2003). On the other hand, the unique nature of handheld and mobile technologies make them excel in supporting learning activities in terms of the mobile nature of human activities and knowledge (Thomas, 2005; Cobcroft et al., 2006). In fact, any handheld device, in addition to mobile phones, can be to some extent used for education purposes, in other words, supporting mobile learning. Hence the conception of mobile learning technologies and devices are not limited to the use of mobile phones. Many handhelds, such as iPod, MP3 player, Personal digital assistant and E-book reader, have unique pros and cons for mobile learning implementation (Corbeil & Valdes-Corbeil, 2007). Based on the research findings of Corbeil & Valdes-Corbeil (2007), we listed a number of the most common handheld devices and discussed their features for mobile learning use, which can be summarized as follows:

- iPod

The iPod enables students to download podcasts of educational materials, such as audio and video lectures. It can be used to present e-books. Also, it can be used as a calendar and a mass-storage device. In addition, students can use the iPod to exchange files, and collaborate on the work even in a distant place. Note that the iPod can utilize iTunes to download a wide spectrum of learning materials. By February 2009, over 100,000 educational audio and video files supporting mobile learning had already been available in iTunes U. However, there are some disadvantages of the iPod, such as high price, one-way communication and small screen sizes.

- MP3 Player

The MP3 player is compact and light. Students can use MP3 players to listen to podcasts, audio lectures, and books. Also, some devices with the voice recording function can be used to record information, such as a lecture. However, the MP3 player can be replaced by another device with the audio playing function. Also, it is time-consuming to transfer files. No interactive communication is offered.

- Personal Digital Assistant (PDA)

Compared to other devices, PDAs have a relatively large screen size and convenient input methods, such as screen keyboard, a stylus or external peripherals. The PDA enables students with many new possibilities to access education, including (i) playing audio, video and flash files; (ii) presenting and editing text and word documents; (iii) accessing e-mail and web resources; (iv) instant interactive communication and learning; (v) serving as a mass storage device; (vi) video recording functions, which can be used to record lectures; (vii) the GPS function, which can be used to support research on geography and the environment. However, the PDA is relatively bulky and expensive. Note that previous functions of the PDA are increasingly embedded in common cellular phones along with technology advances. Hence, differences between PDAs and cellular phones are getting blurred.

- USB Drive

A USB drive is light and small, which can be used to store ad transfer files. Students can use it to save, share and submit their works. However, the function offered by a USB drive is quite limited while other devices may also serve as a mass storage device.

- E-book Reader

E-book readers have large screens which make reading comfortable. It can be used to download, store and play text-based learning materials. Magnification, highlighting, bookmark, and full-text search functions make it easy to be used. However, its functions are limited and only serve for the book reading purpose with limited computing power.

- Laptop/Tablet PC

A laptop/tablet PC is the most complete and functional device among all the devices introduced. It has nearly all the functions that a PDA has, but also offers a big screen and keyboard, facilitating easier operation experiences. Students can easily start their work using laptops at any place and time they want. However, they are relatively expensive and cumbersome to be carried when traveling. Also, it is nearly impossible to use it when walking.

- New Devices (especially designed for mobile learning purposes)

Recent years have seen a number of new devices especially designed for mobile learning purposes. These devices are used for varied purposes and therefore embed different handheld technologies. For instance, in tourist attractions, such as in Louvre Museum and the palace of Versailles, a number of new handhelds were employed and rented to tourists to offer audio guidance. In this case, mobile learning not only enhances tourists' knowledge on the masterpieces presented in museums, but also generates a new source of revenue for the tourist industry. The One Laptop Per Child Association, Inc. (OLPC), as a non-profit organization, developed a new low-cost laptop, which is known as the $100 Laptop, in order to offer children in the developing world with content and software designed for collaborative, joyful, and self-empowered learning. In China, a series of handheld digital learning devices are especially developed for mobile learning purposes. According to CCID Consulting (2009), 6.2 million educational electronic devices were sold in China in 2008 and the figure is expected to reach 7.3 million in 2011.

Note that, in general, there are many technological challenges that mobile learning faces, such as lack of data input capability, low storage, low bandwidth, limited processor speed, short battery life, lack of standardization, limited interoperability, compatibility issues, low screen resolution and small screen size (Maniar et al., 2008). Additionally, usability problems are frequently reported in current mobile learning research, since most mobile learning activities are based on the use of the devices that are not designed for educational use (Kukulska-Hulme, 2007). Hence the devices with special consideration on mobile learning usability issues offer a new approach to facilitate learners' adoption of mobile learning services. It can be predicted that future mobile learning industry tends to rely more on these handhelds that have education in mind.

Utilizing different handheld technologies and devices, a wide spectrum of mobile learning applications has been developed in recent years. As shown in Table 1, Liu & Li (2009) summarized 24 kinds of mobile learning initiatives and classified them into four broad categories from the perspective of functionality, which include:

- **Informal learning:** applications facilitate the learning activities outside predesigned educational establishments.
- **Administration function:** applications are used to administrate the learning process and organize learning activities.
- **Social network:** applications facilitate peer communication as well as instructor-student interactions.

Table 1. A summarization of current mobile learning initiatives (Liu & Li, 2009, p. 7)

Categories	Mobile learning services
Informal learning	Extracurricular study; Searching answers in, for instance, Google on wireless Internet;
Administration function	Sending reminders for examinations or assignment; Informing schedules or coordinating schedules; Calendars; Collecting feedback; Recording attendance or test taker; Recording lectures; Recording information of patients; Retrieving school-related information, such as timetable; Library services; Digital dictionaries, translators; Environmental detectives or recorders; Collecting and analyzing the data of the learning process;
Social network	Interaction between instructor and students, or between peer students; Learning collaboration, such as virus game; Mobile 'blogging'; Accessing online communities, discussion boards and chat rooms via mobile phones;
Learning material utilization	Situated learning, such as learning in a museum, watching birds in open air and mobile excursion games; Displaying lecture videos and courseware; Podcasting lectures; Playing quizzes; Mobile learning in language studying, and mathematics.

- **Learning material utilization:** handheld devices are used to store and display learning materials, such as presenting e-books and lecture videos.

In general, mobile learning applications currently available in tertiary education are mostly for administration and social network purposes, while chief commercial mobile learning applications are for tourism use, such as mobile learning in museums. Note that the mobile learning industry adopts different ways of development in different countries. For instance, mobile learning applications in China are mostly initiated by business communities for students in basic education, while in Europe, mobile learning applications are generally developed by government and educational organizations for students in tertiary education or adults.

Theoretical Underpinnings for Realizing Mobile Learning

Different from traditional education approaches, mobile learning is built on the use of mobile technologies, which brings it a number of unprecedentedly new features. In concert with the unique nature of mobile technologies, these new features can be illustrated as shown in Table 2.

The new features of mobile learning brought by the mobile technologies also bring it new challenges to establish its theoretical underpinnings. Note that most theories of pedagogy fail to capture the unique nature of mobile learning, as they are mostly based on the assumption that learning takes place in a classroom environment, controlled by teachers. Compared with previous education methods, mobile learning is a learner-centered approach. It typically takes place in an unstructured environment and seeks to tailor service for personal needs. This gap leads to a long dearth of proper theories in mobile learning research (Muy-

Table 2. Convergence between learning and technology (Sharples, Taylor & Vavoula, 2005, p. 3)

New Learning	New Technology
Personalized	Personal
Learner-centered	User-centered
Situated	Mobile
Collaborated	Networked
Ubiquitous	Ubiquitous
Lifelong	Durable

inda, 2007). In this light, Sharples, Taylor, & Vavoula (2005) proposed a list of criteria against which a mobile learning theory could be tested. These criteria also offer an important foundation for developing new theoretical underpinning, which are:

- Is it significantly different from current theories of classroom, workplace or lifelong learning?
- Does it account for the mobility of learners?
- Does it cover both formal and informal learning?
- Does it theorize learning as a constructive and social process?
- Does it analyze learning as a personal and situated activity mediated by technology?

Five Mobile Learning Theoretical Underpinnings Proposed by Naismith et al.

Currently theoretical underpinnings of mobile learning research are mostly based on the work of Naismith et al. (2004), who compared new mobile learning practices against existing learning theories, which are behaviorist, constructivist, situated, collaborated, informal and lifelong learning.

Behaviorist Learning Theory

Behaviorist learning emphasizes learning experiences gained as a change in observable actions with proper stimulus and response. With the advance of mobile technologies, mobile learning makes it possible to form a 'drill and feedback' mechanism complied with the behaviorist learning theory. Specifically, mobile learning can give learners content specific questions, then gather their responses in a rapid manner and provide instant feedback, which fits with the behaviorist learning paradigm.

Constructivist Learning Theory

The constructivist theory emphasizes gaining learning experience in a way that learners actively build new ideas or concepts based on both their previous and current knowledge. With a mobile phone, a learner can construct his/her own knowledge and share it freely with peers regardless of time and place. Specifically, an easy way for mobile learning to enable an immersive constructivist learning experience is to offer edutainment (e.g. handheld games).

Situated Learning Theory

Situated learning emphasizes learning activities that take place within authentic contexts where the environment itself appears to be a part of education resources. For situated learning, the environments can be pre-organized, such as studying in a museum (Chang, Chang, & Hen, 2007), or naturally developed, such as watching birds in open air (Chen, Kao, & Sheu, 2003). Specifically, situated learning experience can be realized via three manners, namely problem-based learning, case-based learning, and context-aware learning.

Collaborated Learning Theory

Collaborated learning experiences are promoted as a learning process with proper social interaction. The increasing availability of wireless networks in personal devices not only makes it much easier to communicate and share data, files and messages with partners, but also makes learning collaboration easier to initiate and to respond to. Taking into consideration the recent popularity of the Really

Simple Syndication (RSS) as well as open source software, learning collaboration on a large scale appears to be more socialized and self-initiated.

Informal and Lifelong Learning Theories

Informal and lifelong learning emphasizes the learning activities that take place outside a dedicated learning environment, such as a predetermined curriculum. Informal learning can be intentional with intensive and deliberate learning efforts, or it can be accidental, such as through conversations, TV and newspapers (Naismith et al., 2004). To the extent that mobile devices facilitate instant information acquisition in a seamless and unobtrusive way, mobile learning is especially suitable for offering informal and lifelong learning experience.

Learner-Centered Andragogy: Self-Directed Learning Theory

Considering the learner-centered nature of mobile learning, Liu & Li (2009) sought to use one of the andragogy theories to explain mobile learning activities, which is a self-directed learning theory (SDL). This theory has long been stressed and applied in problem-based, lifelong, and distance learning settings (Fisher, King, & Tague, 2001; Stewart, 2007). In its broadest meaning, "self-directed learning describes a process in which individuals take the initiative, with or without the help of others, in diagnosing their learning needs, formulating learning goals, identifying human and material resources for learning, choosing and implementing appropriate learning strategies, and evaluating learning outcomes" (Knowles, 1975, p. 18). The theory indicates that the level of control that learners are willing to take over their own learning will depend on their abilities, attitude, and personality characteristics (Fisher, King, & Tague, 2001). A common aim for SDL research is to assist individuals in developing the requisite skills for engaging in self-directed learning such as planning, monitoring, and evaluating their own learning (Reio & Davis, 2005), which are also important capabilities to facilitate successful mobile learning implementation.

Liu & Li (2009) applied the self-directed learning theory to mobile learning contexts, and utilized it to explain the success and failure of current mobile learning initiatives. They suggested that mobile learning activities are typically initiated outside a pre-organized learning environment while learners are mostly physically separated from both teachers and peer students. Therefore mobile learning initiates a heightened need for proper self-direction and self-management. To help students finish a mobile learning course that, for instance, takes tens of hours, it is important to sustain their learning desire and help them to effectively self-control and manage the learning process (Liu & Li, 2009). Further, they proposed:

- Education is not inherently a gratification process; anxiety initiated either by education or by lacking social interaction will impede learners in the pursuit of mobile learning. Hence there is a need to sustain students' learning desire.
- Success in mobile learning initiates a requirement for SDL capability, but not all the learners have a proper SDL capability for mobile learning; hence technology and services should help learners to organize their learning process and to evaluate their learning outcomes.
- The misuse of mobile phones in a classroom happens naturally since young students inherently have a limited capability of self-management and self-direction.
- Great autonomy and freedom placed on learners do not guarantee effective mobile learning as well as positive academic outcomes.
- An unstructured learning environment tends to be the typical environment for mobile learning; this type of environment

may cause anxiety for learning and lead to arbitrary learning.

- For those with a low SDL capability, solutions to reduce the requirement for SDL capability are essential; otherwise students may not use mobile learning or discontinue the use after starting to use it.
- From an SDL viewpoint, there are four alternative solutions to implement a successful mobile learning system:
  a) To provide learning environments with proper guidance particularly for situated mobile learning.
  b) To reduce the autonomy and freedom offered to an appropriate level that most learners feel comfortable with.
  c) To help learners manage their learning process using, for instance, SMS reminders and distance instruction.
  d) To motivate students and alleviate learning pressure using more personalized communication and a social network. (Liu & Li, 2009, p. 6)

In essence, different learning theories seek to offer different mobile learning experiences and picture mobile learning from different aspects. It is the inherent nature of mobile learning that lends itself well to motivate learners intrinsically by offering versatile learning experiences. Hence these learning experiences should be integrated and combined instead of being separated. If leveraged appropriately, mobile learning makes it possible to form a learning space which is socialized, personal and digital, trusted, pleasant and emotional, creative and flexible, certified, open and reflexive (Punie, 2007). Similarly, Naismith et al. (2004) stated that mobile learning would initiate a sort of "highly situated, personal, collaborative and long term; in other words, truly learner-centered learning" (p. 36).

ENABLING MOBILE LEARNING FOR NEW LEARNERS

It is evident that a rapid proliferation of mobile devices expands the reach of education to all social-economic levels. In addition to common school/university students, mobile learning appears to be an ideal conduit to deliver training and education to the learner communities unreachable through conventional education approaches. As they are of great demographic importance, these new learners can not be neglected.

Engaging Problem Teenagers and Illiterates

In most parts of the world, it is undeniable that many teenagers are unsatisfied with classroom-based educational environments and they drop out without pursuing any further training or education. Teen dropouts are in general hard-to-reach by traditional educational approaches and are more likely to be the future illiterates, resulting in many serious social problems. For instance, it is reported that in UK, nearly 10 millions adults lack confidence in using literacy skills (BBC, 2007), while in China, the number of people deemed illiterate jumps from 30 million to 116 million from 2000 to 2005, right after India (China Daily, 2007). Further, there are still about 785 million illiterate adults aged over 15 worldwide in 2009 (Indexmundi, 2009). Early dropout of teenagers from schools would lead to serious problems for the society. It is reported that early dropouts are more prone to be unemployed, in prison, living in poverty, receiving government assistance, in poor health, divorced and single parents (Pytel, 2006).

For these learners, mobile learning appears to be an ideal solution to deliver training and education. Current young people, in particular the "millennial generation" that was born in or after 1982, shows a clear preference for technology applications (Oblinger, 2003; McMahon & Pospisil, 2005). With an information technology mindset and a highly developed skill for multitasking, the millennial generation is described as being focused on "connectedness" and social interaction with a preference for group-based methods in study and social occasions (McMahon & Pospisil, 2005).

To engage millennial learners, in particular teen dropouts, mobile learning has great advantages as it accommodates the unique nature of these new learners in comparison to traditional education approaches. Also, in light of the fact that many learners might never be able to afford a personal computer or enroll into formal education again, a mobile phone, which is increasingly popular among young people, becomes an affordable conduit for delivering education. According to Attewell (2005), there are several advantages to implement mobile learning for problem teenagers and illiterates, including:

- Mobile learning helps learners to improve literacy and numeric skills and to recognize their existing abilities;
- Mobile learning can be used for promoting independent and collaborative learning experiences;
- Mobile learning helps learners to identify where they need assistance and support;
- Mobile learning helps to combat resistance to the use of ICT and can help overcome the divide between mobile phone literacy and ICT literacy;
- Mobile learning helps to remove some of the formality from the learning experience and engages reluctant learners;
- Mobile learning helps to concentrate a learner's attention for longer periods;
- Mobile learning helps to raise self-esteem;
- Mobile learning helps to raise self-confidence.

Supporting the Informal and Lifelong Learning of Employees

As human societies are becoming more and more hectic and knowledge-based, employees have to adopt more learning activities to renew and update their knowledge and skills in order to remain competitive in the workplace, and to adapt to an increasingly technological environment. The growing learning requirements go with problems, as today's workforce is increasingly mobile around the world (Edwards, 2005). For instance, it is predicted that 75% of U.S. workforce and 80% of Japan workforce will become mobile by 2011 (IDC, 2008). IDC (2008) estimated that the worldwide mobile worker population will increase from 758.6 million in 2006 to 1.0 billion in 2011, which accounts for 30.4% of the total workforce. Nonetheless, the time available for employees to stay in a stationary place to learn is becoming limited. In 2003, the average time available for training was less than three days (Hayes, Pathak, & Joyce, 2005). Also, there is little evidence to show that time and resources available for formal training will be increased.

In this regard, mobile learning appears to be a desirable way to provide training and education to an increasingly mobile workforce. Great benefits can be achieved through the use of mobile learning. As Koschembahr (2005) stated, mobile learning can assist enterprises in saving cost, enhancing customer services and offering better selling opportunities. Moreover, mobile learning reflects a potential to improve job satisfaction and to reduce job stress as well as employee turnover (Koschembahr, 2005). Also, it enables employees to utilize previously unproductive time as part of people's increasingly hectic lifestyle (Geddes, 2004). With regard to ICT literacy, as Punie (2007) pointed out, mobile learning promotes ICT skills, digital competence and other new skills, and helps to fight ICT resistance. Kineo and UFI/Learndirect (2007) indicated that mobile learning can help address some challenges faced by businesses as follows:

- Mobile learning enables business entities to provide learning to mobile staff and to distribute learning quickly.
- Mobile learning enables the delivery of key data at the point of need - particularly relevant for workers who need access to updated product specifications, pricing details or other time-sensitive information.
- Mobile learning enables companies to utilize staff downtime, those short periods of time waiting or travelling.

Facilitating the Retraining of Aging People

Population aging is a pervasive phenomenon. In the Asia-Pacific area for instance, people aged 50 and above are expected to take up approximately 31% of the total population by 2025 (Watson, 2006), while in Japan, population ageing seems to be more significant and 28.7% of the population will age 65 and above by 2025 (NIPSSR, 2002). In addition to this, it is predicted that almost one third of the working age population will be aged 50 or over by 2050 in developed countries (UN & DESA, 2007). In this light, population aging impresses people with an ongoing trend—aging people will inevitably become an incremental part of the future workforce. Due to lack of enough qualified employees, ageing people nowadays have already been encouraged to join the workforce in some parts of the world. In Europe, a marked rise has been found in the employment rate of people aged 55-64 from 36.6% in 2000 to 43.6% in 2006 (EurActiv, 2007).

The requirement for the retraining of aging learners is intensified, but research targeted at aging learners is in short supply, also within the context of mobile learning. Unlike young and prime adults, aging learners have unique learning requirements and traits. For instance, ageing individuals need a learning approach that facilitates the review of learning materials, as they incur a biologically-based decline in fluid intelligence, which impairs rapid processing of new information (Niessen, 2006). In addition, older learners may have a lack of confidence and thereby resist trying something new. In this concern, mobile learning gains advantages as it tends to address these problems through bringing training into local areas and offering courses in less formal settings (NIACE, 2005). Also, there is little extra economical and physical effort required for aging people to learn via mobile devices in comparison to the computer-based or classroom-based learning approaches.

CONCLUSION

In sum, it is apparent that the potentials of mobile learning are profound and far-reaching. With a worldwide diffusion and increasingly educational use of mobile devices, mobile learning extends learning opportunities to all social-economic levels, and the people who can benefit from mobile learning are increasing. For both learners and society, mobile learning is particularly cost-effective in terms of its capability to be centrally processed and updated with a fast and economical allocation of educational resource in a 24X7 manner for all mobile phone owners regardless of location. As such, in addition to common students, more attention is needed to learners who are previously hard-to-reach or incompatible with traditional educational approaches so as to realize the full potential of mobile learning.

This chapter in general offers some background knowledge on mobile learning with regard to its theoretical and technological underpinnings and potentials. This basic knowledge is important if one wants to further evaluate and understand the significance of a mobile learning application, its potential and contexts of use, such as open source mobile learning applications.

REFERENCES

Attewell, J. (2005). *Mobile technologies and learning: A technology update and m-learning project summary*. London, UK: LSDA.

Banjanovic, A. (2010). *Special report: Towards universal global mobile phone coverage*. Retrieved February 5, 2010, from http://www.euromonitor.com/Special_Report_To wards_universal_global_mobile_phone_coverage

BBC. (2007). *Basic sums 'stress 13.5m adults'*. Retrieved February 8, 2010, from http://news.bbc.co.uk/2/hi/uk_news/education/7027569.stm

Chang, A., Chang, H., & Hen, J. S. (2007). *Implementing a context-aware learning path planner for learning in museum*. Paper presented at the 6th WSEAS International Conference on EACTIVITIES, Tenerife, Spain.

Chen, Y. S., Kao, T. C., & Sheu, J. P. (2003). A mobile learning system for scaffolding bird watching learning. *Journal of Computer Assisted Learning*, *19*(3), 347–359. doi:10.1046/j.0266-4909.2003.00036.x

China Daily. (2007). *Ghost of illiteracy returns to haunt country*. Retrieved February 8, 2010, from http://english.peopledaily.com.cn/200704/02/eng20070402_363005.html

Cobcroft, R., Towers, S., Smith, J., & Bruns, A. (2006). Mobile learning in review: Opportunities and challenges for learners, teachers, and institutions. In *Proceedings of Online Learning and Teaching Conference 2006*. Brisbane.

Consulting, C. C. I. D. (2009). *Handheld digital learning device market becomes more mature*. Retrieved February 9, 2010, from http://www.ccidconsulting.com/insights/content.asp?Content_id=20882

Corbeil, J. R., & Valdes-Corbeil, M. E. (2007). Are you ready for mobile learning? *EDUCAUSE Quarterly*, *30*(2), 51–58.

Edwards, R. (2005). *Knowledge sharing for the mobile workforce*. Retrieved January 22, 2008, from http://www.clomedia.com/content/templates/ clo_article.asp?articleid=945&zoneid=24

EurActiv. (2007). *European youth misses out on jobs boom*. Retrieved January 22, 2008, from http://www.eur activ.com/en/socialeurope/ europeanyouthmissesjobsboom/article168707

Fisher, M., King, J., & Tague, G. (2001). Development of a self-directed learning readiness scale for nursing education. *Nurse Education Today*, *21*, 516–525. doi:10.1054/nedt.2001.0589

Geddes, S. J. (2004). Mobile learning in the 21st century: Benefit for learners. *Knowledge Tree E-Journal*, *30*(3), 214–228.

Hayes, P., Pathak, P., & Joyce, D. (2005). *Mobile technology in education –a multimedia application*. Paper presented at 6th Annual Irish Educational Technology Users' Conference (EdTech), Dublin.

Hujala, H., Kynäslahti, H., & Seppälä, P. (2003). Mobile learning: 'Creative learning'— mobility in action . In Kynäslahti, H., & Seppälä, P. (Eds.), *Mobile learning* (pp. 111–111). Finland: Edita Publishing Inc.

IDC. (2008). *Worldwide mobile worker population 2007-2011 forecast*. Retrieved February 9, 2010, from http://www.workshifting.com/downloads/documents/ IDC_Mobile Worker_excerpt_0_0.pdf

Indexmundi. (2009). *World demographics profile 2009*. Retrieved February 8, 2010, from http://www.indexmundi.com/world/demographics_profile.html

Kineo & UFI/Learndirect. (2007). *Mobile learning reviewed*. Retrieved on April 10, 2008, from http://www.kineo.com/documents/Mobile_learning_reviewed_final.pdf

Knowles, M. (Ed.). (1975). *Self-directed learning: A guide for learners and teachers*. Chicago, IL: Follett Publishing Company.

Koschembahr, C. (2005). *Optimizing your sales workforce through mobile learning*. Retrieved February 8, 2010, from http://www.neiu.edu/~sdundis/textresources/Mobile_Wireless/Optimizing%20Your%20Sales%20Workforce%20through%20Mobile%20Learning.pdf

Kukulska-Hulme, A. (2007). Mobile usability in educational contexts: What have we learnt? *International Review of Research in Open and Distance Learning, 8*(2), 1–16.

Kukulska-Hulme, A., & Traxler, J. (Eds.). (2005). *Mobile learning: A handbook for educators and trainers*. London, UK: Routledge.

Kynäslahti, H. (2003). Mobile learning: In search of elements of mobility in the context of education . In Kynäslahti, H., & Seppälä, P. (Eds.), *Mobile learning* (pp. 47–47). Finland: Edita Publishing Inc.

Kynäslahti, H., & Seppälä, P. (Eds.). (2003). *Mobile learning*. Finland: Edita Publishing Inc.

Liu, Y., & Li, H. X. (2009). What drives m-learning success?—Drawing insights from self-directed learning theory. In *Proceedings of Pacific Asia Conference on Information Systems 2009*. Hyderabad, India.

Maniar, N., Bennett, E., Hand, S., & Allan, G. (2008). The effect of mobile phone screen size on video based learning. *Journal of Software, 3*(4), 51–61. doi:10.4304/jsw.3.4.51-61

McMahon, M., & Pospisil, R. (2005). Laptops for a digital lifestyle: The role of ubiquitous mobile technology in supporting the needs of millennial students. In *Proceedings of ASCILITE 2005*. Retrieved February 8, 2010, from http://www.ascilite.org.au/conferences/brisbane05/ blogs/proceedings/49_McMahon%20&%20Pospisil.pdf

Muyinda, P. B. (2007). M-learning: Pedagogical, technical and organizational hypes and realities. *Campus-Wide Information Systems, 24*(2), 97–104. doi:10.1108/10650740710742709

Naismith, L., Lonsdale, P., Vavoula, G., & Sharples, M. (2004). *Literature review in mobile technologies and learning: A report for NESTA Futurelab*. Retrieved February 8, 2010, from http://www.futurelab.org.uk/resources/ documents/lit_reviews/Mobile_Review.pdf

NIACE. (2005). *NIACE briefing sheet – 54: Mobile ICT resources for older learners*. Retrieved February 9, 2010, from http://archive.niace.org.uk/information/Briefing_sheets/54-Mobile-ICT-resources-for-older-people.pdf

Niessen, C. (2006). Age and learning during unemployment. *Journal of Organizational Behavior, 27*(6), 771–792. doi:10.1002/job.400

NIPSSR. (2002). *Population projections for Japan: 2001-2050*. Retrieved February 9, 2010, from http://www.ipss.go.jp/pp-newest/e/ppfj02/ppfj02.pdf

O'Malley, C., Vavoula, G., Glew, J. P., Taylor, J., Sharples, M., & Lefrere, P. (2003). *Mobilearn WP4 guidelines for learning/teaching/tutoring in a mobile environment*. Retrieved February 8, 2010, from http://www.mobilearn.org/download/results/guidelines.pdf

Oblinger, D. (2003). Boomers & gen-xers, millennials: Understanding the 'new students.' . *EDUCAUSE Review, 38*(4), 37–47.

Portio Research. (2007). *Worldwide mobile penetration will reach 75% by 2011*. Retrieved February 8, 2010, from http://www.portioresearch.com/next_billion_press.html

Punie, Y. (2007). Learning spaces: An ICT-enabled model of future learning in the knowledge-based society. *European Journal of Education, 42*(2), 185–199. doi:10.1111/j.1465-3435.2007.00302.x

Pytel, B. (2006). *Dropouts give reasons: Why do students leave high school without a diploma?* Retrieved February 8, 2010, from http://educationalissues.suite101.com/article.c fm/dropouts_give_reasons

Reio, T. G., & Davis, W. (2005). Age and gender differences in self-directed learning readiness: A developmental perspective. *International Journal of Self-Directed Learning*, *2*(1), 40–49.

Savill-Smith, C., & Kent, P. (2003). *The use of palmtop computers for learning: A review of the literature*. London, UK: Learning & Skills Development Agency.

Sharples, M., Taylor, J., & Vavoula, G. (2005). *Towards a theory of mobile learning*. Paper presented at the 4th World Conference on Mlearning (MLearning 2005), Cape Town, South Africa.

Stewart, R. A. (2007). Evaluating the self-directed learning readiness of engineering undergraduates: A necessary precursor to project-based learning. *World Transactions on Engineering and Technology Education*, *6*(1), 59–62.

Thomas, S. (2005). *Pervasive, persuasive e-learning: Modeling the pervasive learning space*. Paper presented at the 3rd International Conference on Pervasive Computing and Communications, Hawaii.

Traxler, J., & Kukulska-Hulme, A. (2005). *Evaluating mobile learning: Reflections on current practice*. Paper presented at mLearn 2005: Mobile Technology: The Future of Learning in Your Hands, Cape Town, South Africa.

UN & DESA. (2007). *World economic and social survey 2007: Development in an ageing world*. (p. 24). New York, NY: United Nations publication.

Vavoula, G. N. (2005). *WP4: A study of mobile learning practices*. MOBIlearn deliverable D4.4. Retrieved February 8, 2010, from http://www.mobilearn.org/download/results/ public deliverables/ MOBIlearn_D4.4_Final.pdf

Watson, W. (2006). *Ageing workforce 2006 report*. Retrieved January 5, 2008, from http://www.watsonwyatt.com/images/database_uploads/ageing_ap_06/AP_AgeingWorkforce2006.pdf

KEY TERMS AND DEFINITIONS

Andragogy: The art and science of helping adults learn

Distance Learning: A type of education that seeks to deliver education to the students who are not physically "on site" in a traditional classroom or campus remotely by using electronic communication.

Informal Learning: A type of semi-structured learning, which occurs in a wide spectrum of places and time, such as learning at home, work, etc. It can be accomplished through daily interactions and shared relationships among members of society.

Lifelong Learning: A provision or use of both formal and informal learning opportunities throughout people's lives in order to promote the continuous development and improvement of the knowledge and skills needed for employment and personal fulfillment

Pedagogy: The art, science, or profession of teaching

Self-Direction: Directed or guided by oneself, especially as an independent agent

Self-Management: Methods, skills, and strategies by which individuals can effectively direct their own activities toward the achievement of objectives

Chapter 18
Virtual Environments and Mobile Learning:
A Tale of Two Worlds

Rosa Reis
Porto Polytechnic Institute, Portugal

Paula Escudeiro
Porto Polytechnic Institute, Portugal

ABSTRACT

In recent years, Information Technologies have been used to improve educational environments. These technologies led to new forms of learning, encouraging the use of mobile technologies and three-dimensional virtual worlds, and allowing students a ubiquitous learning environment they can explore to improve their learning experience. Although these learning systems have much to offer to both students and teachers, it is necessary to discuss several research questions: (1) How can these platforms make learning experiences more attractive and motivating, particularly to students who do not have powerful intrinsic motivation to learn yet? (2) How can m-learning and virtual worlds together improve the learner's resources?

As an answer to these questions, this chapter will define the virtual world concept, distinguish the different types of virtual worlds, and make a comparative analysis between them in order to bring out the features aimed at helping teachers to adopt them in their classes. In particular, we will focus our choice of virtual world environments on open source platforms. As the prevalence of mobile learning increases, this chapter also describes the m-learning scope, its contextualisation and advantages, as well as the learning methods. Finally, the relation of those methods with social virtual worlds is also discussed.

DOI: 10.4018/978-1-60960-613-8.ch018

INTRODUCTION

The development of Information and Communication Technology (ICT) has contributed to a significant change in the society in which we live. The change is irreversible. We are in a society that is increasingly dominated by technology where all are directly or indirectly dependent on the interaction of man with ICT (Menezes & Moreira, 2009). Thus, new solutions must be found.

The technologies, which have dominated the world, came to develop two different realities, information and informatics. The merge of these two realities contributed to the appearance of social networks. Young people are increasingly changing their behaviours, their interests, and their objectives. Conventional education is no longer appealing to new students because they can no longer stand still long enough to attend a lecture. Given this fact, it is necessary for education to keep pace with the changes. The new people are more attracted to virtual environments by mobile technologies, which facilitate communication and social networking among them. Therefore, it is important, given the high rate of expansion devices and the growing prospective of virtual worlds as the future of the web, to reflect and research around their potential in educational settings. This is because the society is faced with new virtual environments, associated with virtual reality and a new concept of learning, mobile learning. With mobile learning, teachers can make reviews of small units, update information, send information to parents, and teach classes in their entirety.

The purpose of this chapter is to demonstrate that m-learning and virtual worlds can help promote new concepts, new approaches and new strategies. In education, it is possible to change the teaching/learning paradigm, including the development of autonomous learning.

The chapter is organized as follows: Section 2 presents definitions about virtual worlds, describes some social virtual worlds and makes a comparative analysis between them based on the matrix developed by Manninen (2004). Section 3 describes how virtual worlds can help in exploring new educational techniques or enhance and complement methodologies and techniques already known in the academic communities, allowing a greater involvement of students in the study, creating conditions to optimize resources, and avoiding the need for the displacement of people. On the other hand, we will explore and articulate some key issues as a reflection for readers interested in this subject. Section 4 focuses on the m-learning concepts, their characteristics, what we need to consider, and on practical issues (e.g. relation with virtual worlds and m-learning). Finally, in the conclusions, we discuss the advantages and disadvantages of the social worlds for educational purposes, and their relation with m-learning.

VIRTUAL WORLDS: DEFINING VIRTUAL WORLDS

In order to get a better discernment of the virtual world's meaning, it is necessary to clarify and get an unambiguous definition of "virtual world" because, when searching virtual worlds, it is possible to find many and different definitions that lead the reader to have confused ideas about the concept.

Richard Bartle (Bartle, 2004) in the mid 70s defines the virtual world as an environment whose inhabitants are regarded as being self-contained. This leads us to the concept of world, but not to the way the world becomes virtual. Raph Koster (Koster, 2004) defines this concept as a persistent space, which may be experienced by several participants at the same time and being represented by avatars. With this definition, some features of virtual worlds can be easily observed: Persistence and multi-user. However, it does not specify the technology required for the virtual world. Edward Castronova (Castronova, 2004) introduced the element of technology when describing virtual worlds as places worked by computers, which are designed to accommodate a large number of people.

If we blend the elements and put the emphasis on people, it is possible to obtain a more precise definition of virtual worlds:

A virtual world is a simulated persistent space based on the interaction by computers, inhabited by several users, who are represented by iconic images called avatars and who can communicate with each other and the world in a synchronized way.

This definition allows us to demonstrate that a virtual world is more than a simple virtual environment. The term "virtual world" usually brings to our mind a space similar to the real world where we can live, with details and action of gravity, the surface topography of the different ways to move objects, passage of time, and finally the possibility of actively communicating between the various objects created by the users. This world must not be interrupted; it should continue to exist if the user does not connect. Persistence means that participants are members of a dynamic community, and so they communicate and interact with each other and with the environment. This ensures that participants have the feeling of sharing time, being able to see each other's behaviour, and communicating between them. Although the view is the basis for an actual virtual world, you should allow some kind of communication, mediated by the computer, to occur among the participants. Interactivity tends to be dynamic, and thus communication is in real time. Moreover, the sense of presence, which is related to the individual and to the collective, allows, when we get in a shared space, each participant to become a "virtual person", known as avatar, i.e., a digital representation (graphic or text) in the virtual world.

Based on this definition, we can identify characteristics of the virtual world, which are according to Book (2004):

Shared Space: The world allows many users to participate at once. All the users have the feeling of being in the same place, room, and land. The shared space is a common location where interactions may be occurring. The place can be real or fictional, and this space must have the same characteristics for all the participants.

Immersion and Interactivity: The idea of the virtual world allows the users, in the interaction with the environment, to adjust, develop, build or submit contents. The user must feel immersed in the environment and fully engaged in the activities being undertaken. This is normally achieved through the representation of the user and environment in the world (Freitas, 2008).

Persistence: The world's existence continues regardless of whether individual users are logged in. The persistence of the world leads us to the immediate feedback and synchronous use of the actions that take place in the world.

Immediacy: An immediate virtual world allows real-time interaction between the user and the world. Interaction between users is also real-time. It is a definitional quality of virtual worlds.

Socialization: The world allows and encourages the formation of in-world social groups or communities.

The characteristics presented above represent, in the virtual world, the different categories of information by means of three-dimensional objects; thus, it is possible to develop a world similar to the real life.

SOCIAL VIRTUAL WORLDS

The definition of virtual worlds described allows us to consider two different types of virtual worlds: The social virtual worlds and the Massive Multiplayer Online Games (MMOGS). These categories are based on the necessity to apply the world in educational context.

The worlds oriented by socialization have not pre-defined rules. The objectives of members shall live and prosper by using the social practices that they can find in off-line environments, allowing their experiences to be more realistic. The users do not necessarily win or play a game, but socialize with other users.

The game-oriented worlds, which have a basic structure of video games, with obstacles to overcome and with roles especially visible, combine this structure with the necessity of socialization to achieve major objectives. These persistent virtual worlds are flexibly structured by open narratives, where the players can act freely, talk and exchange information. Furthermore, it is clear about the peculiar combination of the imaginary world (fantasy) with the social realism (Castronova, 2004).

Looking at the different worlds, they all have several technologies in a single platform: Audio, video, webcam, text and voice chat (VoIP), graphical tools, scripting, web browser and avatars – the user's projection in the world. Combining these tools and the social aspects, it opens up a large range of new perspectives and new ideas, which will gradually allow new applications.

From the wide range of tools available in the market, we present below those which are the most addressed by the academic community, for the following reasons:

1. Have a great potential for integrating different technologies, allowing presenting e-learning materials and e-content, narratives based on social interactions, sharing documents and files, holding meetings and events, and providing forums for sharing research findings and meetings with international colleagues.
2. Give users the ability to develop the experiences that could be tricky in the real world.
3. Are safe places for students to learn by doing and to work in partnership; the ability to interact with one another simultaneously provides students with the opportunity to learn concepts not easily learned from a textbook (Gros, 2002).
4. The students are encouraged to engage at a higher-level cognitive thinking, such as interpreting, analyzing, and discovering.

The selected virtual worlds will be explored in more detail to help encourage educators to use them in the learning context, because "virtual worlds provide excellent capabilities for creating effective distance and online learning opportunities through providing unique support for distributed groups (online chat, the use of avatars, document sharing etc.)" (Freitas, 2008; p. 5).

OpenSim

OpenSim is a platform for the development of virtual worlds, created in early 2007 by Darren Guard (OpenSimulator, 2010). OpenSim is compatible with Linux and Windows. The environment allows the creation/edition of content and supports the programming languages LSL, OSSL, C#, JavaScript, Visual Basic to create tools and interactions within the environment (Onyesolu, 2009). Many of the features, functionality, appearance and interface of this environment are similar to Second Life (SL). The user can modify the appearance of avatars and the interface, and create interactivity by interacting with the worlds. Interactivity is achieved by encoding scripts. The avatars can be created to communicate with each other, through the same tools that other virtual worlds use for this purpose, for example, chat, voice, and asynchronous messages.

This platform supports the import of 3D models generated with the tools AutoCad, 3D studio, Wings, and Blender, eliminating restrictions on the import of complex objects. It differs from the SL by not being dependent on a third company. The worlds are stored in its own machines or over machines that have control, and offer this functionality to import and export complete worlds for other OpenSim servers.

Despite the features described above, the platform has some problems when we want to create a virtual world, particularly because it offers few features due to its short life.

Croquet

This platform, implemented by the Boston University licensed open source, is based on a model of communication between users not running on a central server but in each person's local machines (Croquet, 2010).

Croquet is a fully dynamic environment; everything is a collaborative object, which can be modified at all times (Smith et al., 2003). The platform allows the import of few 3D models and it is always necessary to change the scripts. The communication between objects is based on the synchronized message-passing model and their interactions are focused on 3D shared space that is used for context-based collaboration. The interaction between users results only by the sending of a function over the network.

The language used by this platform is Squeak, which is a variant of Smalltalk (Onyesolu, 2009). Squeak incorporates the integration with OpenGL libraries and it is a peer-to-peer network architecture providing an infrastructure for synchronous real-time problem solving within shared simulations, which induces difficulties in the creation of virtual worlds.

The users of this environment may set as many portals as they want. Once the connections between the worlds are an analogue of web page hyperlinks, the users can visit different parts of the worlds from various portals.

The functionalities of Croquet can be summed up in two main tasks: (1) simulation of virtual worlds; (2) importation or exportation of virtual worlds and 3D objects.

Wonderland

Wonderland is a platform that allows the creation of virtual worlds, emphasizing the use of immersion and audio communication, text, application and document sharing as well as the development of user-created content (Ally, 2009). At the beginning, Wonderland was created to allow a multi-user, secure, stable and reliable enough environment to conduct real business (Collins, 2008). Thus, it is possible for the users to do their actual work in the virtual world, therefore avoiding the use of collaboration tools separately. The Collaborative work on this platform wins space, leading co-workers to collaborate and work in the same representation, despite the physical limitations of "working from home".

Wonderland provides a multi-platform client through which all avatars can enter a virtual world and interact with each other and with objects. The worlds built on this platform are defined by a group of cells controlled by the server. The cell is used to be possible to manage the object. As the avatar moves through the world, the server periodically informs the client of which cells are to be loaded or unloaded. There are two types of cells: Static and dynamic. The static cell represents the regions that do not move, while the dynamics represent the graphic objects that can be moved, such as avatars. The contents can be added in the virtual worlds in two ways: through static or programmed files; using Java code. Wonderland uses the X3D data interchange format to read the data content in the world.

Second Life

Second Life (SL) is a 3D virtual platform of socialization, where it is possible to create a virtual character, called avatar, to reproduce daily actions of the real world, to realize fantastic feats, like flying, changing self appearance in seconds and "tele-porting" (Zhu et al., 2007). The platform is an authoring environment allowing the user to build his world as a consistent project. For this purpose, SL contains tools for the design and implementation. The behaviour of the objects and avatars can be controlled using the scripting language of the system, called the Linden Scripting Language (LSL).

SL allows high levels of interaction thanks to its potential for communication, collaboration and the creation of virtual content. Interaction and interactivity are common in virtual worlds, together with persistence and immersion. Communication is done through chat tools, voice, private messages and gestures. For voice communication, SL provides a system for the transmission of sound that reproduces the user's voice speaking to a microphone connected to the client computer at the avatar. The sound is transmitted and reproduced from the coordinated avatar in 3D space; thus only the avatars can hear the voice. Another form of interaction comes through gestures. Gesture animations are able to communicate feelings and to simulate an action. Second Life includes a tool that lets users create their own gestures.

COMPARATIVE ANALYSIS OF SOCIAL VIRTUAL WORLDS

To improve the understanding of how the platforms described in the previous section can help us in creating virtual worlds, it is necessary to focus on the characteristics of the design used to build the world. However, all share the same basic attributes (they are virtual, they mimic the real world or a part of it, and they have many simultaneous participants), making similarities and differences often difficult to identify. For the comparison analysis, it has been decided to analyze the various features of design in OpenSim, Croquet, Wonderland, and Second Life.

The review process was developed in three main phases:

1 - Identification and validation of criteria
2 - Classification of each factor
3 - Evaluation of results

Phase 1: Identification and Validation of Criteria

To compare virtual worlds, we started by identifying a set of criteria based on Manninen's matrix (Manninen, 2004). As learning has a social dimension and its roots are linked to social interaction, the choice of this set of criteria allows us to have learning environments which reflect the possibility of interaction, development and cooperation, besides forming a community of sharing, exposure of individuals' perspectives and the joint initiative. According to Manninen (2004), multi-user environments enable the occurrence of direct or indirect interactions between users, usually held in the form of cooperation and collaboration. In this sense, the evaluation criteria will reflect the idea of promoting the CSCL (Computer Supported Collaborative Learning). It can be defined as an educational strategy in which two or more subjects build their knowledge through discussion, reflection and decisions, where computing resources act (among others) as mediators of the teaching and learning process. If we consider the mix of interaction, function and structure as the principle of the virtual world's design, it is possible to analyze which platforms allow students to take a variety of roles, participate in simulations, practise the skills of real life in virtual space and explore situations which could not let them participate safely and easily in the real world. Therefore, these criteria were grouped in 6 main factors: realism in the world, the user's interface and communication, the avatar's characteristics, scalability, communication, and security. Each factor is constituted by a sub-set of features (Table 1).

After identifying the criteria that best characterize objectively the virtual worlds in the study, weights to each criterion will be assigned. The weight ranges are established from 0 to 3 according to their relevance in the virtual world. The numerical values are assigned a qualitative equivalence:

0 - Not applicable; 1 - Shortly applicable; 2 –Applicable enough; 3 - Very appropriate

Phase 2: Classification of Each Factor

As the main goal of the study is to identify a virtual world for educational purposes, it is important to recognize the main features of each world described in the previous section. At the moment, there won't be a solution that meets all our criteria; however, for a product to be acceptable, it must

Table 1. Factors based on some points of Manninen's matrix (Manninen, 2004)

<table>
<tr><th>1. REALISM OF WORLD</th><th>2. USER INTERFACE</th><th>3. COMMUNICATION</th></tr>
<tr><td rowspan="3">♦ Online interaction
♦ Existence of interactive objects
♦ Physical models
♦ Speed of dynamic objects and the world
♦ Dynamic scenarios
♦ AI in the world
♦ Evolution autonomy
♦ Presence of sociability
♦ Similarity with the real world</td><td>♦ Navigation and control
♦ Control and mouse
♦ Support sound</td><td>♦ Audio
♦ Video
♦ Text</td></tr>
<tr><td>4. AVATAR</td><td>5. SCALABITILY</td></tr>
<tr><td>♦ Complex
♦ Configurable
♦ Development
♦ Interaction
♦ Body language</td><td>♦ Distributed by multiple servers
♦ Limiting the creation of objects by user
♦ Limiting the area of the world
♦ Creating users
♦ Limitation of languages</td></tr>
<tr><td colspan="3">6. SECURITY</td></tr>
<tr><td colspan="2">♦ Right on digital creations
♦ Security for the avatar</td><td>♦ Possibility of Paypal</td></tr>
</table>

have at least 80% satisfactory results in the assessments, being set by the examiner depending on its need (Ondrejka, 2006).

Thus, it is important to identify the virtual worlds' platforms, which allow us to build worlds where educators and students easily create new content and where it is possible to assess students' progress, stimulating collaborative work.

Once the comparison matrix was constructed, a weight was given to each criterion, and its value depends on its relevance in the virtual world to be analyzed (Table 2).

Table 2. Matrix for the relevant value assigned to each virtual world

REALISM OF WORLD	OP	CR	WD	SL	AVATAR	OP	CR	WD	SL
Online interaction	3	2	2	3	Complex	2	2	2	3
Existence of interactive objects	3	2	2	3	Configurable	2	2	2	3
Physical models	2	2	1	3	Development	3	2	2	3
Speed of dynamic objects and the world	2	2	2	3	Interaction	3	3	2	3
Dynamic scenarios	3	2	1	2	Body	3	3	3	3
AI in the world	2	1	1	3					
Evolution autonomy	3	2	2	3	**SCALABITILY**	**OP**	**CR**	**WD**	**SL**
Presence of sociability	3	2	2	3	Distributed by multiple servers	2	2	2	2
Similarity with the real world	2	2	2	2	Limiting the creation of objects by user	2	2	1	2
USER INTERFACE	**OP**	**CR**	**WD**	**SL**	Limiting the area	2	1	1	2
Navigation and control	2	2	2	2	Creating users	3	2	2	3
Control and mouse	2	2	2	2	Limit languages	1	1	1	2
Support sound	2	2	2	2					
COMMUNITATION	**OP**	**CR**	**WD**	**SL**	**SECURITY**	**OP**	**CR**	**WD**	**SL**
Audio	3	2	2	3	Right on digital creations	3	3	3	3
Video	3	2	2	3	Security for the avatar	2	2	2	2
Text	3	3	3	3	Possibility of Paypal	1	0	0	3

Table 3. Percentage obtained for each factor

Factor	Ideal Platform	OpSim%	Croq. %	Wond. %	SL %
1.REALISM WORLD	9 X 3	85,2	63	55,6	92,6
2.USER INTERFACE	3 X 3	66,7	66,6	66,7	66,7
3. COMMUNICATION	3 X 3	100	78,8	78,8	100
4. AVATAR	5 X 3	86,7	80	73,4	100
5. SCALABITILY	6 X 3	73,3	53,4	46,7	73,3
6. SECURITY	3 X 3	66,7	55,6	55,6	88,9

The results have been achieved through an observation of applications in specific domains (science education, e-commerce, entertainment) that exist in these virtual worlds' platforms, developing of small objects as well as adding objects and spaces to customize the virtual world.

Phase 3: Evaluation of results

After assigning weights to each of the requirements that make up each factor, we performed the analysis of results. The results were calculated based on the idea of an ideal platform, according to Escudeiro (2007). The ideal platform is one in which all requirements have the maximum score, i.e., each criterion was fulfilled with the weight of 3 (very appropriate), meaning that all requirements will have a maximum relevance to the factor they belong to (see Table 3).

The percentage of compliance for each factor was obtained by calculating the weighted average. For instance:

$$\mathrm{Re}\,alism\,World(Opsim) = \frac{3*5+2*4}{9*3} = 85,2\%$$

The results (Figure 1) allow us to observe that there is a large homogeneity at the communication and interface levels, due to the nature of socialization and the problem that still exists in usability.

All the 4 worlds tested have weaknesses at the level of scalability, due to restrictions of space and objects in the creation of the world, and also

Figure 1.The results of study

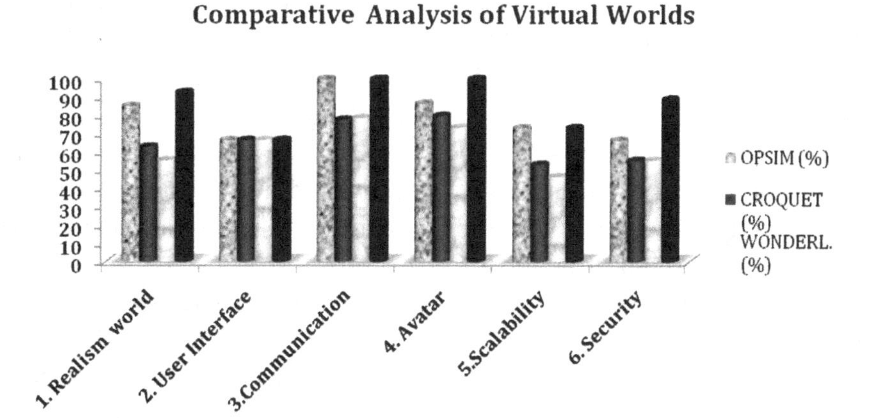

in the scripting language. The world created by the user (avatar) is not as real as it could be.

Distinguishing these 4 environments as the best and worst could be a difficult task as it depends on the user's intentions and expectations. For instance, Second Life has a higher success than other platforms, perhaps due to the existing educational community, to the fact that it allows creating a private space where users can walk around and explore without the inconvenience of dilemmas; besides, it was the first to appear.

On the other hand, the choice of a virtual world platform can be dependent on some key factors, such as:

1. Most students being PC users - Second life won because it is a cross platform and more users could access it.
2. Costs - The academic institutions have a small budget, being necessary to consider how much money the use of a virtual world requires and how much it will be necessary to change the hardware requirements when equipping computer labs. Here, the open source virtual worlds' platforms can win. They give the control over content, access, and updating. We don't need to pay for more regions. It is more an investment than it is a cost because we can generate knowledge in our institution.
3. Management capabilities - In terms of full management capabilities, OpenSim offers more flexibility because there is no limitation on the number of islands it is possible to create, and it allows full access/control of the hosting server.
4. Communication and collaboration - If we need to investigate in terms of communication and collaboration, Croquet is not the best. It works off a peer-to-peer protocol, so we have to work closely with people of Information Technologies.

VIRTUAL WORLDS IN EDUCATION

This study has allowed us to observe that these platforms increase the importance of their use in education. Users can manipulate objects, build and collaborate with each other, discover new information and present information in new and meaningful ways. This allows students and teachers to:

1. Engage in the process of teaching and learning, building up a richer and more dynamic learning experience where they prepare for further study with the content (Gros, 2002).
2. Facilitate the understanding of concepts that are difficult to comprehend and demonstrate in the real world, because these worlds have the potential to be a useful educational tool for teaching and learning.

In these virtual worlds students can have active participation, building their knowledge through interaction between subjects and objects. From this perspective, virtual worlds allow the development of open learning environments. The contents are not pre-defined for the students' actions, giving them the control of the environment. These features allow the students to be:

Participatory: the students learn to work in teams, consolidating the concept of collective and individual, and strengthening their self-respect and the respect of others.

Creative: the students discover and create new knowledge from their experiences.

Innovative: the students have the spirit of challenge.

However, it cannot be affirmed that these virtual worlds provide only benefits. As everything in life, they have disadvantages. For example,

1. We need some time to understand and learn how to use the virtual world efficiently;
2. Learners can easily become distracted;

3. Interaction can lead to conflict because, for all objects to be created, members within the learning process must interact, support and collaborate with fellow members. On the other hand, this could bring benefits as it enhances relevance and, ultimately, learning.

Considering the various features of a virtual world, having a useful collaborative experience goes beyond the mechanics of technology; it requires the use and development of the user's teamwork and social skills, an additional element valuable for collaborative learning (Johnson et al., 1991).

The visibility and easy identification of users provided by virtual worlds facilitate the achievement of users' recognition of their accountability and responsibility towards the group required in effective collaborative learning. A collaborative effort requires all members to work together to achieve a common objective. Thus, all participants get fully involved in the whole process; i.e., there is a group engagement. However, using these tools in the educational context requires a shift in thinking and adjustment in pedagogical methods that will embrace the community. The first hurdle to overcome is to accept that an instructor cannot have total control over a learning space while allowing open participation of students. We need to learn to embrace more educational participation. Therefore, these virtual worlds have the potential to transform the way of working in society:

1. Enable their users to bring a sense of 'self' into their digital space and provide real-time interaction with other avatars in a shared virtual environment;
2. Enable social groups and communities to form around shared interests, among individuals who are dispersed across the globe, and they also have the potential to teach concepts that may be difficult to grasp within a traditional two-dimensional presentation.

The training of teachers becomes necessary. Being a "good" teacher in the traditional method does not guarantee being a good teacher or learning to develop an online educational environment. Therefore, teachers need a training process in which they should go through the experience of being a student online.

Briefly, it has been described how virtual worlds can help in exploring new educational techniques or enhance and complement methodologies and techniques already known in the academic communities. It has been also pointed out how virtual worlds allow greater involvement of students in the study and how they can favour the creation of conditions to optimize resources, avoiding the need for the displacement of people. Further ahead, we will explore and articulate some key issues as a reflection for readers interested in this subject.

VIRTUAL WORLDS IN M-LEARNING

With the rapid growth of information and communication technologies (ICTs), it becomes possible for the emergence of new forms of education. These forms provide new means to combat the shortcomings of traditional teaching. E-learning offers new methods of distance education based on computers and network technologies. Furthermore, m-learning is part of e-learning and therefore part of d-learning (distance-learning), as seen in Figure 2.

These concepts are the most discussed in education or training. Special emphasis was put in m-learning as a new wave of development based on interconnection of devices and infrastructure network. With this new paradigm, there are several attempts to define the term and its essence. M-learning could meet the definition by Georgiev et al. (2004) as a type of e-learning, a method for distance education using computer and Internet technology, which offers education/learning

Figure 2. m -leaning as part of e-learning and d-learning (Georgiev et al., 2004)

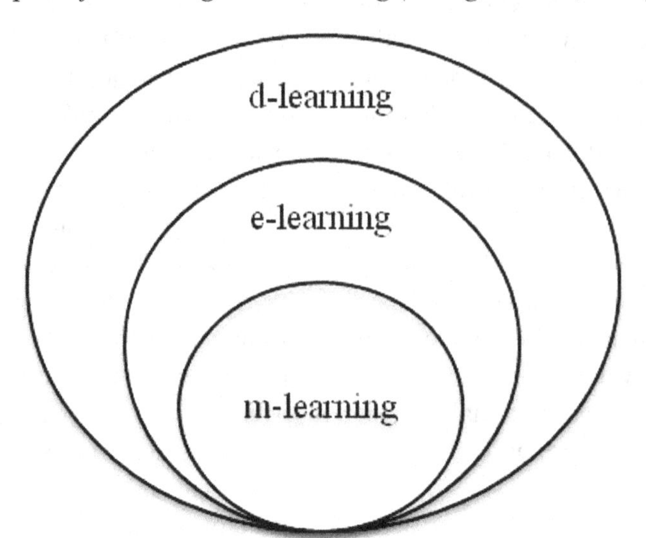

through wireless handheld devices like PDAs, tablet PCs, smart phones and mobile phones.

The essence of m-learning is the access to learning through the use of mobile devices with wireless communication in a transparent manner and with a high degree of mobility. This new paradigm emerges from the advantage of the availability of mobile devices and considering the need for specific education and training (Nyíri, 2002).

M-learning has some main objectives, which promote new methods of education and training, taking advantage of mobile resources and increasing the possibilities of assimilation of the content by the learner; it may have an educational tool that will allow them to accomplish their duties, pushing the boundaries of the classroom so that it is turned into a ubiquitous medium. The most important aspects of virtual environments are:

1. **Communication:** Communication between users is the most important aspect in a virtual environment. Communication tends to be asynchronous as it is necessary to use the Internet. Thus, m-learning reinforces this type of communication because through mobile devices anyone can have access anywhere in the virtual environments. However, it is important to recognize which types of communication are convenient in the virtual learning environments. How much they can make, which resource capabilities they have, and what are the chances of compatibility between users and the resource technology.
2. **Contents:** The organization of contents can be made in different ways: projects, definitions, and thematic units. M-learning promotes the fragmented content similar to when working with learning objects (Ramirez, 2007), advising on the selection of contents in small units, with complete information.
3. **Design of activities and materials:** In virtual environments, many combinations can be performed with educational design, from the defined learning conceptions and objectives. These activities and materials are referred as activities of reading, text and graphics, to describe the instructions. But in m-learning the most used are voice, graphics and animation, whose main aim is to promote learning (Sharma & Kitchens, 2004). However, for the presentation of materials, it is necessary

to overcome the limitations of the device size and the quantity of information to be stored in the memory device.

The integration of three-dimensional virtual environments in mobile phones will allow students to access learning systems with three-dimensional characteristics. The development of equipment in recent years has acquired new functions and extended operational capabilities that allow students to access educational content anywhere in the world, making education closer to individuals. The phone is considered an infrastructure for communication. Moreover, the recent improvements in wireless communication technology are providing greater interoperability of platforms. Nevertheless, defining how learners will interact with them is still a challenge.

This integration is possible by viewing the virtual world on the phone's screen or viewing it through a video projector, which is connected to the phone by Bluetooth connection.

To better understand how mobile devices can use these environments, the academic community has been developing some applications for education. In order to increase the teacher's interest in this research, we will cite some studies that demonstrate the advantage of their applicability in education and have emphasized the main ideas exposed in this chapter:

1. Marçal et al. (2005) developed a framework called VirTraM, whose aim is to develop educational software for mobile devices, intended for training of individuals. This research group implemented a prototype for a similar real science museum to test the framework, where it was observed that, students fixed the theoretical learned. Students can navigate or manipulate objects, interacting only with the phone keys.
2. In 2009, this research group developed an application for teaching hardware. This application simulates a printer on a 3D model and its operation. The application can be accessed via a mobile phone, where the student is allowed to change cartridges through an interactive animation displayed in three dimensions. "The ability to view and interact with the simulated model of the device and use the application virtually anywhere and at anytime presents new opportunities for learning"(Marçal et al., 2009; p.9).
3. The project developed by MIT (Klopfer et al., 2002) uses PDAs with GPS extensions. The main aim is to create a similar real environment that is "polluted", displayed in the map format. This project was developed for children. Children have to take "virtual" probes from the water and/or air in the polluted area or surroundings and analyze the results. The results of different tests and analyses are available for all the colleagues. The project has a specific area on the map where everyone can access. The students are required to collaborate with each other because of time limitations.

These works help us to reinforce the idea that virtual worlds in m-learning help learning. One of the examples is the teaching of how the equipment works. The use of simulations allows the students to view the operation of equipment and resolve technical problems; virtual scenarios with multimedia resources (text, images, video, and sound) help to describe the object under study and in the creation of a three-dimensional area, where the students can test their acquired knowledge, may also provide support, motivate students and improve the understanding of a particular subject.

Learning by using mobile devices can be achieved with the availability of quizzes and formative assessment, where students can respond by sending SMS (text messaging); exercises that include research, lecture and discussion topics leading to the use of forums; small experiments using text, images and video that can be sent to colleagues; strategic games in which one can use the device's keys and finally achieving research exercises, speaking and writing short text as a way of learning a foreign language, sent via SMS to the tutor.

On the other hand, this new technology can be used to inform, for example, to notify teachers when students have just delivered the work, and notify the students of their assessments. Mobile devices offer services that can be used in education, allowing static and dynamic learning.

1. As static learning, students can store/produce images, video, music, quizzes, and games.
2. As dynamic learning, students can produce, upload and download all types of media files, and manipulate the phone for learning purposes.

These services can be highlighted: SMS (text messaging), online publishing and blogging via SMS, MMS, camera, email or web browsers, audio-based learning (MP3 players, podcasting), and media collection via camera-phones. Mobile learning brings to the field of learning challenges not only as a solution but also as a complement to e-learning formats that already exist, since it can integrate all characteristics of collaboration and informal learning (Dias et al., 2008).

It also points to the use of mobile devices as future platforms for virtual worlds using tools such as Croquet, OpenSim, and Second Life. Second Life for 3G services re-creates the online world, which is adjusted for mobile phones. Chris Mahoney, Business Development Manager of Linden Lab said: "This is a great way to Second Life residents to stay connected to friends, business and experiences in the world, wherever they are" (Leominster, 2008). However, the platform of 3G mobile phones should be adapted to be able to re-create the virtual worlds and we need more research in this area.

CONCLUSION

The main purpose of this chapter was to show how 3D virtual worlds and mobile learning can contribute to the quality of teaching and learning in the future. Thus, it described some open-source virtual world platforms and held a comparative analysis between them, based on educational purposes. The results allowed us to reinforce some common features - the existence of a community of users who interact with each other. The possibility of greater heterogeneity in the groups formed in virtual worlds, including cultural, language, and country differences, which can have a transformative role in generating innovation through the exchange of experience and innovation in the future (Ondrejka, 2008).

The mobile technologies as a tool for extrapolating the limits of the surrounding area and extending the possibilities of access to information and the assimilation of content, when combined with 3D virtual worlds can provide students with opportunities to develop competencies, forming the cornerstone of learning. Through three-dimensional virtual worlds, it is possible to represent a wide variety of situations facing various application areas, such as excursions into the real world (e.g. museums, lands, and so on) or the imaginary and representation of objects (cars, machines) or personifications of real or imaginary beings. The peculiar characteristics of three-dimensional virtual worlds, as described at the beginning of the chapter, also lead us to assert that the educational applications for mobile devices with features of virtual reality, where the subject is able to move, hear, see, and manipulate objects as in the real world, represent interesting opportunities available to educators. This combination requires changes at the level of teaching approaches and structuring of three-dimensional virtual applications, which must be adapted to the technology to be used. "Because the use of m-learning is new in education, it is important for educators, researchers, and practitioners to share what works and what does not work in mobile learning so that the field of mobile learning can be implemented in a more timely and effective manner" (Ally, 2009; p.280).

The virtual world technology is today mature, which allows applications to go beyond gaming and entertainment. Thanks to the mobile technology and infrastructure, it is possible to create innovative applications enabling new educational paradigms. In this sense, it is important to ponder: How to produce a 3D scene for devices such as mobile phones? How to enable efficient ways of interaction? How to avoid the limitation of memory and processing of such devices? It is also essential to investigate the impact of mobile technologies on teaching and learning: interface design, model specification and content design in order to establish and develop the best practice for the use of these two concepts that may be strongly linked in the future.

In conclusion, in the near future, these two concepts will be of great importance to the academic community because they will have to adapt their approaches and methods, and the development of new competencies, to the new world of knowledge. Young people today are increasingly linked to new information and communication technologies because they are already part of their life. As educators, we must increasingly adapt to today's world.

REFERENCES

Ally, M. (2009). *Mobile learning: Transforming the delivery of education and training* (2nd ed.). Edmonton, Canada: AU Press, Athabasca University. Retrieved January 25, 2010, from http://www.aupress.ca/index.php/books/120155

Bartle, R. (2004). *Designing virtual worlds*. Indianapolis, IN: New Riders Publishing.

Book, B. (2004). *Moving beyond the game: Social virtual worlds*. Paper presented at the State of Play 2 Conference, New York: New York Law School. Retrieved January 20, 2010, from http://www.virtualworldsreview.com/papers

Castronova, E. (2004). *Synthetic worlds*. Chicago, IL: The University of Chicago Press.

Collins, S., Bentley, K., & Conto, A. (2008). *Virtual worlds in education*. Retrieved January 20, 2010, from net.educause.edu/ir/library/pdf/DEC0801.pdf

Croquet. (2010). *Croquet SDK - introduction*. Retrieved January 28, 2010, from http://www.opencroquet.org/index.php/Croquet_SDK_-_Introduction

Dias, A., Carvalho, J., Keegan, D., Kismihok, G., Mileva, N., Nix, J., & Rekkedal, T. (2008). *Introdução ao mobile learning*. Retrieved January 20, 2010, from http://www.ericsson.com/ericsson/corpinfo/programs/ the_role_of_mobile_learning_in_european_education/products/wp/socrates_wp1_portuguese.pdf

Escudeiro, P. (2007). X-Tec model and QEF model: A case study. In T. Bastiaens & S. Carliner (Eds.), *Proceedings of World Conference on E-Learning in Corporate, Government, Healthcare, and Higher Education 2007* (pp. 258-265). Chesapeake, VA: AACE.

Freitas, S. (2008). *Serious virtual worlds: A scoping study*. Retrieved January 28, 2010, from http://www.jisc.ac.uk/news/stories/2009/01/svw.aspx

Georgiev, T., Georgieva, E., & Smrikarov, A. (2004). M-learning: A new stage of e-learning. In *Proceedings International conference on Computer Systems and Technologies, CompSysTech' 2004, IV (28)* (pp. 1-5).

Gros, B. (2002). Knowledge construction and technology. *Journal of Educational Multimedia and Hypermedia, 11*(4), 323–343.

Johnson, D. W., Johnson, R. T., & Smith, K. A. (1991). *Cooperative learning: Increasing college faculty instructional productivity* (ASHE-FRIC Higher Education Report No.4). Washington, DC: School of Education and Human Development, George Washington University.

Klopfer, E., Squire, K., & Jenkins, H. (2002). Environmental detectives: PDAs as a window into a virtual simulated world. In *Proc. of IEEE International Workshop on Wireless and Mobile Technologies in Education (WMTE 2002)* (pp. 95-98). Sweden: Växjö.

Koster, R. (2004). *Raph Koster's writings on game design*. Retrieved January 20, 2010, from http://www.legendmud.org/raph/gaming/index.html

Leominster, S. (2008). *Second Life News: Second Life© client for mobile phone beta available.*

Manninen, T. (2004). *Rich interaction model for game and virtual environment design*. Finland: Oulu University Press. Retrieved January 2, 2010, from http://herKules.oulu.fi/isbn954272554

Marçal, E., Andrade, R., & Rios, R. (2005). Aprendizagem utilizando dispositivos móveis com sistemas de realidade virtual. In UFRGS (Ed.), *Revista Novas Tecnologias na Educação, 3*(1). Brasil: Porto Alegre.

Marçal, E., Lima, L., Júnior, M., Viana, W., Andrade, R., & Ribeiro, J. (2009). *A utilização de dispositivos móveis com ambientes tridimensionais como ferramentas para favorecer o ensino de hardware. XX Simpósio Brasileiro de informática na Educação*. Brasil: Florianópolis.

Menezes, C. Q., & Moreira, F. L. (2009). In the pursuit of m-learning - first steps in implementing podcast among K12 students in ESL. In *Actas da VI Conferência Internacional de TIC na Educação* (pp. 91–107). Braga: CCUM.

Ondrejka, C. (2006). Escaping the gilded cage: User-created content and building the metaverse . In Balkin, J., & Noveck, B. (Eds.), *The state of play: Law, games, and virtual worlds*. New York, NY: New York University Press.

Ondrejka, C. (2008). Education unleashed: Participatory culture, education, and innovation in Second Life . In Salen, K. (Ed.), *The ecology of games: Connecting youth, games, and learning* (pp. 229–252). Cambridge, MA: The MIT Press.

Onyesolu, O. (2009). Virtual reality laboratories: An ideal solution to the problems facing laboratory setup and management. In *Proceedings of the World Congress on Engineering and Computer Science, vol. 1*. San Francisco.

OpenSimulator. (2010). *Project history*. Retrieved January 28, 2010, from http://opensimulator.org/wiki/History

Sharma, S., & Kitchens, F. (2004). Web services architecture for m-learning. *Electronic Journal of e-Learning, 2*(2). Retrieved January 20, 2010, from http://www.ejel.org/volume-2/vol2-issue1/issue1-art2.htm

Smith, D. A., Kay, A., Raab, A., & Reed, D. P. (2003). Croquet—collaboration system architecture. In *Proceedings of the 1st Conference on Creating, Connecting and Collaborating through Computing* (pp. 2-11). Los Alamitos, CA: IEEE Computer Society Press.

Zhu, Q., Wang, T., & Jia, Y. (2007). Second Life: A new platform for education. In *Proceedings of First International Symposium on Information Technologies and Applications in Education* (pp. 201-204). doi: 10.1109/ISITAE.2007.4409270

KEY TERMS AND DEFINITIONS

Asynchronous Communications: When messages sent by a person are received and answered later by others.

Cognitive Skills: Any mental skills that are used in the process of acquiring knowledge; these skills include reasoning, perception, and intuition.

Collaborative Learning: A set of instructional methods and training that are supported with technology. Students are responsible for their learning and the learning of other students in their group.

Computer Supported Cooperative Learning: An emerging branch of learning that supports collaborative learning using computers and the Internet.

Handheld Device: A mobile device that is a computing device, typically having a display screen.

Open Source: Distribution of free software, which provides the source code. This can be changed.

Social Virtual Words: Socialization spaces that allow the creation of virtual communities.

Chapter 19
Implement Mobile Learning at Open Universities

Harris Wang
Athabasca University, Canada

ABSTRACT

Mobile learning (m-learning) provides convenient access to course materials and other relevant information, especially for learners who are often on the move and cannot afford to spend hours and hours in the classroom, often the case at open universities. As such, implementing mobile learning at open universities makes even more sense. In this chapter we will explore a variety of issues, technologies, and challenges associated with implementing mobile learning at open universities. We will begin with an investigation into open universities' common mandate and their very nature, and then explain the urgency and advantages for implementing mobile learning in their course and program delivery; we then explore the technical requirements of mobile learning, and present some strategies for mobile learning implementation. We will also explore some architectures and technologies for mobile learning systems. We will conclude the chapter by exploring some of the challenges one may have to face when implementing mobile learning at an open university.

INTRODUCTION

Education provided by open universities is mainly for people who wish to pursue higher education on a part-time basis and through distance learning. Hence, students at open universities are often full-time employed or have other commitment such as taking care of kids at home. Because of this very reality, students at open universities are often on the move with fractions of time available for study. They may be traveling on a business trip (Peters, 2006); they may be at a job site in a remote area (Tucker et al., 2009); they may be in a park overseeing kids playing. It is desirable

DOI: 10.4018/978-1-60960-613-8.ch019

for them to utilize the fractions of time they may have in between their normal activities during the day to study. Mobile learning is the solution to make this desire become a reality (Ragus, 2007; Keegan, 2004; Facer et al. 2005; Collins, 2005).

Open Universities' Advantages for Mobile Learning

Mobile learning is learning through mobile devices. Implementing mobile learning requires capable mobile devices, quality learning objects and materials suitable for mobile devices, reliable server services to feed the quality learning objects and materials to mobile devices, and reliable communication network to connect the servers and mobile devices. Fortunately, open universities often have several advantages in implementing mobile learning. In this section, we will explore these advantages in the context of mobile learning implementation.

To implement and deploy mobile learning in open universities such as Athabasca University of Canada, two conditions must be met. Firstly, there must be a significant number of students who have the necessary mobile devices of some sorts to use; the second is that there must be some well designed learning objects accessible for mobile learners to put on their specific mobile devices. Without a large number of potential users, any implementation and deployment of mobile learning would be a waste; without some well designed learning objects available for mobile learners, the potential users of mobile learning won't be using the system.

Generally speaking, open universities (Wikipedia, 2010) are well equipped with resources and expertise for online course production. Benefited from many years of business practice and experience in course production and delivery for distance learners, it is much easier and more convenient for open universities to relocate their personnel and resources to produce courses for e-learning and m-learning. More importantly, compared with traditional universities and colleges, open universities often have a well established business model to deliver creditable degree programs through distance education. This will make their courses and programs developed for delivery through e-learning and mobile learning systems more attractive to potential distance learners.

Compared to traditional universities, another advantage open universities have in implementing e-learning and mobile learning technologies and systems for distance learning is that students at open universities are more motivated and self-disciplined because they are mostly adult learners. They came to a program or course not because someone else wanted them to. They chose to take the program or course for a clear and good reason, and they are determined to succeed in the course and program.

Because students at open universities are mostly adults, they are more likely to have the ability to plan and manage their study. They are more self-disciplined compared to young learners just graduated from high schools. When using mobile devices for mobile learning, they are less likely to be distracted by the fancy and entertainment features such as games. That won't be a concern for educators who develop and deliver mobile learning components in their courses.

At open universities, students are not only adults, but also mostly employed. They either already have mobile devices needed for their work and communication needs, or they have the financial ability to purchase one for their personal use. Therefore, they are better equipped with necessary mobile devices needed for mobile learning.

The Urgency for Implementing Mobile Learning at Open Universities

In the previous section, we have shown that open universities have several advantages when implementing mobile learning, compared to traditional universities. We hence proved the feasibility of implementing mobile learning at open universities.

In this section, we will further prove that there is urgency for open universities to implement mobile learning in their program and course delivery.

Today, traditional universities and colleges not only use e-learning and mobile learning technology and systems to assist their in-class course delivery, they also started to offer distance learning courses through Web-based e-learning and mobile learning systems. This development will definitely have some negative impact on program enrolment and course registration at open universities. Therefore, it is very urgent for open universities to enhance their program and course delivery by utilizing more advanced technologies and systems in their business practice, in order to compete with traditional universities and colleges.

TECHNICAL REQUIREMENTS OF MOBILE LEARNING

In this section, we will explore some of the technical requirements for implementing mobile learning. We will present some details about requirements for capable mobile devices, requirements for Internet access, requirements for learning objects and learning management systems on mobile devices.

Network servers that can store and provide high-quality learning materials to mobile devices, Web servers that can feed the learning objects to different mobile devices

In today's Web-based distance education, network servers are needed to provide distance learners with quality learning materials and services. Among all kinds of network servers, Web servers are usually the front end, through which data and services available on other servers are accessed by learners.

Well designed learning management system to deliver learning objects and coordinate students' learning activities

From learners' perspective, the most important system on a Web server is a well designed learning management system (LMS), or mobile learning management system (mLMS) (Seong, 2006). It is the LMS or mLMS that delivers course contents and services to distance learners; it is the LMS or mLMS that coordinates students' learning activities and mediates collaborations among students in most cases. Even in a peer-to-peer collaboration scenario, an LMS/mLMS on the central Web server is still necessary if the valuable learning objects generated during peer-to-peer collaboration are to be integrated into the course activities and shared by other students in the course.

Learning objects and well organized course materials that can be stored in, and rendered properly on a variety of mobile devices

Education is the transfer of knowledge, and learning objects are the fundamental data units conveying meaningful knowledge that can be transferred to learners. As fundamental data units, learning objects are the building blocks for courses, glued by a well articulated study guide and recommended study plan, and accented by quizzes, assignments and exams. Therefore, good learning objects are very important for any implementation of e-learning and mobile learning. When developing learning objects for mobile learning, special considerations must be given to various limitations of mobile devices and wireless data networks, to ensure that the developed learning objects can be uploaded to, stored and rendered on the mobile devices used by distance learners.

Host computers at home or office needed to communicate with servers directly via available Internet access, and to communicate with mobile devices via USB connections or the like to synchronize learning materials and learners' data

Due to certain limitations of mobile devices and wireless data networks, and because some mobile devices may not have a reliable Internet access or do not have Internet access at all, in most cases, mobile learning can only be used beside mainstream e-learning where desktop and docked laptop computers are used. These desktop or docked laptop computers can be used as host computers for mobile devices. In such a scenario, the hosts directly communicate with the central Web server through reliable high speed Internet access, and relays necessary data to mobile devices by means of synchronization.

Wireless networks that enable mobile devices to communicate with servers and with each other

Wireless network is the key enabling infrastructure for mobile learning. Fortunately, the advances in telecommunication and wireless data network are now providing some much better wireless data services for mobile devices to access the Internet. For iPhones, Blackberries and other Smart Phones, GSM (Global System for Mobile communications) networks are becoming more and more widely available; for mobile devices without phone services such as PDAs, media players and game consoles, there are WiFi (Wireless Fidelity) and WiMAX (Worldwide interoperability for Microwave Access) networks available in corporate buildings, university campuses and even public places such as hotels, restaurants, and airports. There are plans in the U.S. to provide WiFi services free of charge citywide, and some cities are already providing such services on public transit. Enhanced by technologies such as GPRS (General Packet Radio Services) and EDGE (Enhanced Data Rate for GSM Evolution), today the widely available GSM networks are capable of providing wireless data service at 236.8 kbps. Operated at 2.4GHz, a WiFi network can run at as high as 54Mbps. The increasing availability of high quality wireless Internet access has laid a solid foundation for implementing mobile learning at open universities.

Learning objects development and content management system that can be used by professors or subject matter experts to develop and manage learning objects and courses

The content of a course comprises of well developed and carefully chosen learning objects. These learning objects are grouped into sections, units and/or modules. Over the last decade, the importance of learning objects sharable among courses, programs and institutions has been very well recognized, so that some standards have been established and many learning objects are now available from various learning objects repositories around the world on the Internet. However, it is still impossible for a professor to find all the learning objects from the repositories needed for a particular course. Therefore, a learning object development system becomes necessary. Since a professor may simply wish to develop some learning objects during his/her course development, it is natural and convenient to integrate learning object development into a course authoring or content management system (CMS). In any case, learning object development should follow a widely recognized standard such as SCORM (http://www.adlnet.gov/Technologies/scorm/default.aspx) for sharing and portability.

Mobile devices that can store and render various learning objects and course contents, and be used to carry out various course activities and to access learning services provided for the course.

When implementing mobile learning, the availability of capable mobile devices for students is one of the key factors one must consider. Fortunately today, as we mentioned in an earlier section, this may no longer be an issue for open universities, as most of their students should already have a capable mobile device that can be used for mobile learning. If not, it is more likely affordable for them to purchase one from an array of choices including cell phones, personal digital assistants (PDAs), wearable devices, MP3/MP4 players, mobile DVD players, tablet PC and laptop computers.

STRATEGIES FOR IMPLEMENTING MOBILE LEARNING AT OPEN UNIVERSITIES

This section is to explore strategies for implementing mobile learning at open universities. We will answer the following questions through the exploration of implementation strategies. How can distance learners take full advantage of mobile learning? How can educators design learning materials and activities most appropriate for mobile access? How can mobile learning be effectively implemented in both formal and informal learning?

Identify Important Areas Where Mobile Learning Can Enhance Teaching and Learning

When implementing mobile learning at open universities, it is important to identify the areas where mobile learning can enhance teaching and learning, and to determine the priorities of these identified areas. This will provide a better direction as to where to go and where to begin with. This will also ensure that students will benefit the most from mobile learning with limited resources available. These areas may include, for example, course production, learning object development, course delivery, class/course management, instructor-student interaction and student collaboration.

Identify Applications and Systems That Can Be Used in Those Areas

Mobile learning can only be realized through well designed and implemented mobile applications and systems. After the areas have been determined in the previous step, the next step in implementing mobile learning is to find out what applications and systems are needed for each of the areas. In this important process, attention must be paid to priorities of areas, applications and systems, and to the common features and functionalities that can be shared among applications and systems. This system analysis and design task must be carried out in a very systematic manner to ensure better utilization of resources, high efficiency of system implementation, and good integration and collaboration among the applications and systems to be developed.

Identify Key Technologies Needed to Implement Those Applications and Systems

Applications and systems can only be implemented with proper enabling technologies. Fortunately, today's IT industry is providing plenty of options for one to choose when implementing a system. For server-side scripting, there are PHP (http://www.php.net), Java (http://www.java.com), Perl (http://www.perl.com/), and ASP.NET (http://www.asp.net/) with multiple language support; for backend databases, there are commercial ones from Oracle and Microsoft, as well as open source databases such as MySQL (http://mysql.com/) and PostgreSQL (http://www.postgresql.org/). To script for mobile devices, there is Java ME (http://java.sun.com/javame/index.jsp) and Xcode (http://www.apple.com/macosx/developers/#xcode). For applications and systems that would run on mobile devices, one would have to consider the diversity of mobile devices students may have. It is a wise idea to use the technologies supported by most of the mobile devices available on the market.

For Each Course, Identify Learning Objects Including Special Applications and Systems That Can Be Developed for Mobile Learning

The content of a course in e-learning and mobile learning is made of various learning objects including texts, drawings, images, audios, videos, and special application software and systems. These learning objects explain or demonstrate facts, phenomena, principles, theories, or processes, or help students to understand those facts, phenomena, principles, theories, or processes. Therefore, an important step in developing an online course for e-learning and mobile learning is to determine what learning objects are needed for the course and have them developed in a professional way.

Normally, one will take a divide-and-conquer approach in developing learning objects for a course. For example, a course can be divided into units; a unit can be further divided into sections; each section can be further divided into several lessons if deemed necessary. In such a tree-like structure, the leaves are the learning objects including special application software and systems carefully selected and well organized to serve some specific learning objectives. When designing learning objects for mobile learning, one has to consider the limitations mobile devices usually have.

Develop Mobile Games That Can Help Students to Learn Particular Concepts or to Understand Particular Theories, Technologies and Processes

Due to small display, mobile devices don't have the advantages in presenting chunks of text or big images or diagrams for learners to read or view, when compared with big flat panel LCD displays used for desktop or docked laptop computers. The advantages a mobile device has are its mobility and convenience, so that learners can use it on the go even if they have a fraction of time. What can they do in such a short period of time? One of the best things they can do for learning is to play a game or view a demo or simulation, through which particular knowledge or skills can be learned or acquired. In mobile learning, any well designed game, simulation or demonstration for a course is a great addition and will make the course more interesting and attractive to learners (Sanchez, 2008). Therefore, it will be worth investing some money to identify and develop games, simulation and demonstration systems that can be downloaded by distance learners and run on their mobile devices.

Make Mobile Learning Activities Part of Learning Curricula

Since in most cases mobile learning can only be an addition to online course delivery through mainstream e-learning platforms, one must make some great efforts to make mobile learning objects and activities an integral part of entire learning curricula. In order to achieve that, first of all, the knowledge that is conveyed by the mobile learning objects and the skills that learners are expected to acquire through playing the games, simulations and doing other activities on mobile devices must be in the scope of learning outcomes expected from the course; secondly, learning activities carried out on mobile devices and associated learning outcomes should be assessable, and such assessments should be part of the mobile learning assessment for the course.

Integrate Mobile Learning into Mainstream E-Learning

As mentioned in the previous section, in most cases mobile learning is an addition to mainstream e-learning. Hence, it is a wise approach, when implementing mobile learning at open universities, to make mobile learning activities integral part of learning curricula mainly staged on mainstream e-learning platforms (Keegan, 2005). Mobile learning and mainstream e-learning can be two different systems, but they must be able to share data, and then must be able to communicate with each other. For example, learning objects generated on mobile devices by learners should be usable by mainstream e-learning systems such as LMS or mLMS; and a learning management system on mobile devices should be able to communicate with the mainstream partner about student learning profiles and learning assessments.

Reuse Mainstream E-Learning Contents for Mobile Learning

When implementing mobile learning at open universities, a very important task is to have suitable learning objects, applications and systems for learners to use on their mobile devices. Where can we get those learning objects, applications and systems that would serve the learning objectives of a particular course? The most economical approach is to reuse the ones already developed for mainstream learning systems. It may be very difficult to port the applications and systems such as games, demos and simulations, but at least we can reuse the design and programming logic in their mobile versions.

MODES FOR MOBILE LEARNING SYSTEMS

Mobile learning can only be effective and efficient through a well designed mobile learning system. In this section, we will explore some modes that may be adopted for implementing mobile learning at an open university.

Mobile learning can be implemented in two modes: standalone mode, client-server based network mode. In a standalone mobile learning mode, learning objects are often preloaded into mobile devices from a server or host at home or office, or from memory sticks through synchronization or direct USB transfer or other means of data communication.

The system architecture for a standalone mobile learning mode is shown in Figure 1, which consists of a learning object development server, learning object databases, a Web server, host computers at home or office with Internet connections, and mobile devices. The Web server runs the mobile learning management system, serves learning objects to mobile devices, and coordinates students' learning activities.

A standalone learning mode may be adopted for several reasons:

a. Wireless network infrastructure is not widely available for mobile devices, or the bandwidth is too low and the cost is too high for most mobile learners to subscribe. This is particularly true in the 90s.

Figure 1. Standalone mode for m-learn

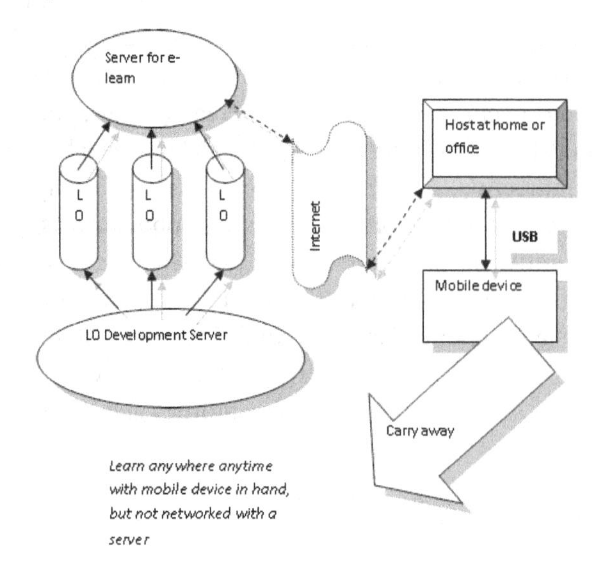

b. Available mobile devices do not have wireless Internet connection. This may be because the devices are not capable, or the users cannot afford to pay for the service.
c. The software system is not there to support the networked mobile learning mode. As we shall see, in order for mobile learners to use a networked mobile learning mode, a well designed learning management system is needed for learners to access the learning resources on a central server directly through wireless Internet connection, and to carry out their learning activities on mobile devices. If such a system is not available or not fully functioning, mobile learning can only be in a standalone mode.

The architecture of a networked mobile learning mode is shown in Figure 2. As can be seen, the main difference between the standalone mode and networked mode is that, in the networked mobile learning mode, mobile devices have direct connection with an e-learning server through a wireless mobile network such as WiFi, WiMAX or GSM, without the mediation of a host at home or office.

In order for mobile devices to utilize the learning resources on the central e-learning server, there must be a mobile learning management system in between to coordinate the activities. Such a system is usually Web-based, and some of the functionalities are implemented on the server side while others can be implemented on the client-side.

In the networked mobile learning mode, since the mobile devices are expected to have access to the Internet via a mobile wireless network, those devices can certainly communicate with each other directly without going through a central e-learning server. In other words, mobile learners in the networked mobile learning mode can collaborate in a peer-to-peer fashion, or as a group, as long as there are suitable software tools available on the mobile devices for the learners to use. In fact, some of these tools are already built into many of the mobile devices today. These include email and instant messenger. Combined with a built-in camera, voice recorder and media player, these tools can be used to foster some good collaboration among students. Instructors can even use such tools to broadcast short lectures.

Figure 2. Networked mode for m-learn

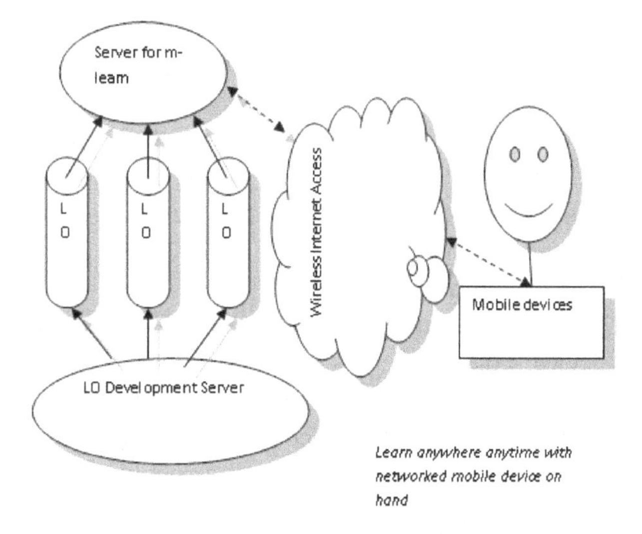

One problem with using these tools to foster collaboration or enhance teaching is that, however, it is hard to make such learning and teaching activities be part of the curricula. For example, it is hard to integrate the learning objects or other relevant data into the course because these tools are not made to communicate with an LMS or mLMS.

ESSENTIAL TECHNOLOGIES FOR MOBILE LEARNING SYSTEMS

Mobile learning can be implemented on a peer-to-peer or client-server basis, or based on an architecture utilizing both peer-to-peer and client-server models. Hence, system design and implementation may be needed on both servers and mobile devices. In this section, we will explore some important technologies that are currently available for implementing mobile learning systems. We will pay particular attention to open source technologies.

Server Design

Network servers for mobile learning are used to store and distribute learning materials to various mobile devices, and to store necessary information about learners and help manage learning through mLMS discussed above.

Servers for mobile learning should have the following functions:

a. Mobile device detection – because of the diversity of mobile devices, servers for mobile learning must be able to detect the type of mobile device, including both hardware and software in order to deliver suitable learning objects to the mobile device. Fortunately, however, this is a very easy thing to do, thanks to the standard HTTP protocol, in which clients are required to reveal these data to the server. In many server- and client-side programming languages, one needs only one statement or two to retrieve these data.
b. Server-side portion of mLMS – as mentioned in the previous section, a learning management system often consists of two portions: a client-side portion and a server-side portion. Functions of mLMS on the server-side may include learners' profile management. For a Web-based mLMS, the best technology for implementing these functions is Ajax (w3schools, 2010). Built on top of HTML and CSS, Ajax is an advanced use of client-side scripting in JavaScript and server-side scripting, mediated by the XMLHttpRequest object and the XML standard. Since the XMLHttpRequest object can request almost any data files from a Web server, including ASP and PHP files, any information on the server side, including those in a database, can be dynamically mapped into a desired area on the display of a mobile device. This makes the interaction between mobile devices and Web servers more powerful and user-friendly.
c. Automatically select or generate learning materials, for a detected mobile device and a specific learner according to his/her status in the course, stored in the profile database. This would make mobile learning context-aware (Berri at al., 2006).
d. The most advanced function a server for mobile learning may have is to generate personalized learning objects according to his/her learning style using adaptive learning technology (Goh et al., 2006; Svetlana et al., 2009).

Learning Object Development

Because of constraints on resources including computing power, size of memory and display, as well as communication bandwidth, learning objects developed for general e-learning cannot be used directly in mobile learning. In some cases, some simple conversion may be sufficient to make general learning objects fit on mobile devices, but in many cases it is better to develop new learning objects especially for mobile devices in order to make mobile learning more effective. These learning objects may include the following:

a. Images and diagrams: because of the small size of display mobile devices usually have, images and diagrams made for mainstream e-learn often become unusable even in a standalone mobile learning mode. Size conversion may be done on the server before the images/diagrams are loaded to mobile devices, but such conversion often produces low quality results due to the loss of resolution. Those large images become even less usable for mobile learners in a networked mode because of the low bandwidth in some wireless network (Zurita & Nussbaum, 2004).
b. Large HTML tables with fixed width: when such tables exist in a learning material, the screen of almost any mobile device will become too small to be practical. One may choose to remove the fixed-width property from such a large table on the server, but this may make the content in a column unreadable.
c. Large trunk of text: although there won't be any problem for any mobile device to display text on a small screen, reading lengthy text on a small screen is not fun, thus may drive learners away from mobile devices. To simply cut a large trunk of text into small pieces may not serve the purpose because a learner still has to read many pieces to comprehend the entire content.

It would be ideal if the learning objects already developed for main stream e-learning platform can be converted and reused on mobile devices, as we have discussed in an earlier section. However, that is not always practical. To make mobile learning a real success, specific learning objects must be developed for mobile devices. These learning objects may include:

a. Concise notes containing condensed info or summary of a lesson or a particular topic

Because of the small areas of display on mobile devices and because learners want to use their mobile devices to learn maybe only to utilize fractions of time available during breaks of their normal activities, they may not prefer to read big chunks of texts on their mobile devices. Instead, they may like to read some short and concise notes that summarize lessons or sections, or define and explain concepts, equations, formulas or theory that will greatly help them to refresh and reinforce what they have learned.

b. Simulations and demos that can run on mobile devices to illustrate a particular concept, principle or theory

To better utilize fractions of time adult learners may have in between their normal activities such as traveling to work or coffee or lunch breaks, another type of learning objects that can be used for mobile learning are simulations and demos. Such a simulation or demo usually runs for a short period of time but can help greatly in explaining a concept, an algorithm or operational procedure, or even demonstrate the principle of a complicated scientific theory. Those simulations and demos can be standalone applications on mobile devices, but it would help in learning profile development and learning assessment if they are made to be an integral part of a mobile learning management system.

c. Games that can play on mobile devices and help learners to learn

In recent years, computer games have been recognized as a good way to transfer knowledge to students. One big reason for that is, a well designed game can express knowledge in an interesting and attractive way, and thus keep learners more engaged in learning activities. Those interesting and attractive features of well designed games on mobile devices can overcome certain limitations of mobile devices, such as small size of display, and make learning on mobile devices more effective. There is no doubt, however, developing good computer games for mobile learning will need big investment.

Learning Object Aggregation and Integration

Equipped with a text messenger, video camera and voice recorder, today's mobile devices can be used easily to produce learning objects in different forms. These mostly learner generated learning objects include texts, images, video and audio clips. To aggregate these learning objects, share them among learners and integrate them into the course will be very beneficial. An effective mechanism and system are needed to make this happen in mobile learning.

Mobile Learning Management System Design and Development

A learning management system is an integrated software system that can be used by learners to access various learning resources and to carry out various learning activities, while a mobile learning management system is the one that can run on mobile devices.

In today's Web dominated world, mobile learning management systems should be Web-based using the HTTP protocol (Berners-Lee, 2007). Although it is technically possible to develop a dedicated client-server system for the sole purpose of mobile learning, the cost would be too high to be practical. Especially given the facts that, although there are so many different mobile devices running different operating systems, they all support Web browsing, why would one not develop a Web-based mobile learning management system that is universally accessible for many different types of mobile devices through a Web browser?

In terms of functionalities, the following should be considered when designing and developing a mobile learning management system:

a. The system should be able to render and present course contents designed for mobile devices in a user-friendly fashion. These contents include texts, diagrams, images, charts, tables, audio clips, video clips, simulation and demo software, as well as computer games for learning purposes.
b. There should be a scheduler in the system to help learners to manage their learning schedules, follow deadlines, and monitor their learning progress in a course and/or program.
c. The system should have mechanisms to support communication and collaboration. Communication and collaboration can be in a peer-to-peer mode between students and instructors, group discussion mode, audio conference and video conference mode. The advancements in technology today have made all these modes of communication possible.
d. The system should support online quizzes and assessment. Online quizzes can be designed like a game, so that students accumulate points for correct answers.
e. The system should be able to harvest learner generated learning objects and send them back to a central server for aggregation and integration.

Utilizing GPS for Location-Based Mobile Learning

Today, many new mobile devices such as iPhones, smart phones and PocketPCs have GPS capability built-in, which means that learners holding such mobile devices will be able to tell their locations. How can we utilize this advanced modern technology in mobile learning?

As have long been realized, one of the problems distance learners at open universities have experienced is a feeling of isolation and being detached from instructors and fellow students. Because students choosing to study at open universities through distance learning are often fulltime employed, distributed and sometime mobile on business trips, it has been difficult to have an in-person meeting for collaboration or group work. However, with the help of GPS, it is now possible for a mobile learner to tell where he or she is and who else are close enough for a meeting in person (Tan et al., 2009).

This can be done in two ways. The first one will be for the student who wants to initiate a meeting to broadcast his or her position among classmates through email, instant messenger or forum, and make a call for meeting. Everyone in the class will know whether he or she is geographically close to the caller, and those located nearby and are interested in the meeting will then respond to the call and join in the meeting.

The second approach is to make all the mobile devices report their GPS locations to the central e-learning server, and then the central server will be able to put students into different groups based on their geographical locations reported by the mobile devices. The central server can even build up a map showing how the students are geographically distributed and send the map back to the mobile devices students are using. By using this map, mobile learners will be able to know who they might be able to meet in person. This approach can be further enhanced by developing and utilizing other advanced algorithms and technologies (Tan et al., 2009).

Typical Functionalities of mLMS for Mobile Learning

Mobile devices are getting more powerful, but still with limitations. As mobile learning is intended for learners on the move with fractions of time in between their normal activities, some applications may be more suitable for mobile learning while others may be not. In this section, we will explore some typical applications that are suitable for mobile learning.

The following are some general applications or functionalities a mobile learning management system should have:

a. **Email**. Like it or not, email has become an important means of communication for almost everyone in the modern society and has become one of the important tools to support students in distance education. There are various email client systems in almost all types of computing platforms, including mobile devices, but to use email for e-learning and mobile learning, it is better to have the email functionality built into the LMS or mLMS to be used by students, so that the data generated in email communication can be shared with other applications such as learning content aggregation.
b. **Search**. From time to time, students will need to search for learning objects as small as a definition and as big as an article. The search functionality should be able to search both within the course and on the Internet, and allow an individual student to annotate the search result and make the search be part of his/her learning profile for the course.
c. **Group forum for discussion**. Forums have been used for years in hosting group discussions. Threaded group discussion is a good means of collaboration on designated topics. It is also a good means of broadcasting.
d. **Peer-to-peer messaging**. Messaging has gained its great popularity in recent years.

It is more effective to keep students engaged and communicating.

e. **Media rendering**. Learning content may come in different media types including text, graphics, charts, audio and video clips, and even movies. So, it is necessary for an mLMS to be able to render all these media types within the system and track students' activities in consuming the learning object.
f. **Learning content downloading and synchronization**. While downloading learning content from a central server or a host computer may be a basic functionality an mLMS should have, synchronizing the learning content with the server or host computer is an advanced function an mLMS should have. Synchronization is a necessary operation to ensure the learning content on a mobile device is updated and current, and consistent with that on the server and the mainstream LMS on a host computer.
g. **Learning object aggregation and integration**. In a course, class participants' contributions through email, p2p messaging and forum discussion are valuable not only to the class, but also to the development and future revision of the course. So, it is necessary for the mLMS to have the functionality to aggregate those user generated learning object and provide assistance in integrating them into the course.
h. **Scheduling and deadline management, including deadline reminder**. Mobile devices are meant to be carried around with the owners. So, they are the ideal platform for scheduling and deadline management. An mLMS should be able to help mobile learners to schedule their tasks, set up and manage deadlines, and remind them about upcoming deadlines.

Other functionalities desirable to have in an mLMS include:

a. **Location detection and location based learning**. Using the GPS capability that came with PDAs, iPhones and other smart phones to detect the location of the learner, and provide localized information and services accordingly. The localized information and services may include local libraries and local offices that provide support to students, providing a list of students in the area for possible in-person meetings and collaboration.
b. **Learning content personalization**. As for all types of learning, it would be ideal if personalized learning content can be provided to each individual learner. For mobile learning, the personalization also includes adaptation to individual devices by generating and providing learning content according to the capability of each individual mobile device.
c. **Learner's portfolio management**. In online distance education, one of the challenges is to know about the learners. This is not only necessary for content personalization, but also needed for learning assessment. A learner portfolio management system should be able to build and manage at least part of a learner's portfolio based on the learner's activities within the mLMS.

CHALLENGES IN IMPLEMENTING MOBILE LEARNING AT OPEN UNIVERSITIES

As we have shown in previous sections, mobile learning provides many promises to researchers and practitioners at open universities to enhance their course and program delivery. There are still some challenges, however. In this section, we will explore the challenges and try to summarize some possible solutions to overcome these challenges.

Practitioners and researchers in mobile learning are still facing many challenges in order to implement and deploy mobile learning in open universities. There challenges include:

a) **Limited processing power of mobile devices**. Over the last few years, central processors built into mobile devices, such as pocket PC and smart phones have evolved greatly (see mobile device review at http://www.mobiletechreview.com). Compared to servers, desktop computers, and even laptop computers, the processing power of mobile devices is still limited. We also have to admit that today's software has become greedier for the CPU and RAM as well as other resources. Users become less patient, too. As such, when implementing and deploying mobile learning at open universities, we have to keep this reality in mind and make sure what we developed for mobile learning is not only usable but also enjoyable. If the learning objects are not deployable, they will become useless; if they are not enjoyable, learners will eventually stay away from mobile learning.

b) **Small size of display**. Small size of display is a common feature of all mobile devices. That may be why they are mobile. For this reason, mobile learning researchers, practitioners, and learners will have to face this challenge for as long as we still utilize mobile learning.

As mentioned in a previous section, many learning objects such as large images, diagrams, tables, and even lengthy textual contents are not suitable for small displays. Simple conversion of these learning objects may only yield unusable results. The best way to overcome this challenge is to redevelop these learning objects especially for mobile devices, although research in new conversion algorithms and technology for this purpose is still worth doing and may generate many good publications.

c) **Low network bandwidth.** In today's world, there are mainly four types of wireless networks available for mobile learning. *Time division multiple access (TDMA)* is a more traditional (or dated) cellular voice communication standard that will eventually be phased out; ***global system for mobile communication (GSM) and general packet radio service (GPRS)*** **are new ones but** will eventually evolve into EDGE, for enhanced data for global evolution, which can be used to transmit email, data, and low-bandwidth learning content; ***carrier division multiple access (CDMA)*** has a bandwidth similar to GSM/GPRS, which has about less than 13Kbps throughput. At such a speed, downloading multimedia learning objects, even just images, would become a suffering.

Wi-Fi wireless local area network is a faster wireless standard providing a bandwidth between 11Mbps and 54 Mbps, but can only beam radio signals up to 300 feet. It is gaining momentum as airports, hotels, malls, and coffee bars start to install Wi-Fi hotspots. Soon, you'll be able to download entire learning simulations on your laptop, tablet or even PDAs and smart phones, if you can afford the cost.

d) **Relatively high cost of data service for mobile devices.** Although mobile communication technology and infrastructure have advanced greatly over the last decade, the cost of gaining Internet access through a wireless carrier is still too high. In Canada, Internet access via a GSM network costs $10 per 10MB data or $90 per month for 25MB with $7/MB for additional data usage. It would cost much more for mobile learners who tend to stay online longer and often with high data usage. For most learners, such high cost is not affordable. As a result, they may choose not to use mobile learning in their study.

e) **High cost of developing quality learning objects for mobile devices**. Because of the small display, low network bandwidth and lower processing power and constraints on other resources, special learning materials must be developed for mobile learning by well trained professionals. This will result in high cost in developing quality learning objects for mobile devices. Will the government be willing to pay?

Mobile learning is definitely a promising area for both researchers and education practitioners to work in, and the challenges we are facing in implementing and developing mobile learning in open universities have made the job even more exciting and productive. Indeed, research is about identifying challenges and problems in an area, and then finding answers and solutions.

REFERENCES

W3schools. (2010). *Ajax introduction*. Retrieved June 10, 2010 from http://www.w3schools.com/ajax/ajax_intro.asp

Berners-Lee, T., et al. (2007). *Hypertext transfer protocol-HTTP/1.0*. Retrieved June 10, 2010 from http://dragon.cs.kent.edu/WEB/resource/http/rfc1945.html

Berri, J., Benlamri, R., & Atif, Y. (2006). Ontology-based framework for context-aware mobile learning. In *Proceedings of the 2006 International Conference on Wireless Communications and Mobile Computing* (pp. 1307-1310). New York, NY: ACM Press.

Collins, T. G. (2005). English class on the air: Mobile language learning with cell phones. In *Proceedings of the Fifth IEEE International Conference on Advanced Learning Technologies* (pp. 402-403). IEEE Computer Society.

Facer, K., Faux, F., & McFarlane, A. (2005). Challenges and opportunities: Making mobile learning a reality in schools. In *Proceedings of MLearn 2005: 4th World Conference on mLearning*. Retrieved June 10, 2010 from http://www.mlearn.org.za/CD/papers/Facer%20-%20Faux%20-%20McFarlane.pdf

Goh, T., & Kinshuk, D. (2006). Getting ready for mobile learning-adaptation perspective. *Journal of Educational Multimedia and Hypermedia, 15*(2), 175–198.

Keegan, D. (2004). Mobile learning: The next generation of learning. In *Proceedings of 18th AAOU Conference, Quality Education for All - New Missions and Challenges Facing Open Universities*. Shanghai, China: TV University.

Keegan, D. (2005). The incorporation of mobile learning into mainstream education and training. In *Proceedings of Mlearn 2005: 4th World Conference on mLearning*. Retrieved June 10, 2010, from http://www.mlearn.org.za/CD/papers/keegan1.pdf

Peters, K. (2006). *Learning on the move mobile technologies in business and education*. Research report of Australian flexible learning framework. Retrieved June 10, 2010 from http://www.flexiblelearning.net.au/projects/resources/2005/Learning%20on%20the%20move_summary.pdf

Ragus, M. (2007). M-learning: A future of learning. *E-Journal of Learning and Innovation, 9*, 22-37. Retrieved June 10, 2010, from http://kt.flexiblelearning.net.au/wp-content/uploads/2006/05/ragus.doc

Sánchez, J., & Salinas, A. (2008). Science problem solving learning through mobile gaming. In *Proceedings of the 12th International Conference on Entertainment and Media in the Ubiquitous Era* (pp. 49-53). New York, NY: ACM Press.

Seong, D. S. (2006). Usability guidelines for designing mobile learning portals. In *Proceedings of the 3rd International Conference on Mobile Technology, Applications & Systems.* New York, NY: ACM Press.

Svetlana, K., & Yonglk-Yoon (2009). Adaptation e-learning contents in mobile environment. In *Proceedings of the 2nd International Conference on Interaction Sciences: Information Technology, Culture and Human* (pp. 474-479). New York, NY: ACM Press.

Tan, Q., et al. (2009). Location-based adaptive mobile learning research framework and topics. In *Proceedings of International Conference on Computational Science and Engineering* (pp. 140-147). IEEE Computer Society.

Tucker, T. G., & Winchester, W. W. (2009). Mobile learning for just-in-time applications. In *Proceedings of the 47th Annual Southeast Regional Conference*. New York, NY: ACM Press.

Wikipedia. (2010). *The open university*. Retrieved June 6, 2010, from http://en.wikipedia.org/wiki/Open_University

Zurita, G., & Nussbaum, M. (2004). A constructivist mobile learning environment supported by a wireless handheld network. *Journal of Computer Assisted Learning, 20*, 235–243. doi:10.1111/j.1365-2729.2004.00089.x

Compilation of References

Abawi, D., Dörner, R., Haller, M., & Zauner, J. (2004). *Efficient mixed reality application development.* In First European Conference on Visual Media Production (pp. 289-294).

Ableson, F., Collins, C., & Sen, R. (2009). *Unlocking Android: A developer's guide*. Greenwich, CT: Manning Publications.

Adaptive Technology Resource Centre. (2010). *A tutor learning content management system*. Retrieved March 21, 2010, from http://www.atutor.ca

Aizawa, K. (2005). *JACET 8000-word list.* Kirihara Shoten.

Alexander, B. (2006, March/April). Web 2.0: A new wave of innovation for teaching and learning? *EDUCAUSE Review*, *41*, 33–44.

Alexander, B., Brown, M., Doogan, V., Johnson, L., Levine, A., & Sabean, R. (2006). *The Horizon report 2006 edition.* Austin, TX: The New Media Consortium.

Alfresco. (n.d.). Retrieved 2010, from http://www.alfresco.com

Allen, M. (2009). *Palm webOS.* Sebastopol, CA: O'Reilly Media, Inc.

Ally, M. (2009). *Mobile learning: Transforming the delivery of education and training* (2nd ed.). Edmonton, Canada: AU Press, Athabasca University. Retrieved January 25, 2010, from http://www.aupress.ca/index.php/books/120155

Amazon Web Services. (2010). *Amazon elastic compute cloud* (Amazon EC2). Retrieved from http://aws.amazon.com/ec2

Ambient Insight. (2008). *The US market for mobile learning products and services: 2008-2013 forecast and analysis*. Retrieved July 16, 2009, from http://www.ambientinsight.com/Resources/Documents/AmbientInsight_2008-2013_US_MobileLearning_Forecast_ExecutiveOverview.pdf

Amuck. (2009). *Welcome to 31 days of iPhone apps.* Retrieved November 6, 2009, from http://www.appsamuck.com

Anderson, C. (2006). *The long tail.* New York, NY: Hyperion.

Andrew, P., Conard, J., Woodgate, S., Flanders, J., Hatoun, G., & Hilerio, I. ... Willis, J. (2005). *Presenting Windows Workflow Foundation.* Indianapolis, IN: Sams.

Andrews, R. (2003). Lrn Welsh by txt msg. *BBC News World Edition*. Retrieved from http://news.bbc.co.uk/2hi/uk_news/wales/2798701.stm

Anido, L. (2006). An observatory for e-learning technology standards. *Advanced Technology for Learning*, *3*(2), 99–108. doi:10.2316/Journal.208.2006.2.208-0876

Apple Computer, Inc. (2009). *Mac OSX reference library*. Retrieved February 5, 2010, from http://developer.apple.com/Mac/library/documentation/Cocoa/Conceptual/ObjectiveC/Introduction/introObjectiveC.html

Apple Computer, Inc. (2010). *iPhone section*. Retrieved February 5, 2010, from http://www.apple.com/iphone

Armbust, M., Fox, A., Griffith, R., Joseph, A. D., Katz, R. H., & Konwinski, A. ... Zaharia, M. (2009). *Above the clouds: A Berkeley view of cloud computing*. Retrieved from http://www.eecs.berkeley.edu/Pubs/TechRpts/2009/EECS-2009-28.html

Asay, M. (2009a). *Up to 24 percent of software purchases now open source*. Retrieved July 16, 2009, from http://news.cnet.com/8301-13505_3-10238426-16.html

Asay, M. (2009b). *IDC: Linux spending set to boom by 21 percent in 2009*. Retrieved July 16, 2009, from http://news.cnet.com/8301-13505_3-10216873-16.html

Attewell, J. (2005). *Mobile technologies and learning: A technology update and m-learning project summary*. London, UK: LSDA.

Attewell, J., & Savill-Smith, C. (2004). Mobile learning and social inclusion: Focusing on learners and learning . In Attewell, J., & Savill-Smith, C. (Eds.), *Learning with mobile devices: Research and development* (pp. 3–12). London, UK: Learning and Skills Development Agency.

Attewell, J. (2005). From research and development to mobile learning: Tools for education and training providers and their learners. In *Proceedings of mLearn 2005*. Retrieved December 20, 2005, from http://www.mlearn.org.za/CD/papers/Attewell.pdf

Attwell, G. (2007). Personal learning environments-the future of e-learning? *eLearning Papers, 2*(1). Retrieved April 24, 2008, from www.elearningpapers.eu

Australian Flexible Learning Framework. (2007). *2017: A mobile learning odyssey*. Australia: Commonwealth of Australia. Retrieved from http://www.flexiblelearning.net.au/flx/go/home/ news/feature/cache/bypass?sector_feature&id_277

Axis. (n.d.). Retrieved 2010, from http://ws.apache.org/axis

Azuma, R. T. (1997). A survey of augmented reality. *Presence (Cambridge, Mass.), 6*(4), 355–385.

Banjanovic, A. (2010). *Special report: Towards universal global mobile phone coverage*. Retrieved February 5, 2010, from http://www.euromonitor.com/Special_Report_To wards_universal_global_mobile_phone_coverage

Barron, J. A., Fleetwood, L., & Barron, A. E. (2004). E-learning for everyone: Addressing accessibility. *Journal of Interactive Instruction Development, 26*(4), 3–10.

Bartle, R. (2004). *Designing virtual worlds*. Indianapolis, IN: New Riders Publishing.

Bauer, M., Bruegge, B., Klinker, G., MacWilliams, A., Reicher, T., & Rib, S. … Wagner, M. (2001). *Design of a component-based augmented reality framework*. In The Second IEEE and ACM International Symposium on Augmented Reality (pp. 45).

BBC. (2007). *Basic sums 'stress 13.5m adults'*. Retrieved February 8, 2010, from http://news.bbc.co.uk/2/hi/uk_news/education/7027569.stm

Berger, S., Mohr, R., Nösekabel, H., & Schäfer, K. (2003). Mobile collaboration tool for university education. In *Proceedings of the Twelfth International Workshop on Enabling Technologies: Infrastructure for Collaborative Enterprises* (p.77).

Berglund, E., & Priestley, M. (2001). Open-source documentation: In search of user-driven, just-in-time writing. In *ACM Special Interest Group for Design of Communication Proceedings of the 19th Annual International Conference on Computer documentation* (pp. 132-141).

Berners-Lee, T., et al. (2007). *Hypertext transfer protocol-HTTP/1.0*. Retrieved June 10, 2010 from http://dragon.cs.kent.edu/WEB/resource/http/rfc1945.html

Berri, J., Benlamri, R., & Atif, Y. (2006). Ontology-based framework for context-aware mobile learning. In *Proceedings of the 2006 International Conference on Wireless Communications and Mobile Computing* (pp. 1307-1310). New York, NY: ACM Press.

Beyers, T. (2009). *Android is even bigger than you think*. Retrieved December 6, 2009, from http://www.fool.com/investing/high-growth/2009/ 11/23/android-is-even-bigger-than-you-think.aspx

Blackboard Website. (n.d.). Retrieved 2010, from http://www.blackboard.com

Bohl, O., Schellhase, J., Sengler, R., & Winand, U. (2002). The sharable content object reference model (SCORM)--a critical review. In *Proceedings of the International Conference on Computers in Education* (pp. 950--951).

Bonita. (n.d.). Retrieved 2010, from http://bonita.objectweb.org

Book, B. (2004). *Moving beyond the game: Social virtual worlds*. Paper presented at the State of Play 2 Conference, New York: New York Law School. Retrieved January 20, 2010, from http://www.virtualworldsreview.com/papers

Bottentuit, J. B., Jr., & Coutinho, C. P. (2007). Virtual laboratories and m-learning: Learning with mobile devices. In *Proceedings of International Multi-Conference on Society, Cybernetics and Informatics (WMSCI)* (pp. 275-278). Orlando, FL, EUA.

Bottentuit, J. B., Jr., & Coutinho, C. P. (2008). The use of mobile technologies in higher education in Portugal: An exploratory survey. In C. Bonk, M. M. Lee & T. Reynolds (Eds.), *Proceedings of E-Learn 2008 Conference on E-Learning in Corporate, Government, Healthcare & Higher Education* (pp. 2102-2107). Las Vegas, NV.

Botturi, L., Derntl, M., Boot, E., & Figl, K. (2006). *A classification framework for educational modeling languages in instructional design.* In 6th IEEE International Conference on Advanced Learning Technologies (ICALT 2006). Kerkrade (The Netherlands).

Breton, E., & Bezivin, J. (2002). Weaving definition and execution aspects of process meta-models. In *Proceedings of the Annual Hawaii International Conference on System Sciences* (pp. 3776-3785).

Brookes, A., & Grundy, P. (Eds.). (1988). *Individualization and autonomy in language learning.* ELT Documents, 131. London: Modern English Publications in association with the British Council (Macmillan).

Bruns, A. (2007, March 21-23). *Beyond difference: Reconfiguring education for the user-led age.* Paper presented at the ICE3: Ideas in cyberspace education: digital difference, Ross Priory, Loch Lomond.

Bull, G., & Garofalo, J. (2002). Proceedings of the Fourth National Technology Leadership Summit: Open resources in education. *CITE Journal, 2*(4), 530-555.

Burgos, D., Tattersall, C., Dougiamas, M., Vogten, H., & Koper, R. (2007). A first step mapping IMS learning design and Moodle. *J. UCS, 13*(7), 924–931.

Burnett, I., Van de Walle, R., Hill, K., Bormans, J., & Pereira, F. (2003). MPEG-21: Goals and achievements. *IEEE MultiMedia, 10*(6), 60–70. doi:10.1109/MMUL.2003.1237551

Burns, A. C., Biswas, A., & Babin, L. A. (1993). The operation of visual imagery as a mediator of advertising effects. *Journal of Advertising, 22*(2), 71–85.

Casado, M., & Mckeown, N. (2005). The virtual network system. In *The Proceedings of the 36th SIGCSE. Technical Symposium on Computer Science Education (SIGCSE'05)* (pp. 76-80).

Castells, M. A. (2004). *Galáxia Internet. Reflexões sobre Internet, negócios e sociedade.* Lisboa, Portugal: Fundação Calouste Gulbenkian.

Castronova, E. (2004). *Synthetic worlds.* Chicago, IL: The University of Chicago Press.

Cawood, S., & Fiala, M. (2007). *Augmented reality: A practical guide.* Raleigh, NC: Pragmatic Bookshelf.

Chang, A., Chang, H., & Hen, J. S. (2007). *Implementing a context-aware learning path planner for learning in museum.* Paper presented at the 6th WSEAS International Conference on EACTIVITIES, Tenerife, Spain.

Chao, L. (2009). *Utilizing open source tools for online teaching and learning: Applying Linux technologies.* Hershey, PA: IGI Global Publishing.

Chen, Y. S., Kao, T. C., & Sheu, J. P. (2003). A mobile learning system for scaffolding bird watching learning. *Journal of Computer Assisted Learning, 19*(3), 347–359. doi:10.1046/j.0266-4909.2003.00036.x

Cheung, W. S., & Hew, K. F. (2009). A review of research methodologies used in studies on mobile handheld devices in K-12 and higher education settings. *Australasian Journal of Educational Technology, 25*(2), 153–183.

Chi, M., & Hausmann, R. (2003). Do radical discoveries require ontological shifts? In Shavinina, L., & Sternberg, R. (Eds.), *International handbook on innovation* (*Vol. 3*, pp. 430–444). New York, NY: Elsevier Science Ltd. doi:10.1016/B978-008044198-6/50030-9

China Daily. (2007). *Ghost of illiteracy returns to haunt country.* Retrieved February 8, 2010, from http://english.peopledaily.com.cn/200704/02/eng20070402_363005.html

Citrix Systems. (2009). *Xen.* Retrieved August 15, 2009, from http://www.xen.org

Claroline Website. (n.d.). Retrieved 2010, from http://claroline.net

Clothier, P. (2005). An introduction to m-learning: An interview with Ellen Wagner. *The E Learning Guild*. Retrieved March 21, 2010, from http://www.elearningguild.com

Cobcroft, R., Towers, S., Smith, J., & Bruns, A. (2006). Mobile learning in review: Opportunities and challenges for learners, teachers, and institutions. In *Proceedings of Online Learning and Teaching Conference 2006*. Brisbane.

Cochrane, T. (2010). Mobilizing learning: Intentional disruption. Harnessing the potential of social software tools in higher education using wireless mobile devices. *International Journal of Mobile Learning and Organisation, 3*(4), 399–419. doi:10.1504/IJMLO.2009.027456

Cochrane, T., & Bateman, R. (2009). Transforming pedagogy using Mobile Web 2.0. *International Journal of Mobile and Blended Learning, 1*(4), 56–83. doi:10.4018/jmbl.2009090804

Cochrane, T., & Bateman, R. (2010b). Smartphones give you wings: Pedagogical affordances of mobile Web 2.0. *Australasian Journal of Educational Technology, 26*(1), 1–14.

Cochrane, T., Flitta, I., & Bateman, R. (2009). Facilitating social constructivist learning environments for product design students using social software (Web2) and wireless mobile devices. *DESIGN Principles and Practices: An International Journal, 3*(1), 15.

Cochrane, T. (2008). Mobile Web 2.0: The new frontier. In *Hello! Where are you in the landscape of educational technology? Proceedings ascilite Melbourne 2008* (pp. 177-186). Retrieved April 24, 2009, from http://www.ascilite.org.au/conferences/melbourne08/procs/cochrane.pdf

Cochrane, T., & Bateman, R. (2010). Smartphones give you wings: Pedagogical affordances of mobile Web 2.0. *Australasian Journal of Educational Technology 26*(1), 1-14. Retrieved June 24, 2010, from http://www.ascilite.org.au/ajet/ajet26/cochrane.html

Cochrane, T., & Bateman, R. (2010a). *Reflections on 3 years of mlearning implementation (2007-2009)*. Paper presented at the IADIS International Conference Mobile Learning 2010.

Cochrane, T., Bateman, R., & Flitta, I. (2009). Integrating mobile Web 2.0 within tertiary education. In *Proceedings of the V-MICTE 2009 - V International Conference on Multimedia and ICT in Education* (pp. 1348-1353).

Cockburn, A., & Williams, L. (2001). *The costs and benefits of pair programming*. Boston, MA: Addison-Wesley Longman Publishing.

Colajanni, M., & Lancellotti, R. (2004). System architectures for Web content adaptation services. *IEEE Distributed Systems Online, 5*(5).

Cole, J. (2005). *Using Moodle: Teaching with the popular open source course management system*. Sebastopol, CA: O'Reilly Media, Inc.

Collins, S., Bentley, K., & Conto, A. (2008). *Virtual worlds in education*. Retrieved January 20, 2010, from net.educause.edu/ir/library/pdf/DEC0801.pdf

Collins, T. G. (2005). English class on the air: Mobile language learning with cell phones. In *Proceedings of the Fifth IEEE International Conference on Advanced Learning Technologies* (pp. 402-403). IEEE Computer Society.

Consulting, C. C. I. D. (2009). *Handheld digital learning device market becomes more mature*. Retrieved February 9, 2010, from http://www.ccidconsulting.com/insights/content.asp?Cont ent_id=20882

Cook, J., Bradley, C., Lance, J., Smith, C., & Haynes, R. (2007). Generating learner contexts with mobile devices. In Pachler, N. (Ed.), *Mobile learning: Towards a research agenda* (pp. 55–73). London, UK: WLE Centre, Institute of Education.

Corbeil, J. R., & Valdes-Corbeil, M. E. (2007). Are you ready for mobile learning? *EDUCAUSE Quarterly, 30*(2), 51–58.

Corter, J. E., Nickerson, J. V., Esche, S. K., Chassapis, C., Im, S., & Ma, J. (2007). Constructing reality: A study of remote, hands-on and simulated laboratories. *ACM Transactions on Computer-Human Interaction, 2*(14), 7–27. doi:10.1145/1275511.1275513

Coutinho, C. P., & Bottentuit, J. B. Jr. (2009a). From Web to Web 2.0 and e-learning 2.0 . In Yang, H. H., & Yuen, S. H. (Eds.), *Handbook of research on practices and outcomes in e-learning: Issues and trends* (pp. 19–37). Hershey, PA: Information Science Reference - IGI Global.

Coutinho, C. (2009b). Using blogs, podcasts and Google sites as educational tools in a teacher education program. In G. Richards (Ed.), *Proceedings of World Conference on E-Learning in Corporate, Government, Healthcare, and Higher Education 2009* (pp. 2476-2484). Chesapeake, VA: AACE. Retrieved July 24, 2009, from http://hdl.handle.net/1822/9984

Coutinho, C. P. (2009a). Challenges for teacher education in the learning society: Case studies of promising practice. In H. H. Yang & S. H. Yuen (Eds.), *Handbook of research on practices and outcomes in e-learning: Issues and trends* (pp. 385-401). Hershey, PA: Information Science Reference/ IGI Global. Retrieved April 24, 2009, from http://hdl.handle.net/1822/9981

Coutinho, C. P. (2009c). E-learning 2.0: Challenges for lifelong learning. In I. Gibson (Ed.). *Proceedings of the 20th International Conference of the Society for Information Technology and Teacher Education, SITE 2009,* (pp. 2768-2773).

Coutinho, C. P., & Bottentuit, J. B., Jr. (2009b). Literacy 2.0: Preparing digitally wise teachers. In A. Klucznick-Toro, et al. (Eds.), *Higher Education, Partnership and Innovation (IHEPI 2009)* (pp. 253-261). Budapest, Hungary: PublikonPublishers/IDResearch, Lda.

Croquet. (2010). *Croquet SDK - introduction*. Retrieved January 28, 2010, from http://www.opencroquet.org/index.php/Croquet_SDK_-_Introduction

Cross, J. (2006). *Informal learning, rediscovering the natural pathways that inspire innovation and performance*. San Francisco, CA: Pfeiffer.

Cruz, S. (2009). Blogue, YouTube, Flickr e Delicious: Software social. In A. A. Carvalho (org.). *Manual de ferramentas da Web 2.0 para professores*. Lisboa, Portugal: Direcção-Geral de Inovação e de Desenvolvimento Curricular do Ministério da Educação.

Curran, K., & Annesley, S. (2005). Transcoding media for bandwidth constrained mobile devices. *International Journal of Network Management, 15*(2), 75–88. doi:10.1002/nem.545

Daden Limited. (2009). *PIVOTE launches an open-source learning system for virtual worlds, the Web and iPhone*. Retrieved February 12, 2010, from http://www.prlog.org/10197856-pivote-launches-an-opensource-learning-system-for-virtual-worlds-the-web-and-iphone.html

DeKoenigsberg, G. (2008). *How successful open source projects work, and how and why to introduce students to the open source world?* In 21st Conference on Software Engineering Education and Training (pp. 274-277).

Denning, P. (2003). Great principles of computing. *Communications of the ACM, 11*(46), 15–20. doi:10.1145/948383.948400

Denning, P. J. (2004). The field of programmers myth. *Communications of the ACM, 47*(7), 15–20. doi:10.1145/1005817.1005836

Denso Wave Incorporated. (2000). *QR code section*. Retrieved March 2, 2010, from http://www.denso-wave.com/qrcode/index-e.html

Dewey, J. (1985). *Democracy and education*. Carbondale, IL: Southern Illinois University Press.

Dias, A., Carvalho, J., Keegan, D., Kismihok, G., Mileva, N., Nix, J., & Rekkedal, T. (2008). *Introdução ao mobile learning*. Retrieved January 20, 2010, from http://www.ericsson.com/ericsson/corpinfo/programs/the_role_of_mobile_learning_in_european_education/products/wp/socrates_wp1_portuguese.pdf

Dillenbourg, P. (2002). *Over-scripting CSCL: The risks of blending collaborative learning with instructional design*. In Three worlds of CSCL. Can we support CSCL (pp. 61-91).

Dolittle, P. E., Lusk, D. L., Byrd, C. N., & Mariano, G. J. (2009). iPods as mobile multimedia learning environments: Individual differences and instructional design. In H. Ryu, & D. Parsons (Eds.), *Innovative mobile learning: Techniques and technologies* (pp. 83-101). New York, NY: Information Science Reference.

Donath, R. (2008). Learning languages in Web.20. *Lend – Lingua e Nuova Didattica, 37*(3), 99-101.

Downes, S. (2005). *E-learning 2.0*. Retrieved March 29, 2010, from http://www.elearnmag.org/subpage.cfm?section=articles&article=29-1

Downes, S. (2006). E-learning 2.0 at the e-learning forum. In *E-Learning Forum*. Canadá: Institute for Information Teachnology. Retrieved April 24, 2008, from http://www.teacher.be/files/page0_blog_entry38_1.pdf

Dumas, M., van der Aalst, W., & Ter Hofstede, A. (2005). *Process-aware information systems: Bridging people and software through process technology*. Hoboken, NJ: Wiley-Blackwell. doi:10.1002/0471741442

Eclipse. (n.d.). Retrieved 2010, from http://www.eclipse.org

Edwards, R. (2005). *Knowledge sharing for the mobile workforce*. Retrieved January 22, 2008, from http://www.clomedia.com/content/templates/ clo_article.asp?articleid=945&zoneid =24

EEC. (2006). *Recommendation of the European Parliament and the Council of 18 December 2006*. Retrieved March 29, 2010, from http://eur-lex.europa.eu/LexUriServ/site/en/oj/2006/l_394/l_39420061230en00100018.pdf

Ellis, R. (1994). *The study of second language acquisition*. Oxford, UK: Oxford University Press.

Enhydra Shark. (n.d.). Retrieved 2010, from http://shark.enhydra.org

Escudeiro, P. (2007). X-Tec model and QEF model: A case study. In T. Bastiaens & S. Carliner (Eds.), *Proceedings of World Conference on E-Learning in Corporate, Government, Healthcare, and Higher Education 2007* (pp. 258-265). Chesapeake, VA: AACE.

EurActiv. (2007). *European youth misses out on jobs boom*. Retrieved January 22, 2008, from http://www.eur activ.com/en/socialeurope/ europeanyouthmissesjobsboom/article168707

Express Computer. (2008). *Computing in the clouds*. Retrieved from http://www.expresscomputeronline.com/20080218/technology01.shtml

Faber, B. (2002). *Educational models and open source: Resisting the proprietary university*. In SIGDOC'02 (pp. 31-39) Toronto, Ontario, Canada.

Facer, K., Faux, F., & McFarlane, A. (2005). Challenges and opportunities: Making mobile learning a reality in schools. In *Proceedings of MLearn 2005: 4th World Conference on mLearning*. Retrieved June 10, 2010 from http://www.mlearn.org.za/CD/papers/Facer%20-%20Faux% 20-%20McFarlane.pdf

Farmer, J., & Dolphin, I. (2005). *Sakai: eLearning and more*. In EUNIS 2005-Leadership and Strategy in a Cyber-Infrastructure World.

Feizabadi, S. (2007). *History of Java*. Retrieved August 6, 2009, from http://ei.cs.vt.edu/book/chap1/java_hist.html

Fernández-Alemán, J. L. (2009). Deducing loop patterns in CS1: A comparative study. In *Proceedings of the 2009 Ninth IEEE International Conference on Advanced Learning Technologies* (pp. 247-248).

Ferretti, S., Roccetti, M., Salomoni, P., & Mirri, S. (2009). Custom e-learning experiences: Working with profiles for multiple content sources access and adaptation. *Journal of Access Services*, *6*(1&2), 174–192. doi:10.1080/15367960802301093

FFMPEG. (2010). *FFmpeg multimedia systems*. Retrieved March 21, 2010, from http://ffmpeg.org

Figueiredo, P., & Refkalefsky, E. (2009). Twitter: Uma nova forma de se comunicar? In *XXXII Congresso Brasileiro de Ciências da Comunicação. Sociedade Brasileira de Estudos Interdisciplinares da Comunicação*: Curitiba. Retrieved April 13, 2010, from http://www.intercom.org.br/papers/nacionais/2009/resumos/R4-1316-1.pdf

Fisher, M., King, J., & Tague, G. (2001). Development of a self-directed learning readiness scale for nursing education. *Nurse Education Today*, *21*, 516–525. doi:10.1054/nedt.2001.0589

Fisler, J., & Schneider, F. (2008). *Creating, handling and implementing e-learning courses and content using the open source tools OLAT and eLML at the University of Zurich*. In ISPRS Conference (pp. 3-11).

Flowerdew, J., & Miller, L. (2005). *Second language listening: Theory and practice*. New York, NY: Cambridge University Press.

Freie Universität Berlin. (2010). *Saros - distributed collaborative editing and pair programming*. Retrieved January 17, 2010, from https://www.inf.fu-berlin.de/w/SE/DPP

Freitas, G. P. (2008). Configurações da fotografia contemporânea em comunidades virtuais. In *XXXI Congresso Brasileiro de Ciências da Comunicação*. Natal. Retrieved December, 8 2009 from http://www.intercom.org.br/papers/nacionais/2008/resumos/R3-1125-1.pdf

Freitas, S. (2008). *Serious virtual worlds: A scoping study*. Retrieved January 28, 2010, from http://www.jisc.ac.uk/news/stories/2009/01/svw.aspx

fring. (2009). *fring*. Retrieved December 6, 2009, from http://www.fring.com

FSim. (2007). *The MS Flight Simulator website*. Retrieved June 5, 2007, from http://www.fsinsider.com/Pages/default.aspx

Gaia Reply. (2010). *Gaia Image Trascoder* (GIT). Retrieved March 21, 2010, from http://gaia-git.sourceforge.net

Gárcia-Mateos, G., & Fernández-Alemán, J. L. (2009). A course on algorithms and data structures using online judging. *ACM SIGCSE Bulletin, 41*(3), 45–49. doi:10.1145/1595496.1562897

García-Zubia, J., Orduña, P., López de Ipiña, D., & Alves, G. (2009). Addressing software impact in the design of remote labs. *IEEE Transactions on Industrial Electronics, 56*(12), 4757–4767. doi:10.1109/TIE.2009.2026368

García-Zubia, J., Irurzun, J., Angulo, I., Orduña, P., Ruiz-de-Garibay, J., & Hernández, U. ... Sancristobal, E. (2010). SecondLab: A remote laboratory under Second Life. In *Proceedings of the IEEE Engineering Education Conference 2010* (pp. 351-356).

Gardner, D., & Miller, L. (1997). *A study of tertiary level self-access facilities in Hong Kong*. Hong Kong: City University.

Gardner, H. (1999). *Intelligences reframed: Multiple intelligences in the 21st century*. New York, NY: Basic Books.

Gartner Research Inc. (2009). *Gartner says 80 & of active Internet users will have a "second life" in the virtual world by the end of 2011*. Retrieved on June 12, 2010, from http://www.gartner.com/it/page.jsp?id=503861

Geddes, M., & Sturtridge, G. (1982). *Individualization*. London, UK: Modern English Publications.

Geddes, S. J. (2004). Mobile learning in the 21st century: Benefit for learners. *Knowledge Tree E-Journal, 30*(3), 214–228.

Georgiev, T., Georgieva, E., & Smrikarov, A. (2004). M-learning: A new stage of e-learning. In *Proceedings International conference on Computer Systems and Technologies, CompSysTech' 2004, IV (28)* (pp. 1-5).

Georgieva, E. (2006). *A comparison analysis of mobile learning systems*. International Conference on Computer Systems and Technologies - CompSysTech' 2006. Retrieved August 6, 2009, from http://ecet.ecs.ru.acad.bg/cst06/Docs/cp/sIV/IV.17.pdf

Gerdes, J., & Tilley, S. (2007). *A conceptual overview of the virtual networking laboratory*. In SIGITE'07 (pp. 75-82).

Glisic, S., & Lorenzo, B. (2009). *Advanced wireless networks: Cognitive, cooperative & opportunistic 4G technology* (2nd ed.). Indianapolis, IN: Wiley. doi:10.1002/9780470745724

Goh, T., & Kinshuk, D. (2006). Getting ready for mobile learning-adaptation perspective. *Journal of Educational Multimedia and Hypermedia, 15*(2), 175–198.

Google Press Center. (2000). Retrieved July 15, 2009, from http://www.google.com/press/pressrel/pressrelease51.html

Google, Inc. (2010). *Nexus One section*. Retrieved March 7, 2010, from http://www.google.com/phone

Google. (2009). *Introducing the Google Chrome OS*. Retrieved January 1, 2010, from http://googleblog.blogspot.com/2009/07/ introducing-google-chrome-os.html

Goyal, R., Lai, S., Jain, R., & Durresi, A. (1998). *Laboratories for data communications and computer networks*. In 1998 FIE Conference (pp. 1113-1119).

Gravier, C., Fayolle, J., Bayard, B., Ates, M., & Lardon, J. (2008). State of the art about remote laboratories paradigms - foundations of ongoing mutations. *International Journal of Online Engineering, 4*(1). Retrieved from http://online-journals.org/index.php/i-joe/article/view/480.

Gros, B. (2002). Knowledge construction and technology. *Journal of Educational Multimedia and Hypermedia, 11*(4), 323–343.

Guo, J., Xiang, W., & Wang, S. (2007). Reinforce networking theory with OPNET simulation. *Journal of Information Technology Education, 6*, 191–198.

Gustavsson, I., Zackrisson, J., Håkansson, L., Claesson, I., & Lagö, T. (2007). The VISIR project - an open source software initiative for distributed online laboratories. In *Proceedings of the Remote Eng. And Virtual Instrumentation Conference 2007*.

Hameed, K., & Shah, H. (2009). *Mobile learning in higher education: Adoption and siscussion criteria.* Paper presented at the IADIS International Conference on Mobile Learning 2009, Barcelona, Spain.

Harp, S. F., & Mayer, R. E. (1998). How seductive details do their damage: A theory of cognitive interest in science learning. *Journal of Educational Psychology, 90*, 414–434. doi:10.1037/0022-0663.90.3.414

Harte, L., & Bowler, D. (2003). *Introduction to mobile telephone systems, 1G, 2G, 2.5G, and 3G technologies and services*. Fuquay Varina, NC: Althos.

Hartig, K. (2009). What is cloud computing? *Cloud Computing Journal.* Retrieved from http://cloudcomputing.sys-con.com/node/579826

Harumoto, K., Nakano, T., Fukumura, S., Shimojo, S., & Nishio, S. (2005). Effective Web browsing through content delivery adaptation. *ACM Transactions on Internet Technology, 5*(4), 571–600. doi:10.1145/1111627.1111628

Harward, V. J., del Alamo, J. A., Lerman, S. R., Bailey, P. H., Carpenter, J., & DeLong, K. … Zych, D. (2008). The iLab shared architecture: A Web services infrastructure to build communities of Internet accessible laboratories. *Proceedings of the IEEE, 96*(6), 931-950.

Hatamura, H., Ikemura, T., & Togo, T. (2006). *Goi no imi-gakushu ni hatsuon wa donoyouni riyo sareruka: Oninriyo to gakushu-konnansei no yoin nitsuite no kento. (How pronunciation is useful for vocabulary learning? Investigation on the relationship between phonemes and learning difficulty, translated by the author)*. Paper presented at Spring Conference of Japan Association of Language Education and Technology, Japan.

Hayes, P., Pathak, P., & Joyce, D. (2005). *Mobile technology in education –a multimedia application.* Paper presented at 6th Annual Irish Educational Technology Users' Conference (EdTech), Dublin.

Helps, C. R. (2006). *Instructional design theory provides insights into evolving information technology technical curricula.* In SIGITE'06 (pp. 129-135). Minneapolis, Minnesota.

Henri, F., Charlier, B., & Limpens, F. (2008). Understanding PLE as an essential component of the learning process. In J. Luca & E. R. Weippl (Eds.), *Proceedings of the 20th World Conference on Educational Multimédia Hypermedia & Telecommunications, EDMEDIA 2008* (pp. 3766-3770). Vienna, Austria: University of Vienna.

Henrysson, A., Billinghurst, M., & Ollila, M. (2005). *Face to face collaborative AR on mobile phones.* In International Symposium on Augmented and Mixed Reality (pp. 80-89).

Hernandez-Leo, D., Asensio-Perez, J. I., & Dimitriadis, Y. (2005). Computational representation of collaborative learning flow patterns using IMS Learning Design. *Journal of Educational Technology & Society, 8*(4), 75–89.

Hernandez-Leo, D., Asensio-Perez, J. I., & Dimitriadis, Y. (2004). IMS learning design support for the formalization of collaborative learning patterns. In *Proceedings of the IEEE International Conference on Advanced Learning Technologies* (pp. 350-354).

Herrington, J., & Oliver, R. (2000). An instructional design framework for authentic learning environments. *Educational Technology Research and Development, 48*(3), 23–48. doi:10.1007/BF02319856

Herrington, A., Herrington, J., & Mantei, J. (2009). Design principles for mobile learning . In Herrington, J., Herrington, A., Mantei, J., Olney, I., & Ferry, B. (Eds.), *New technologies, new pedagogies: Mobile learning in higher education* (pp. 129–138). Wollongong, Australia: Faculty of Education, University of Wollongong.

Herrington, A., & Herrington, J. (2007). *Authentic mobile learning in higher education*. Paper presented at the AARE 2007 International Educational Research Conference, Fremantle, Australia.

Herrington, A., & Herrington, J. (2007). *Authentic mobile learning in higher education.* Paper presented at the AARE 2007 International Educational Research Conference, Fremantle, Australia.

Herrington, J., Herrington, A., Mantei, J., Olney, I., & Ferry, B. (Eds.). (2009). *New technologies, new pedagogies: Mobile learning in higher education.* Wollongong, Australia: Faculty of Education, University of Wollongong. Retrieved May 13, 2010, http://ro.uow.edu.au/newtech

Herrington, J., Mantei, J., Herrington, A., Olney, I., & Ferry, B. (2008). *New technologies, new pedagogies: Mobile technologies and new ways of teaching and learning.* Paper presented at the ASCILITE 2008, Deakin University, Melbourne, Australia.

Heuser, W. (2009). *LINUX on the road - the first book on mobile Linux*. Scotts Valley, CA: CreateSpace.

Hnatyshin, V., & Lobo, A. (2008). Undergraduate data communications and networking projects using OPNET software. In *Proceedings of the 39th SIGCSE Technical Symposium on Computer Science Education* (pp. 241-245).

Ho, C., Raha, S., Gehringer, E., & Williams, L. (2009). *Sangam - a distributed pair programming plug-in for Eclipse*. Retrieved January 15, 2010, from http://collaboration.csc.ncsu.edu/laurie/Papers/Sangam.pdf

Hollingsworth, D. (1995). *The workflow reference model. WfMC. TC-1003*. Workflow Management Coalition.

Honey, P., & Mumford, A. (1984). Questions and answers on learning styles questionnaire. *Industrial and Commercial Training, 24*(7), 1–10.

Hughes, L. (1989). Low-cost networks and gateways for teaching data communications. In *SIGCSE '89: Proceedings of the twentieth SIGCSE Technical Symposium on Computer Science Education* (pp. 6–11).

Hujala, H., Kynäslahti, H., & Seppälä, P. (2003). Mobile learning: 'Creative learning'— mobility in action . In Kynäslahti, H., & Seppälä, P. (Eds.), *Mobile learning* (pp. 111–111). Finland: Edita Publishing Inc.

IDC. (2008). *Worldwide mobile worker population 2007-2011 forecast*. Retrieved February 9, 2010, from http://www.workshifting.com/downloads/documents/IDC_Mobile Worker_excerpt_0_0.pdf

IEEE. Communications Society. (2010). *IDC, 1.2B mobile phones will be sold in 2010*. Retrieved from http://community.comsoc.org/blogs/ajwdct/idc-market-forecasts-mobile-broadband-and-lte

IEEE/ACM. (2008). *IT 2008, curriculum guidelines for undergraduate degree programs in information technology* (Final Draft). Retrieved January 16, 2009, from http://www.acm.org/education/ curricula/IT2008%20 Curriculum.pdf

iLabs. (2010). *Computer software*. Available from http://ilab.mit.edu/wiki

ImageMagick. (2010). *ImageMagick - convert, edit, and compose images*. Retrieved March 2010, from http://www.imagemagick.org

IMS Global Learning Consortium. (2002a). IMS Learner Information Profile (LIP). Retrieved March 21, 2010, from: http://www.imsglobal.org/specificationdownload.cfm

IMS Global Learning Consortium. (2002b). *IMS learner information package accessibility for LIP*. Retrieved March 21, 2010, from http://www.imsglobal.org/specificationdownload.cfm

Indexmundi. (2009). *World demographics profile 2009*. Retrieved February 8, 2010, from http://www.indexmundi.com/world/demographics_profile.html

Jannach, D., Leopold, K., Timmerer, C., & Hellwagner, H. (2006). A knowledge-based framework for multimedia adaptation. *Applied Intelligence, 24*(2), 109–125. doi:10.1007/s10489-006-6933-0

JISC. (2005). *Multimedia learning with mobile phones. Innovative practices with elearning. Case studies: Anytime, anyplace learning*. Retrieved from http://www/jisc.ac.uk/uploaded_documents/southampton.pdf

JISC. (2009a). Effective practice in a digital age. Retrieved from http://www.jisc.ac.uk/publications/documents/ effectivepracticedigitalage.aspx

JISC. (2009b). *Higher education in a Web 2.0 world.* Retrieved from http://www.jisc.ac.uk/publications/documents/heweb2.aspx

Johnson, D. W., Johnson, R. T., & Smith, K. A. (1991). *Cooperative learning: Increasing college faculty instructional productivity* (ASHE-FRIC Higher Education Report No.4). Washington, DC: School of Education and Human Development, George Washington University.

Jonassen, D. H. (2007). *Computadores, ferramentas cognitivas - desenvolver o pensamento crítico nas escolas.* Porto, Portugal: Porto Editora.

Jurado, F., Redondo, M., & Ortega, M. (2006). Specifying collaborative tasks of a CSCL environment with IMS-LD. *Lecture Notes in Computer Science, 4101*, 311–317. doi:10.1007/11863649_38

Kadota, S., & Ikemura, T. (2006). *Eigo goishido handbook.* Taishukan Shoten.

Kappe, D. (2007). *Pair programming with VNC.* Retrieved August 6, 2009, from http://www.pathf.com/blogs/2007/09/pair-programmin

Kato, H., & Billinghurst, M. (1999). *Marker tracking and HMD calibration for a video-based augmented reality.* In Second IEEE and ACM International Workshop on Augmented Reality (pp. 85–94).

Keegan, D. (2004). Mobile learning: The next generation of learning. In *Proceedings of 18th AAOU Conference, Quality Education for All - New Missions and Challenges Facing Open Universities.* Shanghai, China: TV University.

Keegan, D. (2005). The incorporation of mobile learning into mainstream education and training. In *Proceedings of Mlearn 2005: 4th World Conference on mLearning.* Retrieved June 10, 2010, from http://www.mlearn.org.za/CD/papers/keegan1.pdf

Kent NGfL. (2007). *Simulations – Google Search Kent NGfL website.* Retrieved June 15, 2007, from http://www.kented.org.uk/ngfl/software/simulations/index.htm

Kimura, M., & Shimoyama, Y. (2009). Vocabulary learning contents for use with mobile phones in education in Japan. In I. Gibson et al. (Eds.), *Proceedings of Society for Information Technology & Teacher Education International Conference 2009* (pp. 1922-1929).

Kineo & UFI/Learndirect. (2007). *Mobile learning reviewed.* Retrieved on April 10, 2008, from http://www.kineo.com/documents/Mobile_learning_reviewed_final.pdf

Kirk, B. J. (2009). *Android market surpasses 20,000 applications, T-Mobile picks top apps for customers.* Retrieved January 1, 2010, from http://www.mobileburn.com/news.jsp?Id=8419

Klawonn, F. (2008). *Introduction to computer graphics.* Godalming, UK: Springer-Verlag London Limited.

Klein, N., Carlson, M., & McEwen, G. (Eds.). (2008). *Laszlo in action: Rich Web applications with OpenLaszlo.* Greenwich, CT: Manning Publications Co.

Klopfer, E., Squire, K., & Jenkins, H. (2002). Environmental detectives: PDAs as a window into a virtual simulated world. In *Proc. of IEEE International Workshop on Wireless and Mobile Technologies in Education (WMTE 2002)* (pp. 95-98). Sweden: Växjö.

Kneale, B., Horta, A. Y., & Box, L. (2004). Velnet: Virtual environment for learning networking. *Proceedings of the Sixth Conference on Australasian Computing Education, 30,* (pp. 161-169).

Knowles, M. (Ed.). (1975). *Self-directed learning: A guide for learners and teachers.* Chicago, IL: Follett Publishing Company.

Kogure, Y., Shimoyama, Y., Anzai, Y., Kimura, M., & Obari, H. (2006). *Study of mobile phone market for mobile-learning.* Paper presented at the meeting of the Japan Education Association of Technology, Niigata, Japan.

Kolb, D. (1985). *Learning style inventory.* Boston, MA: McBer and Company.

Koschembahr, C. (2005). *Optimizing your sales workforce through mobile learning.* Retrieved February 8, 2010, from http://www.neiu.edu/~sdundis/textresources/Mobile_Wire less/Optimizing%20Your%20Sales%20Workforce%20through%20Mobile%20Learning.pdf

Koster, R. (2004). *Raph Koster's writings on game design.* Retrieved January 20, 2010, from http://www.legendmud.org/raph/gaming/index.html

Kukulska-Hulme, A., & Traxler, J. (Eds.). (2005). *Mobile learning: A handbook for educators and trainers*. London, UK: Routledge.

Kukulska-Hulme, A., & Shield, L. (2008). An overview of mobile assisted language learning: From content delivery to supported collaboration and interaction. *ReCALL, 20*(3), 271–289. doi:10.1017/S0958344008000335

Kukulska-Hulme, A. (2007). Mobile usability in educational contexts: What have we learnt? *International Review of Research in Open and Distance Learning, 8*(2), 1–16.

Kuts, E., Sedano, C. I., Botha, A., & Sutinen, E. (2007). *Communication and collaboration in educational multiplayer mobile games*. In IADIS International Conference on Cognition and Exploratory Learning in Digital Age (CELDA 2007) (pp. 295-298).

Kynäslahti, H., & Seppälä, P. (Eds.). (2003). *Mobile learning*. Finland: Edita Publishing Inc.

Kynäslahti, H. (2003). Mobile learning: In search of elements of mobility in the context of education . In Kynäslahti, H., & Seppälä, P. (Eds.), *Mobile learning* (pp. 47–47). Finland: Edita Publishing Inc.

Laakko, T., & Hiltunen, T. (2005). Adapting Web content to mobile user agents. *IEEE Internet Computing, 9*(2), 46–53. doi:10.1109/MIC.2005.29

Langelier, L. (2005). *Working, learning and collaborating in a network: Guide to the implementation and leadership of intentional communities of practice*. Quebec City, Canada: CEFIRO (Recherche et Études de cas collection).

Laszlo Systems. (2006-2009). *Official documentation*. Retrieved March 15, 2010, from http://www.openlaszlo.org/documentation

Laurillard, D. (2001). *Rethinking university teaching: A framework for the effective use of educational technology* (2nd ed.). London, UK: Routledge.

Laurillard, D. (2007). Pedagogical forms of mobile learning: Framing research questions . In Pachler, N. (Ed.), *Mobile learning: Towards a research agenda* (*Vol. 1*, pp. 153–175). London, UK: WLE Centre, Institute of Education.

Laurillard, D. (2002). *Rethinking university teaching: A conversational framework for the effective use of learning technologies* (2nd ed.). Londres, UK & Nova Iorque, NY: Routledge Falmer. Retrieved October 25, 2009, from http://www.questiaschool.com/read/103888453

Lawrence, I., & Belem, R. C. L. (2009). *Professional Ubuntu Mobile development*. Indianapolis, IN: Wiley Publishing.

Lawson, E. A., & Stackpole, W. (2006). Does a virtual networking laboratory result in similar student achievement and satisfaction? In *SIGITE '06: Proceedings of the 7th Conference on Information Technology Education* (pp. 105–114).

Leominster, S. (2008). *Second Life News: Second Life© client for mobile phone beta available.*

Lerdorf, R., Tatroe, K., Kaehms, B., & McGredy, R. (2002). *Programming PHP*. Sebastopol, CA: O'Reilly & Associates, Inc.

Levy, P. (1997). *Collective intelligence: Mankind's emerging world in cyberspace*. New York, NY: Plenum Publishing Corporation.

Levy, M., & Kennedy, C. (2005). Learning Italian via mobile SMS . In Kukulska-Hulme, A., & Traxler, J. (Eds.), *Mobile learning: A handbook for educators and trainers* (pp. 76–83). London, UK: Taylor and Francis.

LiMo Foundation. (2009). *Welcome to LiMo*. Retrieved November 6, 2009, from http://www.limofoundation.org

Linn, M. C., & Burbules, N. C. (1993). Construction of knowledge and group learning . In Tobin, K. (Ed.), *The practice of constructivism in science education* (pp. 91–119). Hillsdale, NJ: Lawrence Erlbaum Associates.

Lisbôa, E. S., Bottentuit Junior, J. B., & Coutinho, C. P. (2009). Uso do Flickr em contexto educativo. In: VI Congresso Brasileiro de Ensino Superior a Distância (ESuD), 2009, São Luís. *Anais do VI Congresso Brasileiro de ensino Superior a Distância (ESuD).* São Luís - Maranhão: Universidade Estadual do Maranhão.

Liu, C., Tao, S., & Nee, J. (2008). Bridging the gap between students and computers: Supporting activity awareness for network collaborative learning with GSM network. *Behaviour & Information Technology, 27*(2), 127–137. doi:10.1080/01449290601054772

Liu, Y., & Li, H. X. (2009). What drives m-learning success?—Drawing insights from self-directed learning theory. In *Proceedings of Pacific Asia Conference on Information Systems 2009*. Hyderabad, India.

Looser, J. (2007). *AR magic lenses: Addressing the challenge of focus and context in augmented reality*. Doctoral Dissertation, University of Canterbury, New Zealand.

López-de-Ipiña, D., García-Zubia, J., & Orduña, P. (2006) Remote control of Web 2.0-enabled laboratories from mobile devices. In *Proceedings of the 2nd IEEE International Conference on e-Science and Grid Computing 2006* (p. 123).

Luckin, R., Clark, W., Garnett, F., Whitworth, A., Akass, J., Cook, J., et al. (2008). *Learner generated contexts: A framework to support the effective use of technology to support learning.* Retrieved 5 November, 2008, from http://api.ning.com/files/Ij6j7ucsB9vgb11pKPHU6LK-MGQQkR- YDVnxruI9tBGf1Q-eSYUDv- Mil6u-WqX4F1jYA1PUkZRXvbxhnxuHusyL1lRXVrBKnO/LGCOpenContextModelning.doc

MacIntyre, B., Gandy, M., Bolter, J., Dow, S., & Hannigan, B. (2003). *DART: The Designer's Augmented Reality Toolkit*. In The Second International Symposium on Mixed and Augmented Reality (pp. 329-339).

Maemo. (2009). *Maemo basics*. Retrieved January 1, 2010, from http://wiki.maemo.org/Maemo_basics

Maniar, N., Bennett, E., Hand, S., & Allan, G. (2008). The effect of mobile phone screen size on video based learning. *Journal of Software, 3*(4), 51–61. doi:10.4304/jsw.3.4.51-61

Manninen, T. (2004). *Rich interaction model for game and virtual environment design*. Finland: Oulu University Press. Retrieved January 2, 2010, from http://herKules.oulu.fi/isbn954272554

Marçal, E., Lima, L., Júnior, M., Viana, W., Andrade, R., & Ribeiro, J. (2009). *A utilização de dispositivos móveis com ambientes tridimensionais como ferramentas para favorecer o ensino de hardware. XX Simpósio Brasileiro de informática na Educação*. Brasil: Florianópolis.

Marçal, E., Andrade, R., & Rios, R. (2005). Aprendizagem utilizando dispositivos móveis com sistemas de realidade virtual. In UFRGS (Ed.), *Revista Novas Tecnologias na Educação, 3*(1). Brasil: Porto Alegre.

Marino, O., Casallas, R., Villalobos, J., & Correal, D. (2007). *E-learning networked. Environments and architectures: A knowledge processing perspective* (pp. 27–59). New York, NY: Springer-Verlag. doi:10.1007/978-1-84628-758-9_2

Marshall, D. (2009). *Microsoft releases Hyper-V Linux drivers as open source*. Retrieved February 9, 2009, from http://www.infoworld.com/d/virtualization/microsoft-releases-hyper-v-linux-drivers-open-source-205

Martins, E., Gomes, I., & Santos, L. (2009). O Twitter como ferramenta no ensino e atuação de profissionais de publicidade e propaganda. In *XXXII Congresso Brasileiro de Ciências da Comunicação*. Sociedade Brasileira de Estudos Interdisciplinares da Comunicação: Curitiba. Retrieved October 13, 2009, from http://www.intercom.org.br/papers/nacionais/2009/resumos/R4-3861-1.pdf

Mason, R. (1998). Models of online courses. *ALN Magazine, 2*(2), 1–10.

Mayer, R. E. (2001). *Multimedia learning*. New York, NY: Cambridge University Press.

McAndrew, A. (2008). Teaching cryptography with open-source software. *Special Interest Group Computer Science Education, 40*(1), 325–330.

McGettrick, A., Boyle, R., Ibbett, R., Lloyd, J., Lovegrove, G., & Mander, K. (2005). Grand challenges in computing: Education - A summary. *The Computer Journal, 48*(1), 42–48. doi:10.1093/comjnl/bxh064

McLoughlin, C., & Lee, M. (2008a). Future learning landscapes: Transforming pedagogy through social software. *Innovate: Journal of Online Education, 4*(5), 7.

McLoughlin, C., & Lee, M. (2007). *Social software and participatory learning: Pedagogical choices with technology affordances in the Web 2.0 era.* Paper presented at the Ascilite 2007, ICT: Providing Choices for Learners and Learning, Centre for Educational Development, Nanyang Technological University, Singapore.

McLoughlin, C., & Lee, M. (2008b). *Mapping the digital terrain: New media and social software as catalysts for pedagogical change.* Paper presented at the ASCILITE Melbourne 2008, Deakin University, Melbourne.

McMahon, M., & Pospisil, R. (2005). Laptops for a digital lifestyle: The role of ubiquitous mobile technology in supporting the needs of millennial students. In *Proceedings of ASCILITE 2005.* Retrieved February 8, 2010, from http://www.ascilite.org.au/conferences/brisbane05/blogs/proceedings/49_McMahon%20&%20Pospisil.pdf

McNicol, T. (2005). Language e-learning on the move. *Japan Media Review*. Retrieved from http://ojr.org/japan/wireless/1080854640.php

Medford, C. (2008). Cell phone market soars despite recession. *Red Herring*. Retrieved July 16, 2009, from http://www.redherring.com/Home/24181

Meneely, A., Williams, L., & Gehringer, E. F. (2008). A repository of education-friendly open-source projects . In *ACM ITiCSE'08* (pp. 7–12). ROSE.

Menezes, C. Q., & Moreira, F. L. (2009). In the pursuit of m-learning - first steps in implementing podcast among K12 students in ESL . In *Actas da VI Conferência Internacional de TIC na Educação* (pp. 91–107). Braga: CCUM.

Mennila, L. (2007). Novices' progress in introductory programming courses. *Informatics in Education, 6*(1), 139–152.

Milrad, M. (2006). How should learning activities using mobile technologies be designed to support innovative educational practices? In M. Sharples (Ed.), *Big issues in mobile learning* (pp. 28- 30). Report of a workshop by the Kaleidoscope Network of Excellence Mobile Learning Initiative. University of Nottingham.

Minch, R., & Tabor, S. (2003). Networking education for the new economy. *Journal of Information Technology Education, 2*, 191–217.

Mishra, P., Koehler, M. J., & Zhao, Y. (Eds.). (2007). *Faculty development by design: Integrating technology in higher education.* Charlotte, NC: Information Age Publishing.

Mitchell, A., & Savill-Smith, C. (2004). *The use of computer and video games for learning: A review of the literature.* London, UK: Learning and Skills Development Agency. Retrieved July 16, 2009, from http://www.m-learning.org/archive/docs/The%20use%20of%20computer%20and%20video%20games%20for%20learning.pdf

Mobile Learning Engine. (n.d.). Retrieved 2010, from http://mle.sourceforge.net

Moblin. (2009). *Create the mobile Internet future*. Retrieved November 6, 2009, from http://moblin.org

Moblin. (2009). *Moblin*. Retrieved July 16, 2009, from http://www.moblin.com/moblinen/AboutUs.aspx

Möhring, M., Lessig, C., & Bimber, O. (2004). *Video see-through AR on consumer cell phones.* In IEEE and ACM International Symposium on Augmented and Mixed Reality (pp. 252-253).

Moll, L. C. (1990). *Vygotsky and education: Instructional implications and applications of sociohistorical psychology*. New York, NY: Cambridge University Press.

Moodle Pty Ltd. (2010). *Moodle*. Retrieved March 21, 2010, from http://moodle.org

Moodle. (2009). *Moodle statistics*. Retrieved August 6, 2009, from http://moodle.org/stats

MORFEO Project. (2007). *MyMobileWeb Project*. Retrieved March 21, 2010, from http://mymobileweb.morfeo-project.org

Morgan, C., Erlinger, M., Davoli, R., & Goldweber, M. (2007). *Environments for a networking laboratory*. In 20th Annual Conference of the National Advisory Committee on Computing Qualification (pp. 53-59).

Moura, A., & Carvalho, A. (2007). Das tecnologias com fios ao wireless: Implicações no trabalho escolar individual e colaborativo em pares . In Dias, P., Freitas, C. V., Silva, B., Osósio, A., & Ramos, A. (Eds.), *Actas da V Conferência Internacional de Tecnologias de Informação e Comunicação na Educação: Challenges 2007*. Braga: Universidade do Minho.

Moura, A., & Carvalho, A. (2008a). *Mobile learning: Teaching and learning with mobile phone and podcasts*. In 8th IEEE International Conference on Advanced Learning Technologies, 2008 (ICALT 2008) (pp. 631-633). Santander, Spain.

Moura, A., & Carvalho, A. (2008b). Mobile learning with cell phones and mobile Flickr: One experience in a secondary school. In I. Arnedillo-Sánchez, & P. Isaías (Eds.), *Proceedings IADIS Conference Mobile Learning 2008* (pp. 216-220). Algarve, Portugal.

Moura, A., & Carvalho, A. (2009). Peddy-paper literário mediado por telemóvel. *Educação, Formação & Tecnologias, 2*(2), 22-40. Retrieved May 29, 2010, from http://eft.educom.pt

MPEG MDS Group. (2003). *MPEG-21 multimedia framework, part 7: Digital item adaptation*. (ISO/MPEG N5845). Retrieved March 21, 2010, from http://www.chiariglione.org/mpeg/working_documents/ mpeg-21/dia/dia_fcd.zip

MPEG Requirements Group. (2002). *MPEG-21 overview*. (ISO/MPEG N4991).

Mulloni, A. (2007). *A collaborative and location-aware application based on augmented reality for mobile devices*, Master Thesis, Facoltà di Scienze Matematiche Fisiche e Naturali, Università degli Studi di Udine.

Muyinda, P. B. (2007). M-learning: Pedagogical, technical and organizational hypes and realities. *Campus-Wide Information Systems, 24*(2), 97–104. doi:10.1108/10650740710742709

Nagappan, N., Williams, L., Ferzli, M., Wiebe, E., Yang, K., Miller, C., & Balik, S. (2003). Improving the CS1 experience with pair programming. *ACM SIGCSE Bulletin, 35*(1), 359–362. doi:10.1145/792548.612006

Nah, K. C., White, P., & Sussex, R. (2008). The potential of using a mobile phone to access the Internet for learning EFK listening skills within a Korean context. *ReCALL, 20*(3), 331–347. doi:10.1017/S0958344008000633

Naismith, L., Lonsdale, P., Vavoula, G., & Sharples, M. (2004). *Mobile technologies and learning*. Retrieved August 6, 2009, from http://www.futurelab.org.uk/resources/publications-reports-articles/literature-reviews/Literature-Review203

Naismith, L., Lonsdale, P., Vavoula, G., & Sharples, M. (2004). *Literature review in mobile technologies and learning*. Report 11: FUTURELAB SERIES.

Naismith, L., Lonsdale, P., Vavoula, G., & Sharples, M. (2004). *Literature review in mobile technologies and learning: A report for NESTA Futurelab*. Retrieved February 8, 2010, from http://www.futurelab.org.uk/resources/documents/lit_reviews/Mobile_Review.pdf

Narian, D. (2009). *ABI research: Mobile cloud computing the next big thing*. Retrieved August 6, 2009, from http://ip-communications.tmcnet.com/topics/ip-communications/articles/59519-abi-research-mobile-cloud-computing-next-big-thing.htm

National Governors Association. (2002). *A governor's guide to creating a 21st-century workforce*. Retrieved January 16, 2009, from http://www.nga.org/cda/files/AM02WORKFORCE.pdf

Nebula. (2007). *The Nebula device*. Retrieved June 13, 2007, from http://nebuladevice.cubik.org

Negroponte, N. (1996). *Ser Digital*. Lisboa: Editorial Caminho.

Nelson, D., & Ng, Y. M. (2000). Teaching computer networking using open source software. In *ITiCSE' 00: Proceedings of the 5th Annual SIGCSE/SIGCUE ITiCSE Conference on Innovation and Technology in Computer Science Education* (pp. 13-16).

NIACE. (2005). *NIACE briefing sheet – 54: Mobile ICT resources for older learners*. Retrieved February 9, 2010, from http://archive.niace.org.uk/information/Briefing_sheets/ 54-Mobile-ICT-resources-for-older-people.pdf

Niessen, C. (2006). Age and learning during unemployment. *Journal of Organizational Behavior, 27*(6), 771–792. doi:10.1002/job.400

Nimbuzz. (2010). *About Nimbuzz Mobile*. Retrieved January 18, 2010, from http://nimbuzz.com/en/mobile

NIPSSR. (2002). *Population projections for Japan: 2001-2050*. Retrieved February 9, 2010, from http://www.ipss.go.jp/pp-newest/e/ppfj02/ppfj02.pdf

Norbrook, H., & Scott, P. (2003). Motivation in mobile modern foreign language learning . In Attewell, J., Da Bormida, G., Sharples, M., & Savill-Smith, C. (Eds.), *MLEARN 2003: Learning with mobile devices* (pp. 50–51). London, UK: Learning and Skills Development Agency.

NuSOAP. (n.d.). Retrieved 2010, from http://sourceforge.net/projects/nusoap

O'Malley, M. J., & Chamot, A. U. (1990). *Learning strategies in second language acquisition*. New York, NY: Cambridge University Press.

O'Malley, C., Vavoula, G., Glew, J. P., Taylor, J., Sharples, M., & Lefrere, P. (2003). *Mobilearn WP4 guidelines for learning/teaching/tutoring in a mobile environment*. Retrieved February 8, 2010, from http://www.mobilearn.org/download/results/guidelines.pdf

O'Reilly, T. (2005). *What is Web 2.0? Design patterns and business models for the next generation of software*. Retrieved from http://www.oreillynet.com/pub/a/oreilly/tim/news/2005/09/30/what-is-web-20.html

Oblinger, D. (2003). Boomers & gen-xers, millennials: Understanding the 'new students.'. *EDUCAUSE Review*, *38*(4), 37–47.

Ohloh. (2010). *Android-vnc-viewer*. Retrieved January 17, 2010, from http://www.ohloh.net/p/android-vnc-viewer

OMA. (2010a). *OMA data synchronization* V1.2.1. Retrieved from http://openmobilealliance.org/ Technical/release_program/ds_v12.aspx

OMA. (2010b). *Device management Working Group*. Retrieved from http://openmobilealliance.org/Technical/DM.aspx

OMTP Bondi. (n.d.). Retrieved 2010, from http://bondi.omtp.org/1.1

Ondrejka, C. (2006). Escaping the gilded cage: User-created content and building the metaverse . In Balkin, J., & Noveck, B. (Eds.), *The state of play: Law, games, and virtual worlds*. New York, NY: New York University Press.

Ondrejka, C. (2008). Education unleashed: Participatory culture, education, and innovation in Second Life . In Salen, K. (Ed.), *The ecology of games: Connecting youth, games, and learning* (pp. 229–252). Cambridge, MA: The MIT Press.

Onyesolu, O. (2009). Virtual reality laboratories: An ideal solution to the problems facing laboratory setup and management. In *Proceedings of the World Congress on Engineering and Computer Science, vol. 1*. San Francisco.

Open Handset Alliance. (2009). *Android*. Retrieved November 6, 2009, from http://www.openhandsetalliance.com/ android_overview.html

Open Mobile Alliance. (2007). *User agent profile v. 2.0 approved enabler*. Retrieved March 21, 2010, from http://www.openmobilealliance.org/Technical/ release_program/uap_v2_0.aspx

Open Source Education Foundation. (2009). *Welcome to the Open Source Education Foundation website*. Retrieved August 6, 2009, from http://www.osef.org

Open Source for America. (2009). *Open Source for America*. Retrieved August 6, 2009 from http://www.opensourceforamerica.org

Open Source Initiative. (2006). *Frequently asked questions*. Retrieved February 16, 2009, from http://www.opensource.org/docs/osd

Open, G. L. (2010). *The OpenGL – the industry standard for high performance graphics*. Retrieved March 13, 2010, from http://www.opengl.org

OpenCourseWare Consortium. (2009). *OpenCourseWare Consortium*. Retrieved August 6, 2009, from http://www.ocwconsortium.org/home.html

OpenSimulator. (2010). *Project history*. Retrieved January 28, 2010, from http://opensimulator.org/wiki/History

Orduña, P., García-Zubia, J., Irurzun, J., Sancristobal, E., Martín, S., & Castro, M. … González. J. M. (2009). Designing experiment agnostic remote laboratories. In *Proceedings of the Remote Engineering and Virtual Instrumentation Conference 2009*.

Owen, C., Tang, A., & Xiao, F. (2003). *ImageTclAR: A blended script and compiled code development system for augmented reality*. In The International Workshop on Software Technology for Augmented Reality Systems (pp. *23-28*).

Oxford, R. (1990). *Language learning strategies: What every teacher should know*. Boston, MA: Heinle and Heinle Publisher.

Paelke, V., Reimann, C., & Stichling, D. (2004). *Foot-based mobile interaction with games.* In ACM SIGCHI International Conference on Advances in computer entertainment technology (pp. 321-324).

Palm. (2009). *Palm Pre.* Retrieved November 6, 2009, from http://www.palm.com/us/products/ phones/pre/ index.html

Pandey, V., Ghosal, D., & Mukherjee, B. (2004). Exploiting user profiles to support differentiated services in next-generation wireless networks. *IEEE Network, 18*(5), 40–48. doi:10.1109/MNET.2004.1337734

Patterson, D. A. (2006). Computer science education in the 21th century. *Communications of the ACM, 29*(3), 27–30. doi:10.1145/1118178.1118212

PBXes. (2009). *PBXes.* Retrieved December 6, 2009, from https://www1.pbxes.com/index_e.php

Pęcherzewska, A., & Knot, S. (2007). *Review of existing EU projects dedicated to dyslexia, gaming in education and m-learning.* WR08 Report to CallDysMProject.

Perez-Rodriguez, R., Caeiro-Rodriguez, M., & Anido-Rifon, L. (2009). Enabling process-based collaboration support in Moodle by using aspectual services. In *Proceedings of 9th IEEE International Conference on Advanced Learning Technologies* (pp. 301-302). Riga, Latvia.

Peters, K. (2006). *Learning on the move mobile technologies in business and education.* Research report of Australian flexible learning framework. Retrieved June 10, 2010 from http://www.flexiblelearning.net.au/ projects/resources/ 2005/Learning%20on%20the%20 move_summary.pdf

Pfaffenberger, B. (2000). Linux in higher education: Open source, open minds, social justice. *Linux Journal.* Retrieved March 21, 2009, from http://linuxjournal.com/ article.php?sid=5071

PhoneGap. (n.d.). Retrieved 2010, from http://phonegap. com

Piaget, J. (1963). *The psychology of intelligence.* Paterson, NJ: Littlefield, Adams.

Piaget, J. (1976). *The grasp of consciousness.* Cambridge, MA: Harvard University.

Piaget, J. (1973). *To understand is to invent.* New York, NY: Viking Press.

Pilet, J. (2008). *Augmented reality for non-rigid surfaces.* Doctoral Dissertation, Ecole Polytechnique federale de Lausanne, France.

Pincas, A. (2004). Using mobile support for use of Greek during the Olympic Games 2004. In *Proceedings of M-Learn Conference* 2004. Rome, Italy.

Pinelle, D., Gutwin, C., & Greenberg, S. (2003). Task analysis for groupware usability evaluation: Modeling shared-workspace tasks with the mechanics of collaboration. *ACM Transactions on Computer-Human Interaction, 10*(4), 281–311. doi:10.1145/966930.966932

Portio Research. (2007). *Worldwide mobile penetration will reach 75% by 2011.* Retrieved February 8, 2010, from http://www.portioresearch.com/next_billion_press.html

Prensky, M. (2002). *Digital game-based learning.* New York, NY: The twitchspeed.com. Retrieved June 13, 2007, from http://www.twitchspeed.com/site/news.html

Punie, Y. (2007). Learning spaces: An ICT-enabled model of future learning in the knowledge-based society. *European Journal of Education, 42*(2), 185–199. doi:10.1111/j.1465-3435.2007.00302.x

Punithavathi, R., & Duraiswamy, K. (2008). *An optimized solution for mobile computing environment.* Paper presented at the International Conference on Computing, Communication and Networking, St. Thomas, VI.

Pytel, B. (2006). *Dropouts give reasons: Why do students leave high school without a diploma?* Retrieved February 8, 2010, from http://educationalissues.suite101.com/ article.c fm/dropouts_give_reasons

Ragus, M. (2007). M-learning: A future of learning. *E-Journal of Learning and Innovation, 9,* 22-37. Retrieved June 10, 2010, from http://kt.flexiblelearning.net.au/wp-content/uploads/2006/05/ragus.doc

Rasmussen, L. (2009, May 28). *Google Wave developer preview at Google I/O 2009* [Video file]. Retrieved from http://www.youtube.com/watch?v=v_UyVmITiYQ

Redmine. (2009). *Redmine.* Retrieved December 6, 2009, from http://www.redmine.org

Reese, G., Yarger, R. J., & King, T. (2002). *Managing and using MySQL*, (2nd ed.). Sebastopol, CA: O'Reilly Media, Inc.

Reio, T. G., & Davis, W. (2005). Age and gender differences in self-directed learning readiness: A developmental perspective. *International Journal of Self-Directed Learning, 2*(1), 40–49.

Rennison, P., & Andrews, A. (2008). *RealVNC goes mobile*. Retrieved January 17, 2010, from http://www.realvnc.com/company/news/mobile.html

Research, A. B. I. (2010). *5B devices in 2010*. Retrieved from http://www.abiresearch.com/press/1684- Worldwide+Mobile+Subscriptions+Forecast+to+Exceed+Five+Billion+by+4Q-2010

Riekki, J., Davidyuk, O., Forstadius, J., Sun, J., & Sauvola, J. (2005). Context-aware services for mobile users. In *Proceedings of IADIS Distributed and Parallel Systems and Architectures Conference as Part of IADIS Virtual Multi Conference on Computer Science and Information Systems (MCCSIS 2005)*.

Ringstaff, C., & Kelly, L. (2002). *The learning return on our technology investment: A review of findings from research*. San Francisco, CA: WestEd RTEC. Retrieved March 20, 2009, from http://www.wsetedrtec.org

Robins, A., Rountree, J., & Rountree, N. (2003). Learning and teaching programming: A review and discussion. *Computer Science Education, 13*(2), 137–172. doi:10.1076/csed.13.2.137.14200

Roblyer, M. D. (1990). The glitz factor. *Educational Technology, 30*(10), 34–36.

Roblyer, M. D., & Doering, A. H. (2010). *Integrating educational technology into teaching*. Boston, MA: Allyn & Bacon.

Rohs, M., & Gfeller. B. (2004). Using camera-equipped mobile phones for interacting with real-world objects. *Advances in Pervasive Computing,* 265-271.

Roschelle, J. (2003). Unlocking the learning value of wireless mobile devices. *Journal of Computer Assisted Learning, 19*(3), 260–272. doi:10.1046/j.0266-4909.2003.00028.x

Rosen, A. (2006). Technology trends: e-Learning 2.0. *The e-Learning Guild's Learning Solutions E-Magazine*. Retrieved May 24, 2008, from http://www.readygo.com/e-learning-2.0.pdf

Salomon, G. (2000). *It's not the tool but the educational rationale that counts*. Retrieved March, 16, 2010, from http://www.aace.org/conf/edmedia/00/salomonkeynote.htm

Salomoni, P., Mirri, S., Ferretti, S., & Roccetti, M. (2008). A multimedia broker to support accessible and mobile learning through learning objects adaptation. *ACM Transactions on Internet Technology, 8*(2), 9–23. doi:10.1145/1323651.1323655

Sánchez, J., & Salinas, A. (2008). Science problem solving learning through mobile gaming. In *Proceedings of the 12th International Conference on Entertainment and Media in the Ubiquitous Era* (pp. 49-53). New York, NY: ACM Press.

Sanderson, D. (1997). Virtual communities, design metaphors, and systems to support community groups. *SIGGROUP Bulletin, 18*(1), 44–46.

Santos, O., Boticario, J., Raffenne, E., & Pastor, R. (2007). Why using dotLRN? UNED use cases. In *Proceedings of FLOSS International Conference* (pp. 88-107).

Saponas, T., Lester, J., Froehlich, J., Fogarty, J., & Landay, J. (2008). *iLearn on the iPhone: Real-time human activity classification on commodity mobile phones*. (University of Washington CSE Tech Report UW-CSE-08-04-02).

Sauer, F. (2009, October 19). *The enterprise apps in your pocket*. [Web log post]. Retrieved from http://googlewebtoolkit.blogspot.com/2009/10/enterprise-apps-in-your-pocket.html

Savill-Smith, C., & Kent, P. (2003). *The use of palmtop computers for learning: A review of the literature*. London, UK: Learning and Skills Development Agency. Retrieved July 16, 2009, from http://www.m-learning.org/docs/the_use_of_palmtop_computers_for_learning_sept03.pdf

Scheucher, B., Belcher, J., Bailey, P. H., dos Santos, P., & Gütl, C. (2009b). Evaluation results of a 3D virtual environment for Internet-accessible physics experiments. In *Proceedings of the International Conference of Interactive Computer Aided Learning 2009*.

Scheucher, T., Bailey, P. H., Gütl, C., & Harward, V. J. (2009a). Collaborative virtual 3D environment for Internet-accessible physics experiments. In *Proceedings of the International Conference of Remote Engineering and Virtual Instrumentation 2009*.

SCS. (2007). *Simulation: Transactions of the Society for Modeling and Simulation International website*. Retrieved June 3, 2007, from http://www.scs.org/pubs/simulation/simulation.html

Seong, D. S. (2006). Usability guidelines for designing mobile learning portals. In *Proceedings of the 3rd International Conference on Mobile Technology, Applications & Systems*. New York, NY: ACM Press.

Sharma, S., & Kitchens, F. (2004). Web services architecture for m-learning. *Electronic Journal of e-Learning, 2*(2). Retrieved January 20, 2010, from http://www.ejel.org/volume-2/vol2-issue1/ issue1-art2.htm

Sharples, M. (2002). Disruptive devices: Mobile technology for conversational learning. *International Journal of Continuing Engineering Education and Lifelong Learning, 12*(5/6), 504–520. doi:10.1504/IJCEELL.2002.002148

Sharples, M., Crook, C., Jones, I., Kay, D., Chowcat, I., & Balmer, K. (2009). *CAPITAL year one final report (Draft)*. University of Nottingham.

Sharples, M. (2000). The design of personal mobile technologies for lifelong learning. *Computers & Education, 34,* 177-193. Retrieved April 20, 2010, from http://www.eee.bham.ac.uk/sharplem/Papers/handler%20comped.pdf

Sharples, M. (2005). Learning as conversation: Transforming education in the mobile age. In *Proceedings of Conference on Seeing, Understanding, Learning in the Mobile Age* (pp. 147-152). Budapest, Hungary.

Sharples, M., Taylor, J., & Vavoula, G. (2006). *A theory of learning for the mobile age*. Retrieved October 13, 2009, from http://kn.open.ac.uk/public/document.cfm?docid=8558

Sharples, M., Taylor, J., & Vavoula, G. (2005). *Towards a theory of mobile learning*. Paper presented at the 4th World Conference on Mlearning (MLearning 2005), Cape Town, South Africa.

Siemens, G. (2008a). *Connectivism: A learning theory for today's learner*. Retrieved March 11, 2009, from http://www.connectivism.ca/about.html

Siemens, G. (2008b). *Learning and knowing in networks: Changing roles for educators and designers*. Paper presented at the IT Forum. Retrieved October 10, 2009, from http://it.coe.uga.edu/itforum/Paper105/Siemens.pdf

SimBusiness. (2007). *Aspyr - The Sims™ 2 Open for Business website*. Retrieved June 15, 2007, from http://www.aspyr.com/product/info/3

SimCity. (2007). *SimCity societies*. Retrieved June 15, 2007, from http://simcity.ea.com

Sims. (2007). *The Sims official site*. Retrieved: June 15, 2007, from http://thesims.ea.com

Sipdroid. (2009). *Sipdroid*. Retrieved January 18, 2010, from http://sipdroid.org

Skehan, P. (1989). *Individual differences in second-language learning*. London, UK: Edward Arnold.

Skype lite. (2009). *Skype launches on Android platform and more than 100 Java-enabled mobile phones*. Retrieved January 18, 2010, from http://about.skype.com/2009/01/skype_launches_on_android_plat.html

Smart, C. (2009). Linux will regain lost market share, thanks to Moblin. *Linux Magazine*. Retrieved November 6, 2009, from http://www.linux-mag.com/cache/7559/1.html

Smith, D. E., & Nair, R. (2005). The architecture of virtual machines. *Computer, 38*(5), 32–38. doi:10.1109/MC.2005.173

Smith, D. A., Kay, A., Raab, A., & Reed, D. P. (2003). Croquet—collaboration system architecture. In *Proceedings of the 1st Conference on Creating, Connecting and Collaborating through Computing* (pp. 2-11). Los Alamitos, CA: IEEE Computer Society Press.

Song, Y. (2009). Handheld educational application: A review of the research . In Ryu, H., & Parsons, D. (Eds.), *Innovative mobile learning: Techniques and technologies* (pp. 302–323). New York, NY: Information Science Reference.

SourceForge. (2009). *SourceForge*. Retrieved May 6, 2009, from http://sourceforge.net

Spohrer, J. C., & Soloway, E. (1989). Novice mistakes: Are the folk wisdoms correct? In Soloway, E., & Spohrer, J. C. (Eds.), *Studying the novice programmer* (pp. 401–416). Hillsdale, NJ: Lawrence Erlbaum.

Steimann, F., Gößner, J., & Thaden, U. (2009). *Proposing mobile pair programming*. Retrieved December 6, 2009, from http://www.projectory.org/publications/MobilePair-Programming.pdf

Stetz, P. (1999). *The cell phone handbook: Everything you wanted to know about wireless telephony*. Middletown, RI: Aegis Publishing Group.

Stewart, R. A. (2007). Evaluating the self-directed learning readiness of engineering undergraduates: A necessary precursor to project-based learning. *World Transactions on Engineering and Technology Education, 6*(1), 59–62.

Stockwell, G. (2007). Vocabulary on the move: Investigating an intelligent mobile phone-based vocabulary tutor. *Computer Assisted Language Learning, 20*(4), 365–383. doi:10.1080/09588220701745817

Stockwell, G. (2008). Investigating learner preparedness for and usage patterns of mobile learning. *ReCALL, 20*(3), 253–270. doi:10.1017/S0958344008000232

SuperKids. (2007). *SuperKids software review of RockSim – model rocket design and simulation software*. Retrieved June 5, 2007, from http://www.superkids.com/aweb/pages/reviews/science/06/rocksim/merge.shtml

Svetlana, K., & Yonglk-Yoon (2009). Adaptation e-learning contents in mobile environment. In *Proceedings of the 2nd International Conference on Interaction Sciences: Information Technology, Culture and Human* (pp. 474-479). New York, NY: ACM Press.

Swantz, M. L. (2008). Participatory action research as practice . In Reason, P., & Bradbury, H. (Eds.), *The SAGE handbook of action research: Participative inquiry and practice* (2nd ed., pp. 31–48). London, UK: SAGE Publications.

TAG. (2008). *An introduction to mobile technology: Information and marketing in 2008*. Retrieved July 16, 2009, from http://www.tagonline.org/articles.php?id=269

Talagi, S. (2008). Learning goes mobile. In *Western Leader* (p. 4).

Tan, Q., et al. (2009). Location-based adaptive mobile learning research framework and topics. In *Proceedings of International Conference on Computational Science and Engineering* (pp. 140-147). IEEE Computer Society.

Taylor, R. P., & Gitsaki, C. (2005). Using mobile phones in English education in Japan. *Journal of Computer Assisted Learning, 21*, 217–228. doi:10.1111/j.1365-2729.2005.00129.x

Teten, D. (2007). *Web 3.0: Where are we headed*? Retrieved January 13, 2010, from www.teten.com/assets/docs/Teten-Web-3.0.pdf

The Eclipse Foundation. (2009). *Home*. Retrieved December 6, 2009, from http://www.eclipse.org

Third Generation Partnership Project. (2010). *Third Generation Partnership Project*. Retrieved from http://www.3gpp.org

Third Generation Partnership Project 2. (2010). *Third Generation Partnership Project 2*. Retrieved from http://www.3gpp2.org

Thomas, S. (2005). *Pervasive, persuasive e-learning: Modeling the pervasive learning space*. Paper presented at the 3rd International Conference on Pervasive Computing and Communications, Hawaii.

Thornton, P., & Houser, C. (2006). Using mobile phones in English education in Japan. *Journal of Computer Assisted Learning, 21*(3), 217–228. doi:10.1111/j.1365-2729.2005.00129.x

Tomorrow's Professor Listserv. (2002). *Message #289: Mobile learning*. Retrieved from http://sll.stanford.edu/projects/tomoprof/newtomprof/postings/289.html

Torresan, P. (2008). *Intelligenze e didattica delle lingue*. Bologna, Italy: EMI.

Torrisi-Steele, G. (2006). *The making of m-learning spaces*. Paper presented at the mLearn 2005, Cape Town, South Africa. Retrieved May 13, 2009, from https://olt.qut.edu.au/udf/OLT2006/gen/static/papers/Torrisi-Steele_OLT2006_paper.pdf

Torvalds, L., & Diamond, D. (2001). *Just for fun: The story of an accidental revolution*. New York, NY: Harper Business.

Traxler, J. (2007). Defining, discussing and evaluating mobile learning: The moving finger writes and having writ. *The International Review of Research in Open and Distance Learning, 8*(2). Retrieved January 23, 2009, from http://www.irrodl.org/index.php/irrodl/article/view/346

Traxler, J., & Kukulska-Hulme, A. (2005). *Evaluating mobile learning: Reflections on current practice*. Paper presented at mLearn 2005: Mobile Technology: The Future of Learning in Your Hands, Cape Town, South Africa.

Trifonova, A. (2003). *Mobile learning - review of the literature*. (Technical Report DIT-03-009, Informatica e Telecomunicazioni, University of Trento). Retrieved March 9, 2010, from http://eprints.biblio.unitn.it/archive/00000359

Truphone. (2010). *Truphone*. Retrieved January 18, 2010, from http://www.truphone.com

Tucker, T. G., & Winchester, W. W. (2009). Mobile learning for just-in-time applications. In *Proceedings of the 47th Annual Southeast Regional Conference*. New York, NY: ACM Press.

Ubuntu. (2009). *Ubuntu Mobile Internet Device (MID) edition*. Retrieved December 9, 2009, from http://www.ubuntu.com/products/mobile

Ubuntu. (2009). *Ubuntu*. Retrieved December 6, 2009, from http://www.ubuntu.com

Uden, L. (2007). Activity theory for designing mobile learning. *International Journal of Mobile Learning and Organisation, 1*(1), 81–102. doi:10.1504/IJMLO.2007.011190

UN & DESA. (2007). *World economic and social survey 2007: Development in an ageing world*. (p. 24). New York, NY: United Nations publication.

Valentim, H. (2009). *Para uma compreensão do mobile learning: Reflexão sobre a utilidade das tecnologias móveis na aprendizagem informal e para a construção de ambientes pessoais de aprendizagem*. Unpublished Master Dissertation, Universidade Nova de Lisboa, Portugal.

Vamosi, R. (2009). *The pros and cons of iPhone security*. Retrieved January 1, 2010, from http://news.zdnet.co.uk/security/ 0,1000000189,39287778,00.htm

van der Aalst, W. (1998). The application of Petri nets to workflow management. *Journal of Circuits Systems and Computers, 8*, 21–66. doi:10.1142/S0218126698000043

Van Es, R., & Koper, R. (2006). Testing the pedagogical expressiveness of IMS LD. *Journal of Educational Technology & Society, 9*(1), 229–249.

Vaughan-Nichols, S. J. (2007). *2007 desktop Linux survey results revealed*. Retrieved December 1, 2007, from http://www.desktoplinux.com/news/ NS8454912761.html

Vavoula, G. (2007). *Learning bridges: A role for mobile technologies in education*. Paper presented at the M-Learning Symposium. Retrieved from http://www.wlecentre.ac.uk/cms/index.php? option=com_content&task=view&id=105&Itemid=39

Vavoula, G. N. (2005). *WP4: A study of mobile learning practices*. MOBIlearn deliverable D4.4. Retrieved February 8, 2010, from http://www.mobilearn.org/download/results/ public deliverables/MOBIlearn_D4.4_Final.pdf

Vavoula, G., Sharples, M., Scanlon, E., Lonsdale, P., Jones, A., & Sharples, M. (2005). Report on literature on mobile learning, science and collaborative activity. *Kaleidoscope*. Retrieved October 25, 2009, from http://hal.archives-ouvertes.fr/hal-00190175/en

Vetro, A. (2004). MPEG-21 digital item adaptation: Enabling universal multimedia access. *IEEE MultiMedia, 11*(1), 84–87. doi:10.1109/MMUL.2004.1261111

VISIR. (2010). *Computer software*. Retrieved from http://openlabs.bth.se

VMware. (2009). *VMWare*. Retrieved August 6, 2009, from http://www.vmware.com

Voorhoeve, M., & van der Aalst, W. (1997). Ad-hoc workflow: Problems and solutions. In *Proceedings of the 8th International Workshop on Database and Expert Systems Applications* (pp. 36 - 40).

Vygotsky, L. S. (1978). *Mind in society*. Cambridge, MA: Harvard University Press.

W3schools. (2010). *Ajax introduction*. Retrieved June 10, 2010 from http://www.w3schools.com/ajax/ajax_intro.asp

Wadsworth, B. J. (1979). *Piaget's theory of cognitive development*. New York, NY: Longman.

Wagner, D., & Schmalstieg, D. (2009). Making augmented reality practical on mobile phones, part 1. *IEEE Computer Graphics and Applications*, *29*(3), 12–15. doi:10.1109/MCG.2009.46

Wagner, D., & Schmalstieg, D. (2003). *First steps towards handheld augmented reality*. In the 7th International Conference on Wearable Computers (pp. 127-135).

Wagner, D., & Schmalstieg, D. (2007). *ARToolKitPlus for pose tracking on mobile devices*. In 12th Computer Vision Winter Workshop (pp. 139-146).

Wagner, D., Pintaric, T., Ledermann, F., & Schmalstieg, D. (2005). *Towards massively multi-user augmented reality on handheld devices*. In Third International Conference on Pervasive Computing (pp. 208-219).

Wallen, J. (2009). *10 reasons why open source makes sense on smart phones*. Retrieved August 6, 2009, from http://blogs.techrepublic.com.com/10things/?p=808

Ward, J., & Carroll, P. (1998). *Can the use of a computer simulation game enhance mechanical reasoning ability: An exploratory study*. (Working Paper Series MCE-0998), (pp. 1-20). Retrieved June 10, 2007, from http://citeseer.ist.psu.edu/431172.html

Watson, W. (2006). *Ageing workforce 2006 report*. Retrieved January 5, 2008, from http://www.watsonwyatt.com/images/database_uploads/ ageing_ap_06/AP_AgeingWorkforce2006.pdf

WebLab-Deusto. (2010). *Computer software*. Retrieved from http://www.weblab.deusto.es

Wenger, E., White, N., Smith, J., & Rowe, K. (2005). Technology for communities. In Langelier, L. (Ed.), *Working, learning and collaborating in a network: Guide to the implementation and leadership of intentional communities of practice* (pp. 71–94). Quebec City, Canada: CEFIRO.

Wenger, E., White, N., & Smith, J. (2009). *Digital habitats: Stewarding technology for communities*. Portland, OR: CPsquare.

Wheeler, D. (2007). *Why open source software / free software (OSS/FS, FLOSS, or FOSS)? Look at the numbers!* Retrieved May 5, 2009, from http://www.dwheeler.com/oss_fs_why.html

Wikipedia. (2007). *Simulation*. Retrieved June 1, 2007, from http://en.wikipedia.org/wiki/Simulation

Wikipedia. (2010). *High Level Shader Language*. Retrieved March 1, 2010, from http://en.wikipedia.org/wiki/High_Level_Shader_Language

Wikipedia. (2010). *The open university*. Retrieved June 6, 2010, from http://en.wikipedia.org/wiki/Open_University

Wikipedia. (2010a). *Ajax (programming)*. Retrieved from http://en.wikipedia.org/wiki/Ajax_(programming)

Wikipedia. (2010b). *Appcelerator Titanium*. Retrieved from http://en.wikipedia.org/wiki/Appcelerator_Titanium

Wikipedia. (2010c). *Crowdsourcing*. Retrieved from http://en.wikipedia.org/wiki/Crowdsourcing

WiMAX Forum. (2010). *WiMAX Forum*. Retrieved from http://www.wimaxforum.org

Wimberly, T. (2009). *Android closing in on BlackBerry, taking share from iPhone*. Retrieved January 1, 2010, from http://androidandme.com/2009/08/news/ android-closing-in-on-blackberry-taking-share-from-iphone

Winters, N. (2006). What is mobile-learning? In M. Sharples (Ed.), *Big issues in mobile learning* (pp. 5-9). Report of a workshop by the Kaleidoscope Network of Excellence Mobile Learning Initiative. University of Nottingham.

Wohed, P., Russell, N., ter Hofstede, A., Andersson, B., & van der Aalst, W. (2009). Patterns-based evaluation of open source BPM systems: The cases of jBPM, OpenWFE, and Enhydra Shark. *Information and Software Technology*.

Wollen, K., Weber, A., & Lowry, D. (1972). Bizarreness versus interaction of mental images as determinants of learning. *Cognitive Psychology*, *2*, 518–523. doi:10.1016/0010-0285(72)90020-5

Wong, K., Wolf, T., & Gorinsky, S. (2007). Teaching experiences with a virtual network laboratory . In *Special Interest Group Computer Science Education, 2007* (pp. 481–485). Turner. J.

Workflow Management Coalition. (n.d.). Retrieved 2010, from http://www.wfmc.org

World Wide Web Consortium. (2004). *Composite Capability/Preference Profiles (CC/PP): Structure and vocabularies* 1.0. Retrieved March 21, 2010, from http://www.w3.org/TR/2004/REC- CCPP-struct-vocab-20040115

World Wide Web Consortium. (2007). *Content selection for device independence* (DISelect) 1.0. Retrieved March 21, 2010, from http://www.w3.org/TR/cselection

World Wide Web Consortium. (2008a). *Device description repository core vocabulary*. Retrieved March 21, 2010 from http://www.w3.org/TR/ddr-core-vocabulary

World Wide Web Consortium. (2008b). *Device description repository simple API*. Retrieved March 21, 2010, from http://www.w3.org/TR/DDR-Simple-API

World Wide Web Consortium. (2008c). *Synchronized multimedia integration language* 3.0. Retrieved March 21, 2010 from http://www.w3.org/TR/2008/REC-SMIL3-20081201

World Wide Web Consortium. (W3C). (2008). *Extensible Markup Language* (XML) 1.0 (Fifth edition). Retrieved March 19, 2010, from http://www.w3.org/TR/REC-xml

Wright, J., Carpin, S., Cerpa, A., & Gavilan, G. (2007). *An open source teaching and learning facility for computer science and engineering education* (pp. 368–373). In FECS.

WURFL. (2010). *Wireless universal resource file library*. Retrieved March 21, 2010, from http://wurfl.sourceforge

Yate. (2010). *Yate – the next-generation telephony engine home page*. Retrieved June 1, 2010, from http://yate.null.ro

Zebra. (2009). *GNU Zebra—routing software*. Retrieved June 7, 2009, from http://www.zebra.org

Zhu, Q., Wang, T., & Jia, Y. (2007). Second Life: A new platform for education. In *Proceedings of First International Symposium on Information Technologies and Applications in Education* (pp. 201-204). doi: 10.1109/ISITAE.2007.4409270

Zurita, G., & Nussbaum, M. (2004). A constructivist mobile learning environment supported by a wireless handheld network. *Journal of Computer Assisted Learning*, *20*, 235–243. doi:10.1111/j.1365-2729.2004.00089.x

About the Contributors

Li Chao is currently a professor of math and computer science in the school of Arts and Sciences at University of Houston – Victoria, USA. He received his Ph. D. from the University of Wyoming, USA, and he is certified as Oracle Certified Professional and Microsoft Solution Developer. His current research interests are data analysis and technology-based teaching. Dr. Chao is also the author of over a dozen of research articles in data analysis and math modeling, and books in the development of computer labs and database systems.

***

Leila Alem is a senior research scientist at CSIRO ICT Centre. Her group is conducting research in human factors in tele-collaborative settings. Leila is interested in formalising and measuring user's experience when engaged in synchronous computer mediated collaboration, and more specifically, users' sense of presence.

Peta Ashworth is a research scientist at CSIRO ICT Centre. She is managing a small team of social researchers that are investigating stakeholder attitudes to a range of complex issues of strategic importance to Australia. Her research activity is focused on social research.

Thomas Cochrane is an Academic Advisor (elearning and Learning Technologies) with Unitec (March 2004 to present). His role at Unitec includes providing support for elearning and learning technologies for Unitec teaching staff, and pushing the boundaries of educational technology for enhancing teaching and learning at Unitec. His research interests include mobile learning, Web 2.0, and communities of practice. He is currently implementing mobile learning trials for his PHD thesis: "Mobilizing Learning: Transforming teaching and learning in higher education in New Zealand."

Clara Pereira Coutinho, Ph.D., is Associate Professor at the Department of Curriculum and Educational Technology at the Institute of Education and Psychology, University of Minho, Portugal. She teaches Educational Technology, Multimedia Software and Research Methods in Education in graduate and post graduate teacher education programs. She has already published more than 70 papers in proceedings and specialized journals as well as a book. The areas of current research and interest are educational multimedia, Web 2.0 technologies in teacher education programs and research methods in online education.

Paula Maria de Sá Oliveira Escudeiro, Ph.D. in Computer Engineering / Systems and Information Technology in Education / Multimedia Systems, Professor at Polytechnic Institute of Porto, School of Engineering, Department of Computer Engineering, where she teaches regular courses on Software Engineering (Undergraduate Computer Engineering) and Multimedia Learning Systems (MSc Computer Engineering - Graphic Systems and Multimedia). Sub-director of the Computer Engineering Department, Director of the Multimedia Laboratory, Department of Computer Science, she was a member of the Steering Committee of the Computer Science Department, Information Systems Director of the Institute for Technological Development, Vice President of Pedagogical Council, Director of the Center for Development of Multimedia Products for the National Institute of Administration, Element of the Subcommittee on Quality Evaluation for the Polytechnic Institute of Porto, School of Engineering. Researcher in GILT-Graphic & Interaction Learning Technologies Research Group, collaborates in training courses and other activities in collaboration with National and International Institutions.

David Furio Ferri is a Computer Engineer at Universidad Politecnica de Valencia. He earned his MD from Universidad Politecnica de Valencia in 2008. Before he was a senior technician, Dr. Furio Ferri was an Assistant Scholar in the Learning collaboration scholarship in a computer graphics project at Universidad Politecnica de Valencia, from 2008-2009.

Miguelón Giménez is the head of the Summer School of the Technical University of Valencia. His research activity has been focused on new tools for being included as activities at the Summer School.

Diego López-de-Ipiña is research leader in the MORElab. He received his PhD from the University of Cambridge, UK in the topic of Sentient Computing. His main research areas are semantic service middleware, mobile-mediated user-environment interaction, and embedded reasoning techniques for system proactivity and reactivity. He is taking part in the MUGGES FP7 project and the uService ITEA2 project. He is the technical coordinator of the CBDP CELTIC project which deals with mobile mediated user-environment service negotiation and recommendation.

Jaime Irurzun is Computer Engineer (2009) and MSc in Development and Integration of Software Solutions (2010) by the University of Deusto in Bilbao, Spain. He has been working as research intern at the WebLab-Deusto Research Group since year 2007, mainly involved in the development and administration of the WebLab-Deusto Remote Lab.

M. Carmen Juan earned a PhD Computer Science. Her current position is Senior Lecturer at the Technical University of Valencia (Spain). She belongs to the Computer Graphics and Multimedia research group of the ai2 institute. Her research activity has been focused on computer graphics. At the present time, her main research interest is augmented reality.

Midori Kimura is a professor at Tokyo Women's Medical University, Ph.D. in TESOL. Her research interests include learning strategies, learning style, and CALL / MALL, with a special focus on the usage of mobile phones. Her most recent paper is "Development and Effectiveness of Vocabulary Learning Contents for Use with Mobile Phones in Education in Japan," published in *Research Highlights in Technology and Teacher Education 2009*, pp.151-160.

Alvin C.M. Kwan received his doctoral degree in computer science from the University of Essex in the United Kingdom in 1997. He had worked for the computer industry and two universities in Hong Kong before joining the University of Hong Kong in 2001. Currently, he works as a Teaching Consultant at the university.

C.H. Leung received the B.Sc. (Eng.) and Ph.D. degrees from the University of Hong Kong, and the M.Eng. degree from McGill University, all in Electrical Engineering. He has been teaching at the University of Hong Kong since 1986 and is currently an Associate Professor. His research interest is mainly in pattern recognition.

Hongxiu Li is a PhD student of Information Systems Science Institute of Turku School of Economics, Turku University and Turku Centre for Computer Science in Finland. She researches in the areas of e-commerce, e-services, and mobile service. She has together about 30 publications published in international conference proceedings and international academic journals.

Mr. Zexian Liao obtained his Master degree from the School of Communication and Information Engineering, the University of Electronic Science and Technology of China. His current research interest mainly focuses on mobile ad-hoc networks.

Yong Liu is currently a PhD Student of the Institute for Advanced Management Systems Research (School of Computing) in the Åbo Akademi University, TUCS, in Finland. His research interests are in advanced mobile services. His research papers have appeared in journals such as Computers & Education, and Industrial Management & Data Systems, and in conferences, such as PACIS, AMCIS and ECIS.

Silvia Mirri is an assistant professor at the Department of Computer Science of the University of Bologna. In October 2002, she received the Italian Laurea degree in Computer Science from the University of Bologna (in Cesena). In 2007, she received the Ph.D. degree in Computer Science from the University of Bologna. Her research interests include the following ones: multimodal interaction, accessibility of e-learning environments, Web content accessibility and evaluation tools, content transcoding, and multimedia content adaptation.

Chris Moya studied Computer Engineering at the Universitat Ramon Llull (URL), Barcelona and achieved a Postgraduate in ITC and Society by the Universitat Oberta de Catalunya (UOC). He joined the UOC in September 2005 at The Office of Learning Technologies as Head of Projects, in which he made changes to various educational materials to develop them with SCORM technology. Chris also developed various projects using the OpenLaszlo Framework tool. His career has been always related to the development of open source applications, and lately, the tinkering and study of smart phones has become his hobby.

Pablo Orduña is Computer Engineer in the University of Deusto in Bilbao, Spain, 2007. Nowadays, he is Research Associate at the Ambient Intelligence department of DeustoTech - Tecnológico Fundación Deusto, and PhD student at the University of Deusto; his research is focused on Remote Laboratories. He is the lead software designer and developer of WebLab-Deusto.

Rosa Reis is a Professor at the Engineering School of the Polytechnic Institute of Porto, School of Engineering, and Department of Computer Engineering, where she lectures on Database Technology, Software Engineering and Multimedia Information Systems. In December of 1997, she concluded her MSc degree in Information Systems in Education at the Faculty of Sciences and Technology/University of Coimbra, Portugal. Presently she is a PhD student of Informatics at the University of Trás-dos-Montes e Alto Douro in Portugal. Her research field is technology-enhanced learning, particularly on the application of software engineering techniques to the design of collaborative virtual worlds with a special goal to quality evaluation of virtual worlds in Education.She is researcher in the Graphics, Interaction and Learning Technologies research center (GILT) and she has additional interests in mobile learning, LMS, online communities and the impact of ICT on higher education.

Luis Anido-Rifon has a Telecommunications Engineering degree with honours (1997) in the Telematics and Communications branches and a Telecommunications Engineering PhD with honours (2001) by the University of Vigo. Currently, he is Associate Professor in the Telematics Engineering Department of the University of Vigo. He has received several awards by the W3C, the Royal Academy of Sciences and the Official Spanish Telecommunications Association.

Manuel Caeiro-Rodriguez received his PhD in Telecommunications Engineering from the University of Vigo in 2007. He is currently Assistant Teacher at the Department of Telematics Engineering, University of Vigo. He has received several awards by the W3C, NAE CASEE new faculty fellows, and the IEEE Spanish Chapter of the Education Society.

Roberto Perez-Rodriguez has a Telecommunications Engineering degree from the University of Vigo, Spain, in 2008. He has participated in several research and development projects. He is currently a Research Engineer with the Telematics Department, where he is pursuing his doctorate. His research interests include the use of service-oriented architectures and workflow technologies for the provision of e-learning services.

Paola Salomoni is an associate professor of Computer Science at the Department of Computer Science of the University of Bologna. In October 1992, she received the Italian Laurea degree with honors in Computer Science from the University of Bologna. From 1992 through 1995, she was in the research staff of the Computer Science degree Course of the University of Bologna (in Cesena). From 1995 to 2001, she was an assistant professor of the Department of Computer Science. Her research interests include the following: design and implementation of distributed multimedia systems, integration of multimedia services in wireless networks, design and implementation of e-learning environments, accessibility of e-learning environments, content transcoding, and multimedia aontent adaptation.

Elisa Spadavecchia (www.elisaspadavecchia.it) is a secondary high school teacher and a teacher trainer in Italy. At present, she is working at the Educational Interventions Office of the Provincial School Authority in Vicenza, a town near Venice. She is a member of LTEver, the community of the Educational Technology Laboratory at Florence University (http://www.lte.unifi.it/default.asp). Her main professional interests are research in the e-learning area and application of ICT to education and teacher training.

Hal Steger has extensive global marketing experience in the software industry, including several years in mobile, where he evangelizes the use of open source for many types of applications. Hal is Vice President of Worldwide Marketing at Funambol, a provider of open source mobile cloud sync solutions. Prior to Funambol, Hal co-founded and was Vice President of Marketing of Rubric, where he pioneered the category of e-marketing software. Hal has also held senior marketing roles at Oracle, Uniface/Compuware, and other Silicon Valley companies. He holds an undergraduate degree from the University of Michigan with majors in Computer Science and Economics, and an M.S.I.A. (MBA) from Carnegie-Mellon's Tepper School of Business (formerly G.S.I.A.).

Vincent Tam completed his Ph.D. in computer science from the University of Melbourne, Australia in 1998. Dr. Tam was the winner of Innovative Teaching Award (2000) in the School of Computing, National University of Singapore. He is a Senior Teaching Consultant in the University of Hong Kong, the part-time project advisor of an EMB-funded project on e-learning packages, and also the principal investigators of several teaching development projects.

Sebastian Thalanany is a technology strategist - mobile broadband (4G) evolution - at U.S. Cellular. His activities include thought leadership, advocacy, guidance, and LTE (Long Term Evolution) system standards evolution work, in 3GPP. He is also the Chair of the ERA (Evolution, Requirements and Architecture) working group, in 3GPP2. His experience spans the design of mobile devices, networks, and systems. He continues to be engaged in the emerging wireless IP mobility and services standards, leadership and strategy, across global standards organizations - 3GPP (Third Generation Partnership Project), OMA (Open Mobile Alliance), 3GPP2 (Third Generation Partnership Project 2), and IETF (Internet Engineering Task Force).

Harris Wang is an associate professor in the School of Computing and Information Systems at Athabasca University, Canada's Open University. He received a PhD in computer science from the Australian National University, Australia. His research interests include advanced technology for education.

Lawrence Yeung received his B.Eng. and Ph.D. degrees in Information Engineering from The Chinese University of Hong Kong in 1992 and 1995, respectively. He joined the Department of Electrical and Electronic Engineering, The University of Hong Kong in July 2000, where he is currently a Professor.

Dongqing "Holly" Yuan is an Assistant Professor of the University of Wisconsin-Stout. She is an outstanding instructor who uses real-world experience to present complex computer technologies. Holly holds a Master of Science degree from Georgia State University with a major in Computer Science and Ph.D. degree from Nova Southeastern University with a major in Computer Information Systems. Holly has published many papers in the field of networking and security in national conferences and journals, and has received several research grants from NSF and NASA. She is also a speaker in National Computer Application Security Conference. Holly also has been a Cisco Academy curriculum lead, Cisco lab manager and instructor of Cisco academy classes for various universities. Holly also holds many valuable industry certifications which include Cisco Certified Network Professional (CCNP), Cisco Certified Security Professional (CCSP), Cisco Certified Academy Instructor (CCAI), and Linux+.

Jiling Zhong is an Assistant Professor of Troy University. He holds Ph.D. degree in Computer Science from Georgia State University. Dr. Zhong has published many papers in the field of networking, artificial intelligence, database, and security in national conferences journals. Dr. Zhong has extensive experience in curriculum design and online teaching. Dr. Zhong has also helped in ABET accreditation application. Dr. Zhong also has extensive experience in teaching various computer science courses internationally. He has travelled to various international campuses and helped audit their curriculum.

Javier Garcia-Zubia is head of the Industrial Technologies Department, Faculty of Engineering of the University of Deusto in Bilbao, Spain. Also he is project manager and head of the WebLab-Deusto Research Group, whose main achievement is WebLab-Deusto, an open source experiment-agnostic remote lab based on Web 2.0 technologies.

Index

Symbols

A

B

C

D

E

F

G

H

I

J

K

L

M

N

O

P

Q

R

X

Y

Z